Steroid Hormone Receptors:
Basic and Clinical Aspects

HORMONES IN HEALTH AND DISEASE
Series Editor: V. K. Moudgil

Steroid Hormone Receptors: Basic and Clinical Aspects

V. K. Moudgil
Editor

129 Illustrations

Birkhäuser
Boston • Basel • Berlin

V. K. Moudgil
Department of Biological Sciences
Oakland University
Rochester, Michigan 48309-4401
USA

Library of Congress Cataloging In-Publication Data
Steroid hormone receptors : basic and clinical aspects / V. K. Moudgil,
    editor.
        p.   cm. -- (Hormones in health and disease)
    Includes bibliographical references and indexes.
    ISBN 0-8176-3694-3 (h : alk. paper). -- ISBN 3-7643-3694-3 (h :
alk. paper)
    1. Steroid hormones--Receptors.   I. Moudgil, V. K. (Virinder K.),
1945-   . II. Series.
    [DNLM: 1. Receptors, Steroid--physiology. 2. Gene Expression
Regulation.   WK   150 S83575 1993]
QP12.4'05--dc20
DNLM/DLC   93-33850
    for Library of Congress   CIP

COVER: A three-dimensional model for the intramolecular interaction of the T box of human vitamin D receptor. Courtesy of Dr. Leonard Freedman, Sloan-Kettering Institute.

Printed on acid-free paper.

ISBN 0-8176-3694-3
ISBN 3-7643-3694-3

Camera-ready pages prepared by the Editor on IBM PS2, using WordPerfect 5.1
Printed and bound by Quinn Woodbine Inc., Woodbine, NJ
Printed in the United States of America

9 8 7 6 5 4 3 2 1

Contents

Preface

The past few years have witnessed the emergence of steroid hormones as the wonder molecules which generate as much discussion in the scientific literature as they do in a typical living room. This transition has been a result of the tremendous public and scientific interest in the normal functioning of the hormones as well their suggested involvement in several clinical conditions. In the recent past, notable scientific and technological advances have been made in the areas of contraception and regulation of fertility. Steroid receptors are the indispensable mediators of hormonal responses and are complex protein molecules which appear to exist in association with other, yet undefined, proteins and/or factors. Receptors for vitamin D, retinoic acid and the thyroid hormones share structural similarities with steroid receptors, and the roster of this superfamily is still expanding. While our knowledge of the diversity and magnitude of steroid effects has advanced, the precise mode of steroid hormone action has alluded investigators. This volume brings together an international team of prominent investigators who discuss their most recent work on the basic and clinical aspects of steroid/nuclear receptors. The contributions represent updated versions of the invited presentations made at The Second Meadow Brook Conference on Steroid Receptors in Health and Disease. I am grateful to my colleagues on the Scientific Committee: Etienne Baulieu, Jack Gorski, Benita Katzenellenbogen, David Toft and James Wittliff, who provided the vision and guidance in formulating an outstanding program.

The organizers gratefully acknowledge the financial support received from the conference sponsors, Oakland University, The Meadow Brook Hall, Serono Symposia, USA, William Beaumont Hospital, Schering-Plough Corporation, Eli Lilly & Company, The E.I. Du Pont De Nemours & Company, Affinity Bioreagents, Inc., and The Du Pont Merck Pharmaceutical Company. For the efficient execution of the conference program, I am indebted to my colleagues, Bill Underwood, Art Griggs, Amrita Bhakta and John Shiff. The cooperation from the authors and the excellent secretarial assistance received from Rita Perris made the process of editing this volume an enjoyable and memorable experience.

I am also pleased to extend my warmest wishes and cooperation to the publisher, Birkhäuser Boston, who has decided to introduce a new series, Hormones in Health and Disease, with the present volume as the inaugural issue.

Fall 1993   V.K. Moudgil

Contributors

Iris Alroy, Cell Biology and Genetics Program, Sloan-Kettering Institute, New York, New York 10021, USA

Etienne-Emile Baulieu, INSERM U33, Lab Hormones, 94276 Bicêtre Cedex, FRANCE

Candace A. Beck, Department of Pathology, University of Colorado Health Science Center, Denver, Colorado 90262-0216, USA

Nadine Binart, INSERM U33, Lab Hormones, 94276 Bicêtre Cedex, FRANCE

Chris Bradfield, Department of Pharmacology, Northwestern University Medical School, 303 E. Chicago Avenue, Chicago, Illinois 60611, USA

Tauseef Butt, Smithkline Beecham Pharmaceuticals, 709 Swedeland, P.O. Box 1539, King of Prussia, Pennsylvania 19406, USA

Francoise Cadepond, INSERM U33, Lab Hormones, 94276 Bicêtre Cedex, FRANCE

Timothy Carter, Department of Biological Sciences, St. John's University, Jamaica, New York 11439, USA

Maria Grazia Catelli, INSERM U33, Lab Hormones, 94276 Bicêtre Cedex, FRANCE

Beatrice Chambraud, INSERM U33, Lab Hormones, 94276 Bicêtre Cedex, FRANCE

Orla M. Conneely, Department of Cell Biology, Baylor College of Medicine, One Baylor Plaza, Houston, Texas 77030-3498, USA

Richard N. Day, Department of Medicine, University of Virginia, Health Science Center, Box 511, Charlottesville, Virginia 22908, USA

Larry A. Denner, Department of Molecular Biology, Texas Biotechnology Corporation, 7000 Fannin, Houston, Texas 77030, USA

Jing Dong, Hormone Receptor Laboratory, Brown Cancer Center, University of Louisville, Louisville, Kentucky 40292, USA

Dean P. Edwards, Department of Pathology, University of Colorado Health Science Center, Denver, Colorado, 90262-0216, USA

Carlos A. Encarnation, University of Texas Health Science Center, Divison of Medical Oncology, 7703 Floyd Curl Drive, San Antonio, Texas 77030, USA

Petr Folk, Department of Physiology & Developmental Biology, Charles University, 128 00 Vinicna 7, Prague 2, CZECH REPUBLIC

Leonard P. Freedman, Cell Biology and Genetics Program, Sloan-Kettering Institute, New York, New York 10021, USA

Frank S. French, The Laboratories for Reproductive Biology, Department of Pediatrics, CB #7500 MacNider Building, University of North Carolina, Chapel Hill, North Carolina 27599, USA

Suzanne A. W. Fuqua, University of Texas Health Science Center, Division of Medical Oncology, 7703 Floyd Curl Drive, San Antonio, Texas 77030, USA

Jack Gorski, Department of Biochemisty, University of Wisconsin-Madison, 420 Henry Mall, Madison, Wisconsin 53706-1569, USA

Geoffrey L. Greene, The Ben May Institute, University of Chicago, Chicago, Illinois 60637, USA

Steven A. Harris, Department of Biochemistry and Molecular Biology, Mayo Clinic and Graduate Schools of Medicine, Rochester, Minnesota 55905, USA

Kathyrn B. Horwitz, Department of Medicine and Pathology, University of Colorado Health Sciences Center, 4200 E. Ninth Avenue, B-151, Denver, Colorado 80262, USA

Jill Johnson, Department of Biochemisty and Molecular Biology, Mayo Foundation, Rochester, Minnesota 55905, USA

Benita S. Katzenellenbogen, Department of Physiology and Biophysics, University of Illinois, Urbana, Illinois 61801, USA

James P. Landers, Department of Biochemisty and Molecular Biology, Mayo Clinic and Graduate Schools of Medicine, Rochester, Minnesota 55905, USA

Marie-Claire Lebeau, INSERM U33, Lab Hormones, 94276 Bicêtre Cedex, FRANCE

Kimberley K. Leslie, University of Colorado Health Sciences Center, 4200 E. Ninth Avenue, B-198, Denver, Colorado 80262, USA

Ben F. Luisi, MRC Virology Unit, Church Street, Glasgow G11 5JR, UNITED KINGDOM

Jerry R. Malayer, Department of Biochemistry, University of Wisconsin-Madison, 420 Henry Mall, Madison, Wisconsin 53706-1569, USA

Nelly Massol, INSERM U33, Lab Hormones, 94276 Bicêtre Cedex, FRANCE

Richard A. Maurer, Cell Biology and Anatomy, L215, Oregon Health Science University, Portland, Oregon 97201-3098, USA

V. K. Moudgil, Department of Biological Sciences, Oakland University, Rochester, Michigan 48309-4401, USA

Ann M. Nardulli, Department of Physiology, University of Illinois, Urbana, Illinois 61801, USA

Barbara E. Nowakowski, Molecular Biology Ph.D. Program, University of Iowa, Iowa City, Iowa 52242, USA

Yasuhiko Okimura, Department of Physiology and Biophysics, University of Iowa, Iowa City, Iowa 52242, USA

Bert W. O'Malley, Department of Cell Biology, Baylor College of Medicine, One Baylor Plaza, Houston, Texas 77030-3498, USA

Merry Jo Oursler, Department of Biochemistry and Molecular Biology, Mayo Clinic and Graduate Schools of Medicine, Rochester, Minnesota 55905, USA

Magnus Pfahl, Cancer Center, La Jolla Cancer Research Foundation, 10901 North Torrey Pines Road, La Jolla, California 92037, USA

J. Wesley Pike, Department of Biochemistry, Ligand Pharamaceuticals, Inc., 11149 North Torrey Pines Road, Suite 110, La Jolla, California 92037, USA

Alan Poland, McArdle Laboratory for Cancer Research, University of Wisconsin-Madison, 1400 University Avenue, Madison, Wisconsin 53706, USA

Angelo Poletti, Istituto di Endocrinologia, Via Balzaretti 9, 20133 Milano, ITALY

William B. Pratt, Department of Pharmacology, University of Michigan Medical School, Ann Arbor, Michigan 48109, USA

Christine Radanyi, INSERM U33, Lab Hormones, 94276 Bicêtre Cedex, FRANCE

Gérard Redeuilh, INSERM U33, Lab Hormones, 94276 Bicêtre Cedex, FRANCE

Joseph C. Reese, Program in Molecular Medicine, Department of Biochemistry and Molecular Biology, University of Massachusetts Medical Center, 373 Plantation Street, Suite 309, Worcester, Massachusetts 01605, USA

Jack-Michel Renoir, INSERM U33, Lab Hormones, 94276 Bicêtre Cedex, FRANCE

B. Lawrence Riggs, Department of Medicine, Divison of Endocrinology, Mayo Clinic and Graduate Schools of Medicine, Rochester, Minnesota 55905, USA

Michèle Sabbah, INSERM U33, Lab Hormones, 94276 Bicêtre Cedex, FRANCE

Madhabananda Sar, Department of Cell Biology and Anatomy, University of North Carolina, Chapel Hill, North Carolina 27599, USA

Christine Schaupp, Hormone Receptor Laboratory, Brown Cancer Center, University of Louisville, Louisville, Kentucky 40292, USA

Lawrence C. Scherrer, Department of Pharmacology, University of Michigan Medical School, Ann Arbor, Michigan 48109, USA

Ghislaine Schweizer-Groyer, INSERM U33, Lab Hormones, 94276 Bicêtre Cedex, FRANCE

David J. Shapiro, Department of Biochemistry, B-4 RAL, University of Illinois, 1209 W. California Street, Urbana, Illinois 61801, USA

G. Shyamala, Division of Cellular and Molecular Biology, Lawrence Berkeley Laboratory, Building 74, University of California, Berkeley, California 94720, USA

Carolyn L. Smith, Department of Cell Biology, Baylor College of Medicine, One Baylor Plaza, Houston, Texas 77030-3498, USA

David F. Smith, Department of Pharmacology, University of Nebraska Medical Center, 600 South 42nd St., Omaha, Nebraska 68198-6260, USA

Thomas C. Spelsberg, Department of Biochemistry and Molecular Biology, Mayo Clinic and Graduate Schools of Medicine, Rochester, Minnesota 55905, USA

Malayannam Subramaniam, Department of Biochemistry and Molecular Biology, Mayo Clinic and Graduate Schools of Medicine, Rochester, Minnesota 55905, USA

William P. Sullivan, Department of Biochemistry and Molecular Biology, Mayo Foundation, Rochester, Minnesota 55905, USA

David O. Toft, Department of Biochemistry and Molecular Biology, Mayo Clinic, Rochester, Minnesota 55905, USA

Terri L. Towers, Cell Biology and Genetics Program, Sloan-Kettering Institute, New York, New York 10021, USA

Nancy L. Weigel, Department of Cell Biology, Baylor College of Medicine, Houston, Texas 77030, USA

Elizabeth M. Wilson, The Laboratories for Reproductive Biology, Departments of Pediatrics, and Biochemistry and Biophysics, CB #7500 MacNider Building, University of North Carolina, Chapel Hill, North Carolina 27599, USA

James L. Wittliff, Department of Biochemistry and James Graham Brown Cancer Center, University of Louisville, Louisville, Kentucky 40292, USA

Zhong-xun Zhou, The Laboratories for Reproductive Biology, Department of Pediatrics, CB #7500 MacNider Building, University of North Carolina, Chapel Hill, North Carolina 27599, USA

INTRODUCTION

STEROID/NUCLEAR RECEPTOR SUPERFAMILY: RECENT ADVANCES AND RELATION TO HEALTH AND DISEASE

V. K. Moudgil
Department of Biological Sciences
Oakland University

INTRODUCTION

By the year 1993, the word *steroids* has become a household name. The terms, contraception, pregnancy, and hormone-dependent cancers of the prostate, the uterus, and the breast are as frequently discussed on the social and public platforms as they are in the scientific community and literature. Starting from their roles in the regulation of fetal brain development, attainment of puberty, sexual differentiation and maturation, and reproductive function, steroid hormones are involved in normal and tumor cell proliferation, maintenance of mineral balance, and numerous behavioral alterations. The last two decades have also witnessed a growth in the knowledge and use of antisteroids for interruptions in the hormone-induced processes. Although antiestrogens have been employed in the postmenopausal breast cancer patients, interest in and potential of several newly discovered antiprogestins is growing rapidly.

The receptor concept. For their actions, steroid hormones initially interact with specific intracellular *receptor* proteins in the target cells. The tissues that lack functional receptors or are receptor-deficient fail to respond to the circulating normal levels of hormones. During the 1970s

STEROID HORMONE RECEPTORS
V. K. Moudgil, Editor
© 1993 Birkhäuser Boston

and 1980s significant advances were made in the purification, characterization, immunology and molecular biology of steroid hormone receptors (SRs). The experimental data during the last decade indicated similarities between receptors for steroid hormones, thyroid hormones, vitamin D_3, retinoic acid and v-erb A oncogene. These receptors constitute the *steroid/thyroid* hormone receptor *superfamily*, whose membership is still on the rise (O'Malley, 1990). Added to this excitement was the discovery of a number of proteins which possess distinct homologies in their primary structure with the members of the steroid/thyroid superfamily but have no known ligands or defined functions (O'Malley & Conneely, 1992). These *orphan* receptors bring reservoir of potential in examining regulation and evolution of the members of the superfamily. The most recently, a number of accessory transcription factors have been identified that are required for the activation of steroid-responsive target genes (Yoshinaga et al, 1992). Despite these advances, the precise mechanism of steroid hormone action is only incompletely understood.

Since the publication of the first volume on the general aspects of steroid action in health and disease (Moudgil, 1988), a vast literature has accumulated on a number of emerging topics dealing with cellular, molecular and clinical aspects of steroid hormone action. While the purpose of this chapter is to provide a general view of the more widely recognized concepts in steroid hormone action, a brief review of the advances that are detailed in the accompanying chapters was attempted to provide a ready reference source for the new investigators.

Identification of Steroid Receptors: A Historical Perspective

Studies in early 1960s by Elwood Jensen and co-worker employed radioisotopic estrogen to established the target tissue concept due to a selective accumulation of the steroid in responsive tissues. An initial characterization of specific [3H]estradiol binding in the rat uterus was accomplished by mid 1960s (Toft & Gorski, 1966). These and a number of other related studies from the laboratories of Jensen and Gorski helped form the concept of a *two-step model* of steroid hormone action (Jensen et al, 1968; Gorski et al, 1968). Accordingly, the effects of steroid hormones on specific protein synthesis are mediated via specific receptor proteins, which exist in the cytoplasm of target cells. Under physiological conditions, the binding of a steroid hormone to its cytoplasmic receptor results in some yet undefined conformational changes in the SR-complex (SRc), which are collectively termed as *activation*. The term *transformation* has also been used to represent the same process(es).

Activation/transformation leads to *translocation* of SRs into the nucleus where the complexes interact with DNA and chromatin to influence cellular protein synthesis. Since 1984, the two step model has undergone reevaluation and some of its basic principles have been modified.

Cellular location of SRs. Unoccupied SRs may be extracted from the nucleus or cytoplasm depending upon whether the tissue extracts are concentrated or diluted with buffer (Sheridan et al, 1979). The results of immunocytochemical analysis and enucleation of hormone-free intact cells demonstrated that a majority of the unoccupied SRs could be found in the nuclei or the nucleoplasts of the enucleated cells (King & Greene, 1984; Welshons et al, 1984). The nuclear localization of unoccupied receptors for other steroids has also been reported (Gasc et al, 1984; Welshon et al, 1985; Perrot-Applant et al, 1985). Cellular localization and distribution of glucocorticoid receptors (GR) is not established with certainty as many studies have demonstrated the presence of unliganded GR both in the cytoplasm and in the nucleus (Antakly & Eisen, 1984; Wikstrom et al, 1987).

The recovery of SRs in the cytosol fraction of tissue or cell homogenates may represent that population of receptor which is loosely associated with the nucleus. The hormone binding may cause a tighter binding of the SRc with the nuclear sites. The conformational changes, earlier referred to as activation or transformation, would simply represent intranuclear event(s) in the modified model of steroid hormone action.

Activation or Transformation of Steroid Receptors

Effect of ligand binding. Upon homogenization, the SRs are recovered in the high speed supernatant fraction (cytosol) where they can be incubated at low temperatures with hormonal ligands to form *non-transformed* complexes. SRs in such complexes are known to occur in large 7-8 S oligomeric form and exhibit poor affinity for the nuclear sites (Jensen & DeSombre, 1973; Milgrom et al, 1973). The presence of ligand may protect or stabilize steroid binding sites under thermal inactivation conditions (Moudgil, 1987) and/or influence affinity of receptor for DNA (Skafar & Notides, 1985). In addition, the physico-chemical properties of the ligand-free receptor may also be different from liganded SRc and the ligand binding itself may induce alterations in receptor structure, which may precede other modifications prior to a final cellular response to the hormone (Hansen & Gorski, 1985). Although transformation *in vitro* of chicken oviduct or calf uterine progesterone receptor (PR) can be accomplished in the absence of added ligand, the extent of transformation is generally lower (Moudgil & Hurd, 1987; Moudgil et al, 1981, 1985). The work in the author's laboratory had previously demonstrated that the 9S (untransformed) to 4S (transformed) conversion of rat liver GR was absolutely dependent on the occupancy of steroid binding site by ligand (Moudgil et al, 1986a). We noted further that ligand-dependent structural change(s) in the rat liver GR at 0°C are distinct from those caused by transformation (Moudgil et al, 1987). Ligand binding may, therefore, independently induce conformational changes in the receptor which precede or accompany changes induced by other processes in receptor conformation (Hansen & Gorski, 1986).

Transformation of SRs by other agents. Increased nuclear or DNA binding capacity (transformation) can be induced *in vitro* by increasing the ionic strength of the medium, by incubating the receptor preparations at elevated temperatures in the presence of hormone, and by incubation of the cytosol at 0-4°C with 10 mM ATP and other agents (Milgrom et al, 1973; John & Moudgil, 1979; Moudgil & John, 1980a; Moudgil et al, 1981, 1985, 1986, 1987). The transformed 4-5 S receptor exhibits increased affinity for isolated nuclei, chromatin, DNA-cellulose, phosphocellulose, and ATP-Sepharose. The transformation of SRc in intact target cells has also been described under physiological conditions (Munck & Foley, 1979).

Involvement of phosphorylation in SR transformation. Receptor transformation may involve structural and/or conformational alterations induced in the hormone binding proteins or associated components by phosphorylation-dephosphorylation reactions. It has been widely demonstrated that phosphatase inhibitors (molybdate, tungstate, and vanadate) selectively stabilized SRc and blocked their thermal activation to a DNA or ATP-Sepharose binding form(s) (Nishigori & Toft, 1980). The rate of transformation of GR was also reported to be stimulated by incubation with calf uterine alkaline phosphatase (Barnett et al, 1980). These observations formed the basis of the suggestion that transformation of SRs involves dephosphorylation of the receptor itself or of some regulatory component(s) (Leach et al, 1979; Barnett et al, 1980).

Results of some other studies have suggested that actions of molybdate on SRs are direct and may not exclusively be related to their dephosphorylation (Murakami & Moudgil, 1981; Housley et al, 1982). Furthermore, phosphatase inhibitors (levamisole or fluoride), were not effective in blocking receptor transformation, and high concentrations of alkaline phosphatase converted receptor to a transformed form (Nishigori & Toft, 1980; Barnett et al, 1980; Moudgil et al, 1980). The initial clues that molybdate and tungstate had a direct effect on SRc came from studies reported from the authors laboratory (Murakami & Moudgil, 1981, 1982). DNA binding of activated GR was seen to be inhibited by preincubation of receptor with the above transition metal ions, which were effective in extraction of DNA-cellulose-bound or nuclear-bound receptors. Additional work is needed to examine the role of dephosphorylation in SR transformation.

Association of Steroid Receptors With Heat Shock Protein (hsps)

In the hormone-free cellular environment, SRs can be extracted from target tissue cytosols as large protein complexes of around M_r 300,000 and a sedimentation coefficient of 8-10 S. SRs can be stabilized in their native heteromeric structure in the presence of sodium molybdate, an agent shown to block transformation of SRc to their slower sedimenting, DNA binding form (Toft & Nishigori, 1979; Leach et al, 1979; Nishigori & Toft, 1980). These properties of SRs assisted in the isolation of chicken PR (cPR) in its 8 S, molybdate-stabilized forms (Puri et al, 1982; Renoir et al, 1982). PR in these preparations was associated a major 90 kDa protein. Toft and colleagues demonstrated, in these preparations, the existence of two 8 S receptor forms: each associated with a 90 kDa protein (Dougherty and Toft, 1982a). The 90 kDa protein did not bind progesterone and was not associated with the purified PR. In addition, the 90 kDa protein was present in excess over PR-A and B proteins, was absent from receptor preparations purified in the absence of molybdate, and existed only as a component of the 8-10 S SR forms (Dougherty & Toft, 1982; Birnbaumer et al, 1984).

The observations were extended to other systems demonstrating the 90 kDa protein to be a common, non-hormone binding component of the untransformed chick oviduct receptors of four steroid hormones (Joab et al, 1984). The SR-associated 90 kDa protein was later shown to exist in a wide range of receptor systems, tissues and species (Housley et al, 1985; Riehl et al, 1985; Mendel et al, 1986; Renoir et al, 1986) and was identified as a major hsp (Schuh et al, 1985; Sanchez et al, 1985; Catelli et al, 1985). In contrast to the reported behavior of other members of the nuclear receptor superfamily, receptors for thyroid hormone and retinoic acid do not form stable complexes with hsp90 or bind with high affinity to the heterocomplex.

Other receptor-associated proteins. Association of other non-steroid-binding proteins with the hormone-binding components of SRs has been widely reported (Murayama et al, 1980; Kost et al, 1989; Tai et al, 1986; Sanchez, 1990). Recently, presence of a 59 kDa immunophilin was reported in association with SRs (Lebeau, et al, 1992; Tai et al, 1992). A 72 kDa protein that co-purified with the rat liver GR, and which appears to be bound to the transformed DNA-binding form of the receptor, has been identified by Gustafsson and co-workers (Wrange et al, 1986). A 76 kDa protein that co-purified with the human PR has also been reported (Estes et al, 1987). Toft and co-workers (see Chapter 9 in this volume) have recently shown that five nonreceptor proteins specifically co-purify with unactivated cPR; these proteins have molecular masses of ~ 90, 70, 54, 50 and 23 kDa (Smith et al, 1990a,b). The 90 and 70 kDa proteins were previously identified as members of hsp90 and hsp70 protein families (Kost et al, 1989). Except for the 70 kDa protein, the other four proteins could be dissociated from cPR by salt in a process that is molybdate-sensitive. Affinity labeling of hsp70 with azido[^{32}P]-ATP demonstrated PR-associated hsp70 to be an ATP binding protein (Kost et al, 1989).

The recent success in the reported reconstitution of SRs with hsp90 and hsp70 indicate that the phenomenon may be biologically relevant (Smith et al, 1990a,b; Scherrer et al, 1990; Inano et al, 1990). The reconstitution of cPR depleted of hsp70 and hsp90 by treatment with 0.5 M KCl and 10 mM ATP, upon incubation in rabbit reticulocyte lysate at 30°C in the absence of progesterone, was observed to be consistent with the *in vivo* effects of progesterone in promoting hsp dissociation (Smith et al, 1990a,b). Previous studies from this laboratory have shown an interaction between SRs and immobilized ATP (Moudgil & John, 1980; Moudgil et al, 1981). It has also been shown that ATP-Sepharose binding is a property of transformed SRs (Miller & Toft, 1978; Moudgil & John, 1980b). Furthermore, addition of free ATP to cytosol preparations has been shown to transform the 8S non-DNA binding forms of SRs to their DNA binding 4S forms (John & Moudgil, 1979; Moudgil & John, 1980a; Moudgil et al, 1981). It is, however, not clear whether the effects of ATP discussed above are mediated via hsp70 or other receptor-associated or co-purified proteins.

Significance of the association of hsp90 with SRs. Bresnick et al (1989) have presented evidence that hsp90 is necessary for imparting GR a conformation that promotes steroid binding. It has also been suggested that hsp90 is associated with tubulin-containing complexes in L-cell cytosol and in intact PtK cells (Sanchez et al, 1988). Binding of 8 S GR to actin filaments through the hsp90 was also reported recently (Miyata & Yahara, 1991). Hsp90 may be involved in the regulation of SR function by capping its DNA binding site to prevent it from binding to hormone response elements (HREs) and thus block the transcription of regulatory genes (Baulieu, 1989). Yamamoto and colleagues have recently reported that hsp90 is involved in the signal transduction pathway for SRs, and that the latter fail to function optimally if cellular hsp90 levels are reduced (Picard et al, 1990). Hsp90 may be a critical factor in maintaining GR in a nonfunctional state (Codepond et al, 1991).

Baulieu (1989) has proposed that a hormone agonist may induce the dissociation of the receptor oligomer to unmask the functional DNA binding domain. RU 486, the PR antagonist, may stabilize the 8 S form of PR that is unable to dissociate from hsp90 to elicit a hormonal response. Accordingly, transformation may result from dissociation of the 90 kDa protein exposing DNA binding site masked in the 8 S receptor form. This proposal is consistent with the observation made in our laboratory (Moudgil & Hurd, 1987) that the ability of calf uterine PR to undergo 8 S to 4 S transformation *in vitro* is impaired when the receptor is occupied by RU 486. The molybdate-stabilized nontransformed GRc also contains a 90 kDa non-steroid binding phosphoprotein that is lost on transformation (Mendel et al, 1986). The phenomenon of the loss of 90 kDa protein from the heteromeric complex has also been observed in intact cells (Mendel et al, 1986). A number of tyrosine kinases which are products of viral oncogenes, including the Rous Sarcoma virus-transforming tyrosine protein kinase (PK), have been shown to form complexes with hsp90 (Schuh et al, 1985). It is also possible that the association of hsp90 with both viral tyrosine kinases and SRs may render these into inactive complexes. Dissociation of hsp90 from these complexes may convert the complexes into their active DNA binding forms. Alternatively, dissociation of hsp90 and conversion of SRc to 4 S form may not be sufficient to make a SR competent for transactivation, and that additional conformational alteration(s) in the receptor molecule must precede the activation of target genes by PR (Bagchi et al, 1990).

Composition of Steroid Receptors and Significance of Hsps

The heteromeric structure of chicken PR. Two hormone-binding components of cPR have been described and designated PR-A and PR-B proteins based on their differential elution from ion-exchange columns (Schrader & O'Malley, 1972). These two 4S receptor proteins have been isolated and characterized, showing molecular weights of about 79,000 (PR-A) and 108,000 (PR-B) (Schrader et al, 1981). The cloning of partial and entire cDNA sequences of the cPR have been accomplished (Conneely et al, 1986, 1987a; Jeltsch et al, 1986; Gronemeyer et al,

1987). The polypeptide product expressed from the cDNA has been shown to display functional activities characteristic of the classical cPR (Conneely et al, 1987a). PR-A and PR-B are produced from a single mRNA transcript by initiation of translation at alternate sites (Conneely et al, 1987a, 1989; Gronemeyer et al, 1987). Truncated mRNAs lacking the receptor B protein translational signal were shown to be capable of generating the A receptor by initiation of translation at a second internal start site. Thus, PR-A appears to be a truncated version of PR-B; the two proteins differ by an additional 128 amino acids located at the N-terminus of the B protein. Both proteins appear to functionally activate a progesterone-responsive target gene (Carson et al, 1987; Tora et al, 1988; Conneely et al, 1989). Results of recent studies indicate that the PR isoforms may originate from different transcripts, and that both human and cPR isoforms A and B are translated from different mRNAs (Kastner et al, 1990).

PR: an evolving heteromeric complex. The composition of Chicken PR under different experimental conditions is described by Toft and Colleagues (Smith et al, Chapter 9). The avian PR occurs in the complexes along with hsp90, hsp70 and three recently identified proteins, p54, p50 and p23 (Smith et al, 1990a,b). Although association of different proteins with SRs now appears to represent a general phenomenon, the authors caution that it is still speculative whether such association occurs in a cell and is of functional significance.

The avian PR salt-stripped of its associated proteins can be reconstituted with endogenous hsp90 when suspended in rabbit reticulocyte in a time and temperature-dependent process that requires ATP, Mg^{++} and K^+. Under sub-optimal ATP requirements, association *in vitro* of another protein, p60, with PR complex is described. The authors suggest that p60 may be a transient participant in the assembly of PR complex. Assembly of PR and dissociation of its associated proteins has been summarized in a model describing the dynamic process.

Heat shock and immunosuppression. Like hsp90, p59 is a non hormone binding 59 kDa protein which has been reported to occur as a part of the heteromeric structure of untransformed receptors for estrogen, progesterone glucocorticoid and androgen. It does not interact directly with the SRs. Lebeau et al (Chapter 10) have characterized association of p59 with hsp90 in the 8-9 S SRc. The p59 cDNA was cloned and sequenced. Although the deduced sequence of 458 amino acids appeared unique, there was a 55% homology seen in the 100 N-terminal amino acids to peptidyl-propyl isomerase and FK506 binding protein (FKBP); the latter is known to interact with immunosuppressants FK506 and

Rapamycin. By employing hydrophobic cluster analysis, three similar globular domains with an addition tail region that could bind calmodulin were identified. The domains identified were analogous to FKBP-12. The first domain of p59 was superimposable on FKBP-12 and the ligand (immunosuppressant) binding domains were almost identical. The authors propose that p59 is an immunophilin that can associate with hsp90 and have given it a name, hsp-binding immunophilin (HBI). The p59/HBI may be involved in alterations in conformation and trafficking of intracellular proteins such as SRs.

Intracellular trafficking and hsps. The questions of the cellular dynamism of hsps has been addressed by Pratt and Scherrer (Chapter 8). The authors have asked the question as to how the largely cytoplasmic hsps interact with SRs (which are either predominantly nuclear or arrive at the nuclear sites upon hormone binding). It is proposed that in those cells where GR is present in the cytoplasm in the absence of hormone, its subsequent entry into the nucleus and binding to HRE upon hormone binding must involve a trafficking system. The SR may attach to the hsp heterocomplex which may serve as a transport particle for intracellular trafficking.

According to Pratt and Scherrer, SRs remain "docked" to hsp90-containing protein complex until binding of hormone triggers its dissociation from hsp90 component to allow its high affinity interaction with transcriptional machinery. Docking to hsp complex may provide regulation against transcriptional activity in the absence of hormone. The authors propose that unliganded GR may remain cytoplasmic prior to its hormone-driven transport to the nuclear sites. Since hsp70 is not present in the cytoplasmic GR heteromer, its association may be facilitated by the nuclear environment.

Its has been proposed that GR contains two nuclear localization signals, both are hormone-dependent in intact cells and could transport the receptor into the nucleus. The receptor appears to be bound to the hsp heterocomplex during its migration through both cytoplasm and the nucleus. It is speculated that SRs bind to hsp90 transport machinery immediately upon their synthesis. In hormone-free conditions, the nuclear signal is "engaged" to facilitate nuclear localization. The GR, which is cytoplasmic, remains docked with hsp90 and requires hormone for freeing it for migration into the nucleus. The chapter provides detailed discussion on receptor trafficking, relationship between the hsp heterocomplex and protein movement to the plasma membrane, bidirectional movement of proteins bound to hsp90 and involvement of cytoskeletal pathways in movement of SRs in cytoplasm and nucleus.

Developmental regulation of hsps. Heat shock response appears to be a general phenomenon, and hsp90 and hsp70 are amongst the most prominent hsps. Shyamala (Chapter 11) has addressed the issues of physiological relevance and significance of SR:hsp interactions. It is hypothesized that steroid hormones and cellular hsp90 may mutually regulate their cellular functions. To test this hypothesis, the author's laboratory has studied the developmental aspects of estrogenic regulation of hsp90 gene expression in uterine and mammary tissues. The developmental induction of hsps appears to be distinct from the one induced by heat shock as is evidenced by the selective estrogenic regulation of uterine hsp90 but not hsp70. This increase in hsp90 appears to be intrinsic to the uterine tissue and was seen primarily in the endometrium.

The mammalian cells express two forms of hsp90 encoded by distinct genes. The murine analogs of the two hsp90 forms are hsp86 and hsp84. The cellular responsiveness to estrogen may be influenced by the relative magnitude of expression of hsp86 and hsp84. For example, hsp86 appears to dominate in the estrogenic environment while hsp84 is more abundant in murine cells in the absence of estrogen. Analysis of hsp90 in human endometrium during the menstrual cycle was also consistent with its estrogenic regulation since a progressive increase in steady state levels of hsp90 was observed during preovulatory phase. Furthermore, mammary glands of pregnant and lactating mice were also noted to express higher levels of hsp90. The development regulation of hsp90, however, was noted to be an estrogen-independent phenomenon. Preliminary data on alterations in the hsp90 levels in randomly selected human breast tumors are provided along with discussions on the significance of hsp90 detection/involvement in the clinical management of estrogen receptor (ER)-dependent mammary and endometrial cancers.

Regulation of Receptor Structure and Function by Phosphorylation

A potential regulation of SR function by phosphorylation was initially recognized by Munck and Brinck-Johnsen (1968) who observed that specific glucocorticoid binding in rat thymus lymphocytes was energy-depended, and that the loss or acquisition of hormone-binding capacity paralleled changes in the levels of intracellular ATP (Ishii et al, 1972; Rees & Bell, 1975). Based on the subsequent observations made on the effects of ATP on hormone binding (John & Moudgil, 1979; Sando et al, 1979) and activation of SRs (John & Moudgil, 1979; Moudgil & John, 1980; Barnett et al, 1980), it was suggested that the steroid binding as well as activation of SRs are influenced by phosphorylation-dephosphory-lation processes. Consequently, phosphorylation of SRs and members of the nuclear receptor superfamily has been reported in a number of systems. It is now generally agreed that SR are phosphoproteins (see reviews: Auricchio, 1989; Moudgil, 1990; Orti et al, 1992).

Amplification of phosphorylation by hormone administration. A role for phosphorylation in steroid hormone action is indicated in cases where this covalent modification enhances ligand binding by SRs (Denner et al, 1987; Sullivan et al 1988a,b; Moudgil, 1990). Phosphorylation of cPR *in vivo* has been widely demonstrated (Dougherty et al, 1982, 1984; Puri et al, 1984; Denner et al, 1987; Sullivan et al, 1988a, b; Nakao & Moudgil, 1989). The three major proteins of cPR (PR-A, PR-B, and hsp90) can be phosphorylated in the absence of progesterone, but hormone administration leads to increased phosphorylation of PR-A and PR-B proteins with a loss of hsp90 (Sullivan et al, 1988a,b; Nakao & Moudgil, 1989). Hormone-dependent phosphorylation also leads to an apparent decrease in the mobility of PR-A protein upon SDS-PAGE, the effect being termed "upshift" (Logeat et al, 1985; Horwitz et al, 1985, 1986). We have observed that the phenomena of upshift of phosphory-lated PR-A and the loss of hsp90 following hormone administration are progesterone specific: the effect cannot be mimicked by estradiol, cortisol or dihydrotestosterone (DHT) (Nakao & Moudgil, 1989). This hormone-dependent phosphorylation has been observed predominantly on serine residues located in the amino terminus of chicken and human PRs (Sullivan et al, 1988a, b; Sheridan et al, 1989; Denner, 1990b).

Other members of the steroid/thyroid nuclear receptor superfamily also undergo hormone-dependent phosphorylation, including GR (Orti et al, 1989, 1993; Hoeck & Groner, 1990), ER (Denton et al, 1992), AR (Van Laar et al, 1991), VDR (Pike & Sleater, 1985; Pan & Price, 1987; Brown & DeLuca, 1990). The significance of a phosphorylation-dependent post-translational maturation step in steroid hormone action is not well understood (Sheridan et al, 1989). While PR from T47D cells may undergo processing or nuclear down regulation (Wei et al, 1987), it has been have shown that transformation and nuclear down-regulation of human PR are independent of phosphorylation, but that nuclear phospho-rylation affects the interaction of receptors with transcription elements (Sheridan et al, 1988). Sequence specific DNA binding of rabbit uterine PR to the uteroglobin gene reportedly is influenced by receptor phospho-rylation (Bailly et al, 1986). Active hormone- and DNA-binding domains of the cPR have been separately expressed in *E. coli* (Eul et al, 1989; Power et al, 1990) and no post-translational modifications which would be unique to eukaryotic cells were required for the cPR to bind progestins with wild-type affinity (Eul et al, 1989). However, hormone-dependent phosphorylation that causes loss of hsp90 may regulate PR function in a eukaryotic target cell (Baulieu, 1989). Indeed, it has been shown that phosphorylation of cPR or other proteins in the transcription complex can modulate PR-mediated transcription *in vivo* (Denner et al, 1990a).

Phosphorylation in vitro by different kinases. It has been shown that SRs can be phosphorylated *in vitro* by a number of PKs (Migliaccio et al, 1984, 1989; Ghosh-Dastidar, 1986; Singh & Moudgil, 1984, 1985; Denner et al, 1987; Boyle & vander Walt, 1988; Hsieh, 1991, 1993; Nakao et al, 1992; Weigel et al, 1992; Huggenvik et al, 1993; Jerutka et al, 1993). The question as to whether SRs possess any intrinsic kinase activity is still a matter of debate. Receptors for progesterone (Garcia et al, 1983; Denner et al, 1987), glucocorticoids (Miller-Diener et al, 1985; Kurl & Jacob, 1984; Singh & Moudgil, 1984), and estrogen (Baldi et al, 1986) all have been reported to have PK activities, which were attributed either to copurified enzymes or considered intrinsic to the receptor protein. In some cases, the enzyme activities were reported to be separable from the SR upon more extensive purification (Denner et al, 1987; Sanchez & Pratt, 1986; Hapgood et al, 1986; Perisic et al, 1987; Garcia et al, 1986). Meggio and coworkers have reported co-purification of type II casein kinase with a major 90 kDa substrate in the rat liver cytosol (Meggio et al, 1985). It was suggested that at physiological ionic strength, the substrate and the kinase might form a complex. This 90 kDa substrate was later identified as the heat-shock protein which binds SRs (Dougherty et al, 1987). A kinase that binds to rabbit uterine PR was identified as casein kinase by Milgrom and co-workers (Logeat et al, 1987). However, there was no enhancement of kinase activity by treatment with the hormone and the phosphorylated receptor did not exhibit the characteristic "upshift" in its electrophoretic mobility. It is claimed that the presence of this PK in highly immunopurified preparations of rabbit PR is neither an artifact of the purification procedure nor of the cell homogenization (Savouret et al, 1989). Results of a recent study have suggested that cAMP-mediated phosphorylation regulates PR transcription activity *in vivo* (Denner et al, 1990a). Furthermore, a progesterone-induced phosphorylation site on cPR has been identified that contains a consensus sequence for the catalytic subunit of cAMP-PK. This site, which is located in the region between DNA and hormone binding domains in one of the two cPR transactivation domains, was preferentially phosphorylated by cAMP-PK *in vitro*. The observed effect of hormone treatment in vivo and possible involvement of an enzymatic activity that uses a SR as a substrate, has implications in understanding the regulation of progesterone action.

Functional analysis of PR phosphorylation. Weigel *et al.* (Chapter 12) have presented data on the phosphorylation of the cPR with an emphasis on its functional aspects. Although, both PR forms A and B contain all four of the phosphorylation sites identified, sites one and two are phosphorylated in the absence of hormone and undergo hyperphosphorylation upon hormone administration. There may be additional phosphorylation sites since the constitutive sites which turn over slowly or sites which represent nuclear function might have lower level of phosphorylation that is not easily detected. In this pursuit, a yeast expression system was developed for the synthesis of PR. In the absence of progesterone, only SER 211 and Ser 260 were phosphorylated as was the case with chicken oviduct slices. In the presence of progesterone an additional phosphopeptide at Ser 530 was detected in B receptor. When ^{32}P-incorporation was examined in the yeast and whole cell extract form chicken oviduct, another phosphopeptide at site Ser 367 was identified. These results raise the possibility that multiple kinases may be involved in the phosphorylation of receptor in the hormone-free and hormonally active cellular environments. Since regions Ser 211 and Ser 260 are dispensable as their deletion does not alter hormone binding of receptor expressed in yeast or mammalian cells, Weigel et al. conclude that cPR does not require phosphorylation for hormone binding.

Phosphorylation enhances transcriptional activation of receptor either directly or via altering activity of a factor. Chicken PR cotransfected into CV_1 cells can be activated by 8-Br cAMP or by okadaic acid in the absence of hormone (Denner et al, 1990a). Addition of progesterone with either 8 Br cAMP or okadaic acid resulted in activity that is higher than seen with progesterone alone. Clearly some members of the SR superfamily are capable of ligand-independent transcriptional activation.

Steroid Receptor Interaction with DNA and Nuclear Components

All SRs are segmented proteins that exhibit common structural and functional properties (Hollenberg et al, 1985; Miesfeld et al, 1986; Greene et al, 1986; Loosfelt et al, 1986; Conneely et al, 1986; Jeltsch et al, 1986; Misrahi et al, 1987; Arriza et al, 1987; Chang et al, 1988; Tilley et al, 1989). A SR is composed of three major domains. The amino-terminal part of the protein is an immunoreactive domain of variable length (Hollenberg et al, 1987). The DNA-binding domain of approximately 70 amino acids is a highly conserved region in the middle of the molecule.

The steroid-binding domain, containing around 250 amino acids, is also highly conserved and is located at the carboxyl-terminal end of the protein (Yamamoto, 1985, Hollenberg et al, 1987; Conneely et al, 1987; Carson et al, 1987a; Evans, 1988). Initially observed in an RNA polymerase III transcription factor, Zn^{++} fingers have been described as a common structural motif in the DNA binding domain of SRs and a number of nucleic acid binding proteins (Weinberger, 1985; Yamamoto, 1985). Binding of the steroid ligand appears to be a prerequisite for specific DNA binding and transcriptional activation *in vivo* (Carson-Jurica, 1990; Yamamoto, 1985).

Occupancy of the steroid binding site by the hormonal ligands confers on SRs the ability to act as *trans*-acting regulatory proteins capable of altering gene expression. The steroid-charged receptor complexes have been shown to interact specifically with promotor *enhancer* elements of target cell genes (Green & Chambon, 1986; Green et al, 1986). The enhancer or the hormone-responsive DNA sequences are located in the 5'-flanking region (Payvar et al, 1981; Pfahl, 1982; Govindan et al, 1982). Far upstream from the initiation site, the DNA binding site(s) have been located within introns of the transcription unit (Payvar et al, 1981; Moore et al, 1985). Accordingly, the transformed SRc bind preferentially to specific DNA sequences in the regions upstream from the trancriptional start site of hormone-regulated genes (Mulvihill et al, 1982; Compton et al, 1983; Karin et al, 1984; Groner et al, 1984; Scheidereit & Beato, 1984; Jost et al, 1984).

Discussions on the structural and functional aspects of DNA binding by steroid/nuclear receptors are presented by Freedman and colleagues in Chapter 2. The authors have described how the structure and function analyses have assisted them to predict characteristics of DNA binding domains of glucocorticoid-, estrogen- and Vitamin D_3 receptors. One can postulate how SRs discriminate by sequence specific contacts between closely related HREs. It is also possible to distinguish symmetry of target sites through protein-protein interactions. Taken together, Freedman et al present an explanation of SR regulation of diverse genes by a limited number of HREs.

Transcription factors and DNA bending. Interaction of SRs with specific sequences on DNA may result into conformational changes which are critical to the mediation of a hormonal signal. It has recently been shown that interaction of transcription factors with their cognate recognition DNA sequences results in the bending of DNA. The latter may be involved in transcriptional activation. Shapiro and colleagues (Chapter 3) have described their work with the transcription activation by

ER. Employing purified ER DNA binding domain expressed in *E. coli*, crude human ER or partially purified ER expressed in yeast, Shapiro et al demonstrate that ER interacts specifically with a series of ERE-containing DNA fragments inducing DNA bending at different angles. The degree of DNA bending appeared to be independent of ligand site occupancy of ER or any ER associated proteins. The demonstration that binding of ER from several sources induce conformational changes in DNA suggests an important role for DNA bending in ER-induced transcription activation.

Regulation of prolactin gene by ER. ER directly interacts with a DNA element(s) near or within the PRL gene to mediate estrogen-dependent stimulation of transcription. The estrogen-regulated element (ERE) identified in rat PRL (rPRL) gene has been located around 1.5 kb upstream from PRL transcriptional start site. Malayer and Gorski (Chapter 4) have described ER-induced transcriptional activation of rPRL gene and discussed mechanisms involved in ER action. The rPRL model system is extremely useful in understanding the estrogen-induced regulation of gene expression since prolactin is a major product of the pituitary gland which expresses it at a high level.

The authors' data suggest that ER binding causes alterations in the topology of DNA at a flanking enhancer site which are critical in estrogen-induced activation of rPRL gene. Although proteins may act to trap non-β DNA structure to regulate topology and influence gene transcription and DNA replication (Frank-Kamen et al, 1989), the authors propose that negative supercoiling may lead to formation of non-β DNA structures which trap ER and/or other nuclear proteins: the resulting thermodynamic energy could be then used for the regulation of DNA topology and chromatin structure. .

Tissue-specific transcription factor and gene expression. It has been recently shown that cellular response to a steroid may require cooperativity between SRs and other transcriptional factors (Scule et al, 1988). Since the number of transcriptional factors appears to be limited, different specific hormone response units could be generated by combination of various SRs with individual factors. Maurer et al (Chapter 5) provide evidence that ER acts in concert with tissue-specific homeodomain transcription factor Pit-1 to activate the PRL gene expression. While Pit-1 alone may be able to activate proximal promotor region under certain conditions, it is not sufficient to activate PRL distal enhancer. To achieve the basal activity of the distal enhancer and the hormone responsiveness, both ER and Pit-1 are required. The requirement for ER interaction with a tissue-specific factor, such as Pit-1, provides insurance against expression of PRL gene in tissues other than the pituitary. As to the mechanism of functional interactions between ER and Pit-1, ER binding sites in PRL gene may exist adjacent to the Pit-1 binding sites and that this proximation may facilitate their interaction with the distal PRL enhancer. ER and Pit-1 may also bind cooperatively to an adapter to allow interaction between distal and proximal regions of PRL gene and this looping may bring the unit in close proximity to the transcription initiation complex.

Vitamin D receptor and gene expression. The vitamin D metabolite 1,25-dihydroxyvitamin D_3 [1,25$(OH)_2D_3$] regulates cell growth and differentiation, immune function and many cellular activities associated with mineral metabolism. Pike (Chapter 6) provides evidence in support of receptor-mediated alteration of gene expression in response to 1,25$(OH)_2D_3$. The latter has been shown to upregulate osteocalcin (OC), a protein of adult bone abundantly expressed in osteoblasts. The OC gene is also regulated by retinoic acid and glucocorticoids, in addition to its regulation by 1,25$(OH)_2D_3$. Although its function is not clear, OC may have a role as an index of bone turnover. Pike and colleagues have delineated the organization of the OC gene and examined its basal functional activity under a variety of experimental conditions including mapping of vitamin D response elements (VDRE) within the OC gene promotor. A direct involvement of vitamin D receptor (VDR) in OC transcription was demonstrated by introducing an expression vector containing the receptor cDNA into VDR-negative cells. The interaction between VDR and DNA elements require the presence of a protein partner; the nuclear accessory factor (NAF). Incidentally, the formation of VDR-NAF heteromer does not require DNA. These observations raise the interesting possibility that thyroid hormone, retinoic acid and vitamin D function *in vivo* may involve heterodimeric complexes contrasting with the proposed homodimeric functional units of sex steroids and GRs. The chapter provides a detailed discussion on the properties of NAF and other SR accessory factors and the demonstration that retinoic acid receptor (RAR) also display NAF activity.

Receptors For thyroid hormone and retinoic acid. Recent success in cloning of specific intracellular receptors for thyroid hormone (T_3) and retinoic acid has significantly advanced our understanding of the molecular mechanisms that govern biological responses to thyroid hormones and vitamin A-derived hormones (e.g. retinoids). While mammalian T_3 receptors (TRs) are encoded by two genes, TRα and TRβ, RARs are encoded by at least six genes. RARs can be divided into two subfamilies that contain the classical RARs and the retinoid X receptors

(RXRs). TRs and both the subfamilies of RARs belongs to the subgroup of the nuclear receptor superfamily which possesses conserved DNA binding domains and also include COUP and VDR.

It has been assumed that SRs and other members of the nuclear receptor superfamily function by binding as homodimers to specific DNA regions (response elements). Pfahl and colleagues (Chapter 7) suggest that RAR/RXR heterodimers are required for specific DNA binding and transcription activity. RXRs act both as auxiliary receptors as well as independently in the presence of specific ligands (e.g. 9 cis retinoic acid). It is suggested that 9-CIS-retinoic acid binding leads to the formation of RXR homodimers that can activate a distinct retinoic response pathway. Interestingly, receptors for T_3 and retinoic acid can regulate transcription in a manner that allows their interactions with other transcription factors without binding to response elements.

The screening of cDNA libraries for RXRα related receptors revealed that the COUP receptors are negative regulators with a potential to restrict retinoic acid signaling to certain genes. Thus, COUP orphan may be involved in the age-related regulation of retinoic acid sensitive programs during development.

Ligand Requirements in DNA Binding and Transcriptional Activation

The functional domains of SRs have been shown to be highly conserved (Green & Chambon, 1986). Human and chicken ER appear to contain in their primary structure six distinct regions; the hydrophobic region E contains the hormone binding domain between amino acids 301 and 552 (Green et al, 1986). Under physiologic conditions, the occupancy of the hormone binding domain is thought to promote the receptor binding to DNA. The receptors which contain large deletions within the hormone binding domain fail to induce transcription suggesting that the hormone binding domain is indispensable for gene activation (Kumar et al, 1986). The function of a steroid binding domain, in the absence of hormone, appears to be to discourage receptor binding to the HRE, thereby preventing transcription.

While a steroid hormone binding may not determine the specificity of GR and PR binding to DNA *in vitro* (Willmann & Beato, 1986; Bailly et al, 1986), the presence of hormone appears to be essential for gene activation *in vivo* (Green et al, 1986). It is possible that the regions in the GR which are necessary for full transcriptional activation are not specifically involved in steroid or DNA binding (Giguere et al, 1986). The steroid-free GR binds specifically to MMTV DNA; and therefore, the function of hormone *in vivo* could be to modulate nuclear partitioning of the receptor (Willmann & Beato, 1986). The protein-DNA interaction in GRE, however, appears to require the presence of the hormone. Hormone-free GR has been noted to be unable to recognize specific response elements of a target gene *in vivo* (Becker et al, 1986). The hormone may, therefore, unmask a preformed DNA binding domain eclipsed in the unoccupied receptor, probably by a non-hormone binding component of SR heteromer.

Ligand requirement and DNA binding. Although ER is widely believed to be localized in the nucleus of target cells in the absence of the hormone, and in some instances the steroid-free receptor can also bind DNA, the exact role of ligand in ER function in the mammalian cells is not clearly understood. Reese and Katzenellenbogen (Chapter 15) show that hormone-free ER is capable of binding to reporter templates within intact cells but the presence of ligand enhances this interaction. The delineation of the role of the ligand in DNA binding becomes even more important when ERs occupied by estradiol or antiestrogen are found capable of DNA binding. The authors have employed a temperature-sensitive ER mutant, C447A, to gather evidence that the observed ligand-independent DNA binding of ER may be responsible for the known functional differences between the ER and other members of the nuclear receptor superfamily.

One clue to the role of ligand in ER function became apparent when the binding affinity of wild type and C447A receptors for estradiol was examined *in vitro* at 37°C. A 30-min incubation at 37°C caused 80-90% loss of specific hormone binding capacity of the mutant receptor while wild type receptor was stable during this period. The authors noted that the thermal instability of the mutant receptor was fully preventable by ligand-site occupancy. The ligand stabilizes the receptor protein or its dimerization within the cells to impart greater stability for ER-DNA interaction. The temperature instability of the mutant receptor observed *in vitro* may be related to observed impaired transactivation in transfected CHO cells discussed in the chapter. The DNA binding by steroid-free ER and the hormone-independent transactivation of weakly acting promoters occurs presumably via the TAF-1 function located in the N-terminus of the receptor. The presence of ligand enhances the above processes through ER binding to TAF-2 function.

Gene activation by ligand dependent/independent pathways. In addition to the ligand-induced promotion of DNA binding, reversible post-translational events such as phosphorylation/ dephosphorylation reactions have also been shown to regulate SR activities. Enhanced phosphorylation of SRs has been observed in the presence of receptor-specific cognate ligands. For example, mutation of a single amino acid 118 of human ER leads to reduced phosphorylation and receptor-mediated transcription. Whereas ligand-dependent activation of SRs may involve multiple phosphorylation steps, ligand-independent activation may be influenced by altered cellular phosphorylation pathways. O'Malley and colleagues have discussed ER activation by ligand-dependent and ligand independent pathways (Smith et al, Chapter 13). The ligand-independent activation, which the authors relate to receptor-dependent target gene transcription, has been discussed for cPR, human RAR-α, human GR, and orphan receptors. Although the exact mechanism is not known, ligand-independent activation, seen with many receptor systems, is not a general phenomenon.

Ligand-independent activation of SRs does not appear to be cell-type specific. Smith et al note that, like cPR, ER-dependent gene expression can be stimulated in CV_1 and HeLa cells by dopamine in the absence of estradiol. Furthermore the ligand-independent activation ability appears to be an intrinsic functional characteristic of ER pertaining to a region in the ligand binding domain. EGF has been shown to mimic estrogen-induced production of lactoferrin, an iron binding growth promoting glycoprotein. O'Malley and coworkers have noted that in EGF receptor-positive HeLa cells transiently transfected with human ER expression vector, ER, but not estrogen, was necessary for mediation of EGF-induced dose-dependent increase in reporter gene expression.

The authors provide insight into the possible mechanism(s) of ligand-independent activation by ER by employing ER antagonists, 4-hydroxy-tamoxifen (4HT) and ICI 164,384. Only the ligand-dependent estrogenic activation was blocked in the presence of 4HT (possibly by its blocking of TAF-12) but the ligand-independent activation could be achieved via TAF-1 located in the amino terminal.

The compound IC 164,384 is a pure antiestrogen and appears to block both the ligand-dependent and the ligand-independent activation of ER-mediated gene transcription. The significance of ligand-independent activation in the brain, the uterus, and the breast has been discussed citing examples that under conditions where the levels of steroids may be undetectable or undesired, transcriptional activity may be regulated by cellular agents (e.g. growth factors) capable of modulating cellular signaling pathways. The relevance of the above discussion is crucially important in the evaluation and treatment of estrogen-independent or tamoxifen-resistant tumors.

Steroid Receptors, Antihormones and Clinical Considerations

ER variants and drug resistance. Earlier efforts to correlate receptor presence with successful outcome of hormonal therapy in the breast cancer patients relied heavily on the detection and quantitation of SRs. With the advent of use of molecular biology in the analysis of clinical samples, qualitative analyses of the samples appeared to have become better predictors of the clinical management of breast and endometrial cancers. Presence of ER in the breast cancer biopsies is regarded as one of the most important factors predicting response to endocrine therapy. Whereas patients with positive ER expression in the cancerous tumors have 40-80% chance of response, several variations in ER could lead to resistance to antiestrogenic drugs. Encarnacion and Fuqua (Chapter 17) discuss the structure-function relationship of various ER regions and the occurrence and role of ER variants in breast cancer.

Tamoxifen, has been most widely used antiestrogen for post-menopausal women with breast carcinoma. The drug interacts with ER to block its action but without interfering with receptor dimerization or its interaction with DNA. However, long-term tamoxifen use is associated with drug resistance. This resistance could be either be due to loss or emergence of ER expressing cells or presence of ER RNA variants. The authors do caution that, although ER variants observed in their laboratory and by others, have potential relationship to development of tamoxifen resistance, it has not been definitely proven clinically. Other observations of importance in this regard are that wild type ER may be present along with variants and that the latter represent RNA splicing variants but not mutations in DNA. Variant ER mRNA with substitutions in A/B region, and C domain of ER have been identified. The authors' laboratory has observed an ER variant with a deletion of exon 5 in ER-negative/PR-positive breast cancer biopsies. It is suggest that since the variant mRNA codes for a 40 kDa protein that lacks a functional steroid binding domain, it could explain the basis for the tumors to be ER-negative. Another variant of 3' truncated ER has been identified which lacks exon 7 but is capable of binding to DNA. Regardless of their currently proven role in taxomifen-resistance, the identification and analysis of ER variants in breast cancer biopsies should facilitate approaches to clinical management of breast cancer.

Receptor agonist/antagonists and breast cancer. Many tumors, which are transformed from cells that normally grow under the influence of hormones, continue to be responsive to the circulating levels of these hormones. In many instances tumorigenesis may be accompanied by a loss/alteration of genetic information which results into formation of spontaneous cell variants. Many breast cancer cells become hormone resistance and do not express PR or ER. The success rate of endocrine therapy involving SR antagonists is significantly influenced in hormone-dependent breast cancers. The antagonists bind SRs to generate inactive complexes which fail to induce gene expression in hormone-responsive cells. However, the situation is more complicated when sometimes receptor antagonist start behaving as a SR agonist. Horwitz and Leslie (Chapter 14) have provided a detailed discussion on the possible mechanism and circumstances which influence SR antagonists to function as agonists. In addition to the molecular and cellular heterogeneity of receptors and tumor cells, the authors discuss their relationship to progression and resistance of breast cancer. The success in cloning of cDNA for ER, and delineating the structures of ER gene and the receptor protein has allowed analyses of polymorphic forms of ER gene and identification of alternate mRNA transcripts. Accordingly, ER structure and regulation in normal cells may be tissue-specific. Furthermore, malignant cells also express ER mutants described earlier.

The authors provide experimental data on cellular heterogeneity of PR observed with $T47D_{Co}$ cells, a subline of an ER-negative and PR-positive T47D cell line of human breast cancer. While explaining their tumor cell "remodeling" hypothesis, the authors note that although tamoxifen decreases PR levels in a majority of cells, it paradoxically increases PR levels on a select subpopulation. It is possible that the emergence of subpopulations is related to tumor progression and recurrence of breast cancer. In this context, the chapter provides discussions on the proposed mechanisms to explain acquired resistance to tamoxifen, progestin resistance, and natural PR isoforms. The authors caution about the use of tamoxifen as a chemoprevent in women at high risk of developing breast cancer since the molecular heterogeneity discussed above can lead to inappropriate or undesired responses to endocrine therapy.

RU486 and antiprogestins: basic mechanism(s) and clinical potential. Availability and widespread testing of RU486 (Mifepristone) has opened up new avenues to its use in basic research and clinical practice. Until 1990, it was the only steroid analog available with demonstrable antiprogestin activity. RU486 is now widely recognized as the effective antiprogestin and antiglucocorticoid: the two activities are discernable at difference steroid concentrations (Baulieu, 1989; Ulmann et al, 1990).

Lack of RU486 interaction with cPR. Earlier attempts to use RU486 as a chemical probe for investigating molecular mechanism of progesterone action resulted in some unexpected observations. The cPR, which is one of the more widely used systems employed to study progesterone action at the receptor level, failed to exhibit any affinity for RU486 (Groyer et al, 1985; Moudgil et al, 1986b: Eliezer et al, 1987). RU486 interacted in the chicken cytosol with a molecule which was immunologically distinct than cPR (Eliezer et al, 1987). It was subsequently shown that other recently synthesized antiprogestin, ZK98299, also did not bind cPR (Moudgil et al, 1991). These observations now represent a general phenomenon regarding the lack of interaction between RU486 and the antiprogestins: two additional antiprogestins, ORG 31806 and ORG 31710 also failed to bind cPR specifically (Mizutani et al, 1992). The primary structure analysis of cPR revealed that a substitution of a single amino acid, cysteine 575, by glycine in the hormone binding domain of cPR allows the resultant mutant to bind RU486 (Benhamou et al, 1992). The above findings have potential implications in cases where RU486 fails to induce abortion (nearly 1% women): such patients may possess receptors which bind progestins optimally but are resistant to the antiprogestins due to a possible mutation at position 722 that is required for RU486 binding (Baulieu, 1989; Benhamou et al, 1992).

Mechanism of RU486 action. Initial work from this laboratory and by others suggested that RU486-bound calf uterine PR had impaired ability to undergo 8 S to 4 S transformation (Moudgil & Hurd, 1987). Baulieu suggested that RU486 binding to PR stabilizes the 8 S nontransformed heteromeric complex; the presence in this complex of hsp90 (or other receptor components) precludes an interaction of PRc with its cognate HRE (Baulieu, 1989). We proposed that the inability of RU486-bound PRc to induce a functional response was due to the differential conformational changes induced in the receptor by the agonist and antagonist ligands (Moudgil et al, 1989). This view was supported by the finding that transcriptional inactivation of SRs by RU486 involves alterations in the conformation of receptor in the C-terminus of the steroid binding domain (Allan et al, 1992). There is some evidence that agonist and the antagonist ligands may bind to different regions within or around the hormone binding domain (Vegeto et al, 1992). Milgrom and colleagues have reported that receptors bound to RU486 form abortive complexes with hormone HREs (Guiochon-Mentel et al, 1988). These observations are supported by the finding that the antihormone-bound receptors can undergo transformation and interact with DNA but in a manner that is transcriptionally nonproductive (Pham et al, 1991).

Another clue to RU486 action comes from a study demonstrating that RU486-bound hPR interacts with HREs in a structurally altered form (El-Ashry et al, 1989). Further, Gronemeyer and colleagues have shown the presence of two transcription activation functions (TAFs) located in the N-terminal (TAF-1) or the hormone binding domain (TAF-2). While both the TAFs are active in the presence of agonists, TAF-2 was seen to be inactive in the presence of RU486 (Meyer et al, 1990). These findings indicate that RU486-charged PR is able to bind HRE but that this union does not result into a functional response.

Extra-gestational considerations of RU486. The principal target organs for progesterone action are the uterus, the breast and the brain. Largely known as an effective agent for termination of pregnancy, RU486 is being considered as a potential drug for treatment of a number of clinical conditions related to progestin abundance/influence and hyper-cortisolism. As a glucocorticoid antagonist, its potential is being evaluated for treatment of Cushing's Syndrome. PR-containing non-cancerous tumors of the brain (meningiomas) could also be regulated by RU486. RU486 has been shown to inhibit PR-mediated growth of breast cancer cells, and is not antiproliferative in PR-negative cells (Bardon et al, 1985; for details see review by Horwitz, 1992). Alternatively, depending upon the biological response that is being measured, RU486 may exhibit both the agonist and the antagonist actions in the cancer cells (Horwitz, 1985b).

Androgen receptor and relation to disease. The effects of biological-ly active androgens, testosterone and dihydrotestosterone, on the normal male sexual development are mediated via the androgen receptor (AR). With the recent success in the purification of AR and raising AR antibodies, cloning of human AR cDNA was accomplished by the Wilson and colleagues who have described molecular and biological aspects of human AR with respect to health and disease (Zhou et al, Chapter 16).

It is known that natural mutations in AR gene are associated with congenital and acquired human diseases. Since the AR gene is located on X chromosome, mutations in AR gene would be more detrimental than those that occur in autosomal SR genes. For example, a deletion of AR gene results in androgen insensitivity characterized by a female sex phenotype and absence of sexual hair. The chapter also describes a number of single base mutations associated with complete androgen resistance. Just recently it was discovered that defects in AR could lead to adult onset spinal/Bulbar muscular atrophy characterized by reduced fertility, gynecomastia and neuropathy.

A number of nonandrogenic steroids including some antiandrogens which display moderate binding affinity, cause nuclear transport of AR but addition of androgen causes a strong nuclear localization. Hydroxy-flutamide was the only compound, among the antiandrogens tested, that caused perinuclear localization seen in the absence of hormone. Not only was hydroxyflutamide devoid of agonist activity itself, it also inhibited agonist activity of a strong androgen agonist, R1881.

Although it is not established with certainty yet, but AR gene mutations can occur in prostate cancer cells. The oncogenic expression of mutant form of AR could be directed via different mechanisms. An androgen dependent human prostate cancer cell line LNCAP, described by the authors, expresses AR but not other SRs. Alteration of a single amino acid within the steroid binding domain alters steroid binding specificity and may cause an antagonist to act like an agonist. The latter postulation is supported by the observation that hydroxyflutamide causes increase in the transcriptional activity.

Is estrogen directly involved in osteoporosis? It has been widely believed estrogen action is a key component in the process of maintenance of mineral bone balance. It's role, however, was considered to be indirect until recently when ERs were identified in both bone-forming osteoblasts and bone-resorbing osteoclasts. Spelsberg and colleagues (Chapter 18) present their findings that suggest that estrogen plays a major role in coupling osteoclasts and osteoblasts functions; any alterations in the coupling process might be implicated in osteoporosis. The chapter, in addition to summarizing the author's contribution related to identification of ER in bone tissues, provides an overview of the related work performed by other laboratories.

Bone tissues are known to undergo remodeling at locations termed bone remodeling units and the bone mass requires a balance of remodeling of each of these units. Estrogen regulates this remodeling. During estrogen deficiency, an increase in the frequency of remodeling units occurs causing osteoclasts to construct deeper resorption spaces and inability of osteoblasts to refill these spaces. The functions of osteoblasts and osteoclasts may be coupled and deeper resorption may result in bone perforation that are more pronounced in postmenopausal women.

Employing more specific and sensitive techniques than were used previously, Spelsberg and coworkers have demonstrated the presence of not only ER but of receptors for androgens and glucocorticoids in the human osteoblasts. Although the exact role of estrogen in the bone is still evolving, it appears to act primarily to decrease bone reabsorption and not to influence its formation. Estrogen may act on osteoclasts and regulate c-fos/c-jun gene expression, inhibit lysosome gene expression and activities and cause an overall inhibition of bone resorbing activities of osteoclasts.

Dioxin, the environmental hazards, and the Ah receptor: Halogenated aromatic hydrocarbons are environment pollutants representing potential health hazards. 2, 3, 7, 8 Tetrachlorodibenzo-p-dioxin (TCDD) is a prototype aromatic hydrocarbon which elicits numerous species- and tissue-specific effects by inducing alterations in gene expression. Initial experimental evidence has shown that the aromatic hydrocarbon responsiveness is an inherited simple autosomal dominant trait, *the Ah locus*. The latter is a subject of review by Poland and Bradfield (Chapter 20).

The biological effects of the TCDD are mediated via the aryl hydrocarbon receptor (Ah receptor), an intracellular protein that interacts with TCDD with high affinity. The Ah receptor shares many common structural and functional features with SRs: it is recovered in its unliganded form in the cytosol fraction of target cells as a 280 kDa heterocomplex which include hsp90. Like SRs, the Ah receptor can be stabilized in the presence of sodium molybdate in its heteromeric form that retain greater ligand binding capacity. In addition, upon interaction with TCDD or its agonists, the Ah receptor exhibits tighter association with the nucleus that can be disrupted by high salt environment. The liganded nuclear Ah receptor interacts with dioxin response elements which are similar to the classical enhancer elements noted for SR.

Poland and Bradfield describe a differentiated hepatoma cell line, Hepa 1c7c7, as a model system employed to study Ah receptor-mediated induction of P-4501A1. Unlike the SRs, the liganded nuclear Ah receptor occurs as a heteromeric form comprising the Ah receptor and another protein Arnt (Ah receptor nuclear translocator) suggesting that genes other than those for the Ah receptor are involved in P-4501A1 induction. The purification and cloning of Ah receptor allowed Poland and colleagues to delineate structural analysis of the 100 kDa Ah receptor and the 110 kDa Arnt protein. Upon binding TCDD, the Ah receptor interacts with the Arnt protein to form a heterodimer which then binds to the enhancer sequence. The authors point to the different mechanism by which SRs act as transcriptional activators by recognizing specific DNA enhancer sites by two zinc fingers. The successful cloning of the Ah receptor and the Arnt proteins will facilitate studies on the physiological purpose of the Ah receptor and help search for an endogenous ligand for this receptor.

Steroid Receptors in Emerging Models/Systems

Over the years, many experimental model systems have been employed by investigators to gain insight into the structure-function relationship of ER. Normal mammalian cells express limited amounts of specific SRs, although neoplastic transformed cells express sufficiently large quantities of ER and PR. Recently several SR have been expressed in yeast but it has been difficult to obtain intact biological active SRs expression in *E. coli*. Wittliff and colleagues (Chapter 19) expressed the human ER gene in *E. coli* and yeast. The expressed ER retained its wild-type characteristics and exhibited isoforms similar to those seen in human breast cancer. The authors also observed specific association of the expressed ER with ERE. A distinct advantage of this approach is the ability to generate large quantities of receptor preparations to allow for crystallographic studies and receptor isoform analysis in the presence and absence of hormonal and antihormonal ligands.

SUMMARY, CONCLUSION AND FUTURE DIRECTIONS

Nearly thirty years after the initial characterization of ER, nature and characteristics of other SRs have been revealed. With the advent of the tools of molecular biology, cloning of different SRs has been accomplished. Many systems have been employed to express the cloned receptors and the diversity of expression systems available allows investigators to perform analysis of the different domains of receptors to attribute specific functions. Nearly all the SRs have now been characterized with respect to their cellular location, tissue distribution and alterations during various clinical and experimental conditions. These advances have been made possible due to the availability of both polyclonal and monoclonal antibodies raised against SRs or specific receptor regions. The predictions of 1960s that the detection of SRs human biopsies could aid in the endocrine treatment of hormone-dependent tumors have born fruits. Antihormones, which work at receptor level, have been employed in the treatment of postmenopausal breast cancers. Similar approaches have been used or are being tried in relation the use of antihormones in conditions which seem to be influenced by the circulating levels of androgens, progestins and glucocorticosteroids. An unprecedented social, political and scientific debates about the current use of the antiprogesin-antiglucocorticoid RU486 for fertility regulation has brought to focus the power and potential of steroids. Amongst all these developments and discussions, the central issue of how steroid hormones and other members of the steroid/thyroid hormone superfamily work at molecular level has remained unsolved. The future efforts, therefore, should not only be invested in the learning more about the molecular mysteries of SRs but also in bridging the gaps between the members of the superfamily in the hope that common, simpler mechanisms may be revealed to explain their mode of action.

The chapters in this volume, written by the pre-eminent and acknowledged experts, have provided lively discussions on the important questions: Is hormone limited merely to binding to the receptor, or is it involved in exposing certain domains of ligand-free receptor which are eclipsed by cellular components? Does it merely navigate the SRs to select sequences on the hormone-responsive genome? The authors have described modifications of receptor structure that influence its cellular regulation, thus providing knowledge about the flexibility of the target cell response to reversibly controlled hormone-dependent alterations.

The advances during the last decade followed the functional or mechanistic analyses of individual receptors. The picture is a bit more complicated with the discovery that the functions of SRs may be regulated by their association with a number of cellular proteins, the hsps. These observations are compounded with the revelations regarding the presence of orphan receptors. The search for the biological ligands for the latter must proceed at a rate that is compatible with the investigations on the growth factors and cellular signalling pathways which influence steroid hormone actions. Have all the mediators of chemical signals evolved from a common primordial molecule? Answers to the precedding questions will provide an investigator an overview of the receptors which is inclusive of the knowledge about the identity of the individual members of the superfamily and their applications to the issues related to health and disease.

ACKNOWLEDGMENTS

The work in the authors laboratory is supported by National Institutes of Health grant DK 20893. The excellent secretarial assistance by Ms. Rita Perris in the preparation of this chapter is gratefully acknowledged.

REFERENCES

Allan GF, Leng X, Tsai SY, Weigel NL, Edwards DP, Tsai M-J and O'Malley BW (1992): Hormone and antihormone induce distinct conformational changes which are central to steroid receptor activation. *J Biol Chem* 267: 19513-19520.

Antakly T and Eisen HJ (1984): Immunocytochemical localization of glucocorticoid receptor in target cells. *Endocrinology* 115: 19841989.

Arriza JL, Weinberger C, Carelli G, Glaser TM, Handelin BL, Housman, DE and Evans RM (1987): Cloning of human mineralocorticoid receptor complementary DNA: Structural and functional kinship with the glucocorticoid receptor. *Science* 237: 268-275.

Auricchio F (1989): Phosphorylation of steroid receptors. *J Steroid Biochem* 32: 613-622.

Bagchi MK, Tsai SY, Tsai MJ and O'Malley BW (1990): Identification of a functional intermediate in receptor activation in progesterone-dependent cell-free transcription. *Nature* 345: 547-550.

Bailly A, Le Page C, Rauch M and Milgrom E (1986): Sequence-specific DNA binding of the progesterone receptor to the uteroglobin gene: Effects of hormone, antihormone and receptor phosphorylation. *EMBO J* 5: 3235-3241.

Baldi A, Boyle DM and Wittliff JL (1986): Estrogen receptor is associated with protein and phospholipid kinase activities. *Biochem Biophys Res Commun* 135: 597-606.

Bardon S, Vignon F, Chalbos D and Rochefort H (1985): RU486, a progestin and glucocorticoid antagonist, inhibits the growth of breast cancer cells via the progesterone receptor. *J Clin Endocrinol Metab* 50: 692-697.

Barnett CA, Schmidt TJ and Litwack G (1980): Effects of calf intestinal alkaline phosphatase, phosphatase inhibitors and phosphorylated compounds on the rate of activation of glucocorticoid-receptor complexes. *Biochemistry* 19: 5446-5455.

Baulieu EE (1989): Contragestation and other clinical applications of RU486, an antiprogesterone at the receptor. *Science* 245: 1351-1357.

Becker PB, Gloss B, Schmid W, Strahle U and Schutz G (1986): *In vivo* protein-DNA interactions in a glucocorticoid response element require the presence of the hormone. *Nature* 326: 686-688.

Benhamou B, Garcia T, Lerouge T, Vergezac A, Gofflo D, Bigogne C, Chambon P and Gronemeyer H (1992): A single amino acid that determines the sensitivity of progesterone receptors to RU486. *Science* 255: 206-209.

Birnbaumer M, Bell RC, Schrader WT and O'Malley BW (1984): The putative molybdate-stabilized progesterone receptor subunit is not a steroid-binding protein. *J Biol Chem* 259: 1091-1098.

Boyle DM and van der Walt LA (1988): Enhanced phosphorylation of progesterone receptor by protein kinase C in human breast cancer cells. *J Steroid Biochem* 30: 239-244.

Bresnick EH, Dahlman FC, Sanchez ER and Pratt WB (1989): Evidence that the 90-kDa heat shock protein is necessary for the steroid binding conformation of the L-cell glucocorticoid receptor. *J Biol Chem* 264: 4992-4997.

Brown TA and Deluca HF (1990): Phosphorylation of the 1,25-dihydroxyvitamin D_3 receptor. A primary event in 1,25-dihydroxyvitamin D_3 action. *J Biol Chem* 265: 10025-10029.

Carson-Jurica MA, Schrader WT and O'Malley BW (1990): Steroid receptor family: Structure and functions. *Endocrine Rev* 11: 201-220.

Carson MA, Tsai MJ, Conneely OM, Maxwell BL, Clark JM, Dobson ADW, Elbrecht A, Toft DO, Schrader WT and O'Malley BW (1987): Structure function properties of the chicken progesterone receptor A synthesized from complementary deoxyribonucleic acid. *Mol Endocrinol* 1: 791-801.

Catelli MG, Binart N, Feramisco JR and Helfman DM (1985): Cloning of the chick hsp 90 cDNA in expression vector. *Nucl Acids Res* 13: 6035-6047.

Chang C, Kokontis J and Liao S (1988): Structural analysis of a complementary DNA and amino acid sequences of human and rat androgen receptors. *Proc Natl Acad Sci USA* 85: 7211-7215.

Codepond F, Schweizer-Groyer G, Segard-Maurel I, Jibard N, Hollenberg SM, Giguere V, Evans RM and Baulieu EE (1991): Heat shock protein 90 as a critical factor in maintaining glucocorticoid receptor in a nonfunctional state. *J Biol Chem* 266: 5834-5841.

Compton JG, Schrader WT, O'Malley BW 1983: DNA sequence preference of the progesterone receptor. *Proc Natl Acad Sci USA* 80: 16-20.

Conneely OM, Sullivan WP, Toft DO, Birnbaumer M, Cook RG, Maxwell BL, Zarucki-Schulz T, Greene GL, Schrader WT and O'Malley, BW (1986): Molecular cloning of the chicken progesterone receptor. *Science* 233: 767-770.

Conneely OM, Dobson ADW, Tsai MJ, Beattie WG, Toft DO, Huckaby CS, Zarucki T, Schrader WT and O'Malley BW (1987a): Sequence and expression of a functional chicken progesterone receptor. *Mol Endocrinol* 1: 517-525.

Conneely OM, Maxwell BL, Toft DO, Schrader WT and O'Malley BW (1987b): The A and B forms of the chicken progesterone receptor arise by alternate initiation of translation of a unique mRNA. *Biochem Biophys Res Commun* 149: 493-501.

Conneely OM, Kettelberger DM, Tsai MJ, Schrader WT and O'Malley BW (1989):The chicken progesterone receptor A and B isoforms are products of an alternate translation initiation event. *J Biol Chem* 264: 14062-14064.

Denner LA, Bingman III WE, Greene GL and Weigel NL (1987): Phosphorylation of the chicken progesterone receptor. *J Steroid Biochem* 27: 235-243.

Denner LA, Weigel NL, Maxwell BL, Schrader WT and O'Malley BW (1990a): Regulation of progesterone receptor-mediated transcription by phosphorylation. *Science* 250: 1740-1743.

Denner LA, Schrader WT, O'Malley BW and Weigel NL (1990b): Hormonal regulation and identification of chicken progesterone receptor phosphorylation sites. *J Biol Chem* 265: 16548-16555.

Denton RR, Koszewski NJ and Notides AC (1992): Estrogen receptor phosphorylation. Hormonal dependence and consequence on specific DNA binding. *J Biol Chem* 267: 7263-7268.

Dougherty JJ and Toft DO (1982): Characterization of two 8S forms of chick oviduct progesterone receptor. *J Biol Chem* 257: 3113-3119.

Dougherty JJ, Puri RK and Toft DO (1984): Polypeptide components of two 8 S forms of chicken oviduct progesterone receptor. *J Biol Chem* 259: 8004-8009.

Dougherty JJ, Rabideau DA, Iannotti AM, Sullivan WP and Toft DO (1987): Identification of the 90-kDa substrate of rat liver type II casein kinase with the heat shock protein which binds steroid receptors. *Biochim Biophys Acta* 927: 74-80.

Eliezer N, Hurd C and Moudgil VK (1987): Immunologically distinct binding molecules for progesterone and RU 38486 in the chick oviduct cytosol. *Biochim Biophys Acta* 926: 34-39.

Estes PA, Suba EJ, Lawler-Heavner JL, El-Ashry D, Wei LL, Toft DO, Sullivan WP, Horwitz KB and Edwards DP (1987): Immunological analysis of human breast cancer progesterone receptors. 1. Immuno-affinity purification of transformed receptors and production of monoclonal antibodies. *Biochemistry* 26: 6250-6262.

El-Ashry D, Onate SA, Nordeen SK and Edwards DP (1989): Human progesterone receptor complexed with the antagonist RU486 binds to hormone response elements in a structurally altered form. *Mol Endo* 3: 1545-1558.

Eul J, Meyer ME, Tora L, Bocquel MT, Quirin-Sticker C, Chambon P and Gronemeyer H (1989): Expression of active hormone and DNA binding domains of the chicken progesterone receptor. *EMBO J* 8: 83-90.

Evans RM (1988): The steroid and thyroid hormone receptor superfamily. *Science* 240: 889-895.

Garcia T, Tuohimaa P, Mester J, Buchou T, Renoir JM and Baulieu EE (1983): Protein kinase activity of purified components of the chicken oviduct progesterone receptor. *Biochem Biophys Res Commun* 113: 960-966.

Garcia T, Buchou T, Renoir JM, Mester J and Baulieu EE (1986): A protein kinase copurified with chick oviduct. *Biochemistry* 25: 7937-7942.

Gasc JM, Renoir JM, Radanyi C, Joab I, Tuohimaa P and Baulieu EE (1984): Progesterone receptor in the chick oviduct: An immunohisto-chemical study with antibodies to distinct receptor components. *J Cell Biol* 99: 1193-1201.

Ghosh-Dastidar P, Coty WA, Griest RE, Woo DDL and Fox CF (1984): Progesterone receptor subunits are high-affinity substrates for phosphorylation by epidermal growth factor receptor. *Proc Natl Acad Sci USA* 81: 1654-1658.

Giguere S, Hollenberg SM, Rosenfeld MG and Evans RM (1986): Functional domains of the human glucocorticoid receptor. *Cell* 46: 645-652.

Gorski J, Toft, DO, Shyamala, G, Smith D and Notides A (1968): Hormone receptors: Studies on the interaction of estrogen with the uterus. *Rec Prog Horm Res* 24: 45-80.

Govindan MV, Spiess E and Majors J (1982): Purified glucocorticoid receptor-hormone complex from rat liver cytosol binds specifically to cloned mouse mammary tumor virus long terminal repeats *in vitro*. *Proc Natl Acad Sci USA* 79: 5157-5161.

Groner B, Kennedy N, Skroch P, Hynes NE and Ponta H (1984): DNA sequences involved in the regulation of gene expression by glucocorti-coid hormones. *Biochim Biophys Acta* 781: 1-6.

Green S and Chambon P (1986): A superfamily of potentially oncogenic hormone receptors. *Nature* 324: 615-617.

Green S, Walter P, Kumar V, Krust A, Bornet JM, Argos P and Chambon P (1986): Human oestrogen receptor cDNA: Sequence expression and homology to v-erb-A. *Nature* 320: 134-139.

Greene, G.L., Gilna, P., Waterfield, M., Baker, A., Hort, Y. and Shine, J. (1986): Sequence and expression of human oestrogen receptor complementary DNA. *Science* 231, 1150-1154.

Gronemeyer H, Turcotte B, Quirin-Stricker C, Bocquel MT, Meyer ME, Krozowski Z, Jeltsch JM, Lerouge T, Garnier JM and Chambon P (1987): The chicken progesterone receptor: sequence, expression and functional analysis. *EMBO J* 6: 3985-3994.

Groyer A, Bouc YL, Joab I, Radanyi C, Renoir J-M, Robel P and Baulieu EE (1985): Chick oviduct glucocorticoid receptor: Specific binding of the synthetic steroid RU486 and immunological studies with antibodies to chick progesterone receptor. *Eur J Biochem* 149: 445-451.

Guiochon-Mantel A, Loosfelt H, Ragot T, Bailly A, Atger M, Misrahi, M, Perricaudet M and Milgrom E (1988): Receptors bound to antiprogestin form abortive complexes with hormone responsive elements. *Nature* 336: 695-698.

Hansen JC and Gorski J (1985): Conformational and electrostatic properties of unoccupied and liganded estrogen receptors determined by aqueous two-phase partitioning. *Biochemistry* 24: 6078-6085.

Hansen JC and Gorski J (1986): Conformational transitions of the estrogen receptor monomer. Effects of estrogens, antiestrogen, and temperature. *J Biol Chem* 261: 13990-13396.

Hapgood JP, Sabbatini GP and Holt CV (1986): Rat liver glucocorticoid receptor isolated by affinity chromatography is not a Mg^{2+}- or Ca^{2+}-dependent protein kinase. *Biochemistry* 25: 7529-7534.

Hoeck W and Groner B (1990): Hormone-dependent phosphorylation of the glucocorticoid receptor occurs mainly in the amino-terminal transactivation domain. *J Biol Chem* 265: 5403-5408.

Hollenberg SM, Weinberger C, Ong ES, Cerelli G, Oro A, Lebo R, Thompson EB, Rosenfeld MG and Evans RM (1985): Primary structure and expression of a functional human glucocorticoid receptor cDNA. *Nature* 318: 635-642.

Hollenberg SM, Giguere V, Sequi P and Evans RM (1987): Colocalization of DNA binding and transcriptional activation functions in the human glucocorticoid receptor. *Cell* 49: 39-46.

Horwitz KB (1992): The molecular biology of RU486. Is there a role for antiprogestins in the treatment of breast cancer? *Endocrine Rev* 13: 146-163.

Horwitz KB (1985): The antiprogestin RU38486: Receptor-mediated progestin versus antiprogestin actions screened in estrogen-insensitive T47Dco human breast cancer cells. *Endocrinology* 116: 2236-2245.

Horwitz KB, Francis MD and Wei LL (1985): Hormone-dependent covalent modification and processing of human progesterone receptors in the nucleus. *DNA* 4: 451-460.

Horwitz KB, Wei LL and Francis MD (1986): Structure analyses of progesterone receptors. *J Steroid Biochem* 24: 109-117.Hsieh J-C, Jurutka PW, Galligan MA, Terpening CM, Haussler CA, Samuels DC, Shimizu Y, Shimizu N, Haussler MR (1991): Human vitamin D receptor is selectively phosphorylated by protein kinase C on serine 51, a residue crucial to its trans-acting function. *Proc Natl Acad Sci USA* 88: 9315-9319.

Housley PR and Pratt WB (1983): Direct demonstration of glucocorticoid receptor phosphorylation by intact L-cells. *J Biol Chem* 258: 4630-4635.

Housley PR, Dahmer MK and Pratt WB (1982): Inactivation of glucocorticoid-binding capacity by protein phosphatases in the presence of molybdate and complete reactivation by dithiothreitol. *J Biol Chem* 257: 8615-8618.

Housley PR, Sanchez ER, Westphal HH, Beato M, Pratt WB (1985): The molybdate-stabilized L-cell glucocorticoid receptor isolated by affinity chromatography or with a monoclonal antibody is associated with a 90-92K nonsteroid binding phosphoprotein. *J Biol Chem* 260: 13810-13817.

Huggenvik JI, Collard MW, Kim Y-W and Sharma RP (1993): Modification of the retinoic acid signaling pathway by the catalytic subunit of protein kinase-A. *Mol Endocrinol* 7: 543-550.

Inano K, Haino M, Iwasaki M, Ono N, Horigome T and Sugano H (1990): Reconstitution of the 9 S estrogen receptor with heat shock protein 90. *FEBS Lett* 267: 157-159.

Ishii DN, Pratt WB and Aronow L (1972): Steady-state level of the specific glucocorticoid binding component in mouse fibroblasts. *Biochemistry* 11: 3896-3904.

Jeltsch JM, Krozowski Z, Quirin-Stricker C, Gronemeyer H, Simpson RJ, Garnier JM, Krust A, Jacob F and Chambon P (1986): Cloning of the chicken progesterone receptor. *Proc Natl Acad Sci USA* 83: 5424-5428.

Jensen EV, Suzuki T, Kawashima T, Stumpf WE, Jungblut PW and DeSombre ER (1968): A two-step mechanism for the interaction of estradiol with rat uterus. *Proc Natl Acad Sci USA* 59: 632-638.

Jensen EV and DeSombre ER (1973): Estrogen-receptor interaction. *Science* 182: 126-134.

Jerutka PW, Hsieh J-C, MacDonald PN, Terpening CM, Haussler CA, Haussler MR and Whitfield GK (1993): Phosphorylation of serine 208 in the human vitamin D receptor. The predominant amino acid phosphorylated by casein kinase II, *in vitro*, and identification as a significant phosphorylation site in intact cells. *J Biol Chem* 268: 6791-6799.

Joab I, Radanyi C, Renoir JM Buchou T, Catelli M-G, Binart N, Mester J and Baulieu EE (1984): Common non-hormone binding component in nontransformed chick oviduct receptors of four steroid hormones. *Nature* 308: 850-853.

John JK and Moudgil VK (1979): Activation of glucocorticoid receptor by ATP. *Biochem Biophys Res Commun* 90: 1242-1248.

Jost JP, Seldran M and Geiser M (1984): Preferential binding of estrogen-receptor complex to a region containing the estrogen-dependent hypomethylation site preceding the chicken vitellogenin II gene. *Proc Natl Acad Sci USA* 81: 429-433.

Karin M, Haslinger A, Holtgreve H, Richards RI, Krauter P, Westphal HM, Beato M (1984): Characterization of DNA sequences through which cadmium and glucocorticoid hormones induce human metallothionine in-11A gene. *Nature* 308: 513-519.

Kastner P, Bocquel MT, Turcotte B, Garnier JM, Horwitz KB, Chambon P and Gronemeyer H (1990): Transient expression of human and chicken progesterone receptors does not support alternative translational initiation from a single mRNA as the mechanism generating two receptor isoforms. *J Biol Chem* 265: 12163-12167.

King WJ and Greene GL (1984): Monoclonal antibodies localize oestrogen receptor in the nuclei of target cells. *Nature* 307: 745-747.

Kost SL, Smith DF, Sullivan WP, Welch WJ and Toft DO (1989): Binding of heat shock proteins to the avian progesterone receptor. *Cell Mol Biol* 9: 3829-3838.

Kumar V, Green S, Staub A and Chambon P (1986): Localization of the oestradiol-binding and putative DNA-binding domains of the human oestrogen receptor. *EMBO J* 5: 2231-2236.

Kurl RN and Jacob ST (1984): Phosphorylation of purified glucocorticoid receptor from rat liver by an endogenous protein kinase. *Biochem Biophys Res Commun* 119: 700-705.

Leach KL, Dahmer MK, Hammond ND, Sando JJ and Pratt WB (1979): Molybdate inhibition of glucocorticoid receptor inactivation and transformation. *J Biol Chem* 254: 11884-11890.

Lebeau M-C, Massol N, Herrick J, Faber LE, Renoir J-M, Radanyi C and Baulieu EE (1992): p59, an hsp 90-binding protein. Cloning and sequencing of its cDNA and preparation of a peptide-directed polyclonal antibody. *J Biol Chem* 267: 4281-4284.

Logeat F, LeCunff M, Pamphile R and Milgrom E (1985): The nuclear-bound form of the progesterone receptor is generated through a hormone-dependent phosphorylation. *Biochim Biophys Res Commun* 131: 421-427.

Loosfelt H, Atger M, Misrahi M, Guiochon-Mantel A, Meriel C, Logeat F, Benarous R and Milgrom E (1986): *Proc Natl Acad Sci USA* 83: 9045-9049.

Meggio F, Agostinis P and Pinna LA (1985): Casein kinases and their protein substrates in rat liver cytosol: evidence for their participation in multimolecular systems. *Biochim Biophys Acta* 846: 248-256.

Mendel DB, Bodwell JE, Gametchu B, Harrison RW and Munck A (1986): Molybdate-stabilized nonactivated glucocorticoid-receptor complexes contain a 90K-Da steroid binding phosphoprotein that is lost on activation. *J Biol Chem* 261: 3758-3763.

Meyer M-E, Pornon A, Ji J, Bocquel M-T, Chambon P and Gronemeyer H (1990): Agonistic and antagonistic activities of RU486 on the functions of the human progesterone receptor. *EMBO J* 9: 3923-3932.

Miesfeld R, Rusconi S, Gudowski PJ, Maler BA, Okret S, Wikstrom AC, Gustafsson JA and Yamamoto KR (1986): Genetic complementation of a glucocorticoid receptor deficiency by expression of cloned receptor cDNA. *Cell* 46: 389-399.

Migliaccio A, Rotondi A and Auricchio F (1984): Calmodulin-stimulated phosphorylation of 17B-estradiol receptor on tyrosine. *Proc Natl Acad Sci USA* 81: 5921-5925.

Migliaccio A, Domenico MD, Green S, de Falco A, Kajtaniak EL, Blasi F, Chambon P and Auricchio F (1989): Phosphorylation on tyrosine of in vitro synthesized human estrogen receptor activates its hormone binding. *Mol Endocrinol* 3: 1061-1069.

Milgrom E, Atger M and Baulieu EE (1973): Acidophilic activation of steroid hormone receptors. *Biochemistry* 12: 5198-5205.

Miller JB and Toft DO (1978): Requirement for activation in the binding of progesterone receptor to ATP-Sepharose. *Biochemistry* 17: 173-177.

Miller-Diener A, Schmidt TJ and Litwack G (1985): Protein kinase activity associated with the purified rat hepatic glucocorticoid receptor. *Proc Natl Acad Sci USA* 82: 4003-4007.

Misrahi M, Atger M and Milgrom E (1987): A novel progesterone-induced messenger RNA in rabbit and human endometria. Cloning and sequence analysis of the complementary DNA. *Biochemistry* 26: 3975-3982.

Miyata Y and Yahara I (1991): Cytoplasmic 8 S glucocorticoid receptor binds to actin filaments through the 90-kDa heat shock protein moiety. *J Biol Chem* 266: 8779-8783.

Mizutani T, Bhakta A, Kloosterboer HJ and Moudgil VK (1992): Novel antiprogestins ORG 31806 and 31710: Interaction with mammalian progesterone receptor and DNA binding of antisteroid receptor complexes. *J Steroid Biochem Mol Biol* 42: 695-704.

Moore DD, Marks AR, Buckley DI, Kapler G, Payvar F and Goodman HM (1985): The first intron of the human growth hormone gene contains a binding site for glucocorticoid receptor. *Proc Natl Acad Sci USA* 82: 699-702.

Moudgil VK (1988): Ed. *Steroid Receptors in Health and Disease.* New York: Plenum Press.

Moudgil VK (1990): Phosphorylation of steroid hormone receptors. *Biochim Biophys Acta* 1055: 243-258.

Moudgil VK, Anter MJ and Hurd C (1989): Mammalian progesterone receptor shows differential sensitivity to sulfhydryl group modifying agents when bound to agonist and antagonist ligands. *J Biol Chem* 264: 2204-2211.

Moudgil VK, Eessalu TE, Buchou T, Renoir JM, Mester J and Baulieu EE (1985): Transformation of chick oviduct progesterone receptor *in vitro*: Effects of hormone, salt, heat and adenosine triphosphate. *Endocrinology* 116: 1267-1274.

Moudgil VK, Kruczak, VH, Eessalu TE, Paulose CS, Taylor MG and Hansen JC (1981): Activation of progesterone receptor by ATP. *Eur J Biochem* 118: 547-555.

Moudgil VK, Lombardo G, Eessalu T and Eliezer N (1986a): Hormone dependency of transformation of rat liver glucocorticoid receptor *in vitro*: Effects of heat, salt and nucleotides. *J Biochem* 99: 1005-1016.

Moudgil VK, Lombardo G, Hurd C, Eliezer N and Agarwal MK (1986b): Evidence for separate binding sites for progesterone and RU486 in the chick oviduct. *Biochim Biophys Acta* 889: 192-199.

Moudgil VK and Hurd C (1987): Transformation of calf uterine progesterone receptor: Analysis of the process when receptor is bound to progesterone and RU 38486. *Biochemistry* 26: 4993-5001.

Moudgil VK and John JK (1980a): ATP-dependent activation of glucocorticoid receptor from rat liver cytosol. *Biochem J* 190: 799-808.

Moudgil VK and John JK (1980b): Interaction of rat liver glucocorticoid receptor with adenosine 5'-triphosphate. Characterization of interaction by use of ATP-Sepharose affinity chromatography. *Biochem J* 190: 809-818.

Moudgil VK, Nath R, Bhakta A and Nakao M (1991): ZK98299, a novel antiprogesterone that does not interact with chicken oviduct progesterone receptor. *Biochim Biophys Acta* 1094: 185-192.

Moudgil VK, Nishigori H, Eessalu TE and Toft DO (1980): Analysis of the avian progesterone receptor with inhibitors. In: Roy A, Clark JH, eds. *Gene Regulation by Steroid Hormones*. New York: Springer-Verlag, 103-119.

Mulvihill ER, Le Pennee JP and Chambon P (1982): Chicken oviduct progesterone receptor: Location of specific regions of high affinity binding in cloned DNA fragments of hormone-responsive genes. *Cell* 24: 621-632.

Munck A and Foley R (1979): Activation of steroid hormone-receptor complexes in intact cells in physiological conditions. *Nature* 278: 752-754.

Munck A and Brink-Johnsen T (1968): Specific and nonspecific physicochemical interactions of glucocorticoids and related steroids with rat thymus cells *in vitro*. *J Biol Chem* 243: 5556-5565.

Murakami N and Moudgil VK (1981): Inactivation of rat liver glucocorticoid receptor by molybdate. *Biochem J* 198: 447-455.

Murayama A, Fukai F, Hazato T and Yamamoto T (1980): Estrogen receptor of cow uterus. II. Chracterization of a cytoplasmic factor which binds with native 4 S estrogen receptor to give 8 S estrogen receptor. *J Biochem* (Tokyo) 88: 963-968.

Nakao M, Mizutani T, Bhakta A, Ribarac-Stepic N and Moudgil VK (1992): Phosphorylation of chicken oviduct progesterone receptor by cAMP-dependent protein kinase. *Arch Biochem Biophys* 298: 340-348.

Nakao M and Moudgil VK (1989): Hormone specific phosphorylation and transformation of chicken oviduct progesterone receptor. *Biochem Biophys Res Commun* 161: 295-303.

Nishigori H and Toft DO (1980): Inhibition of progesterone receptor activation by sodium molybdate. *Biochemistry* 19: 77-83.

O'Malley BW (1990): The steroid receptor superfamily: More excitement predicted for the future. *Mol Endocrinol* 4: 363-369.

O'Malley BW and Conneely OM (1992): Orphan receptors: In search of a unifying hypothesis for activation. *Mol Endocrinol* 6: 1359-1361.

Orti E, Mendel DB, Smith LI and Munck A (1989): Agonist-dependent phosphorylation and nuclear dephosphorylation of glucocorticoid receptors in intact cells. *J Biol Chem* 264: 9728-9731.

Orti E, Hu L-M and Munck A (1993): Kinetics of glucocorticoid receptor phosphorylation in intact cells. Evidence for hormone-induced hyperphosphorylation after activation and recycling of hyperphosphorylated receptors. *J Biol Chem* 268: 7779-7784.

Orti E, Bodwell JE and Munck A (1992): Phosphorylation of Steroid Hormone Receptors. *Endo Rev* 13: 105-128.

Pan LC and Price PA (1987): Ligand-dependent regulation of the 1,25-dihydroxyvitamin D_3 receptor in rat osteosarcoma cells. *J Biol Chem* 262: 4670-4675.

Payvar F, Wrange O, Carlstedt-Duke J, Okret S, Gustafsson JA and Yamamoto KR (1981): Purified glucocorticoid receptors bind selectively *in vitro* to a cloned DNA fragment whose transcription is regulated by glucocorticoids *in vivo*. *Proc Natl Acad Sci USA* 78: 6628-6632.

Perisic O, Radojcic M and Kanazir DT (1987): Protein kinase activity can be separated from the purified activated rat liver glucocorticoid receptor. *J Biol Chem* 262: 11688-11691.

Perrot-Applanat M, Logeat F, Groyer-Picard MT and Milgrom E (1985): Immunocytochemical study of mammalian progesterone receptor using monoclonal antibodies. *Endocrinology* 116: 1473-1484.

Pfahl M (1982): Specific binding of the glucocorticoid-receptor complex to the mouse mammary tumor proviral promotor region. *Cell* 31: 475-482.

Pham TA, Elliston JF, Nawaz Z, McDonnell DP, Tsai M-J and O'Malley BW (1991): Antiestrogen can establish nonproductive receptor complexes and alter chromatin structure at target enhancers. *Proc Natl Acad Sci USA* 88: 3125-3129.

Picard D, Khursheed B, Grabedian MJ, Fortin MG, Lindquist S and Yamamoto KR (1990): Reduced levels of hsp90 compromise steroid receptor action *in vivo*. *Nature* 348: 166-168.

Pike JW and Sleator NM (1985): Hormone-dependent phosphorylation of the 1,25-dihydorxyvitamin D_3 receptor in mouse fibroblasts. *Biochem Biophys Res Commun* 131: 378-385.

Power RF, Conneely OM, McDonnell DP, Clark JH, Butt TR, Schrader WT and O'Malley BW (1990): High level expression of a truncated chicken progesterone receptor in *Escherichia coli*. *J Biol Chem* 265: 1419-1424.

Puri RK, Grandics P, Dougherty JJ and Toft DO (1982): Purification of nontransformed avian progesterone receptor and preliminary characterization. *J Biol Chem* 257: 10831-10837.

Puri RK, Dougherty JJ and Toft DO (1984): The avian progesterone receptor: Isolation and characterization of phosphorylated forms. *J Steroid Biochem* 20: 23-29.

Rees AM and Bell PA (1975): The involvement of receptor sulfhydryl groups in the binding of steroids to cytoplasmic glucocorticoid receptor from rat thymus. *Biochim Biophys Acta* 411: 121-132.

Renoir JM, Yang CR and Formstecher P, Lustenberger P, Wolfson A, Redeuilh G, Mester J, Richard-Foy H and Baulieu EE (1982): Progesterone receptor from chick oviduct: purification of molybdate-stabilized form and preliminary characterization. *Eur J Biochem* 127: 71-79.

Renoir JM, Buchou T and Baulieu EE (1986): Involvement of a non-hormone binding 90-kilodalton protein in the nontransformed 8S form of the rabbit uterus progesterone receptor. *Biochemistry* 25: 6405-6413.

Riehl RM, Sullivan WP, Vroman BT, Bauer VT, Pearson GR and Toft DO (1985): Immunological evidence that the non-hormone binding component of avian steroid receptors exists in a wide range of tissues and species. *Biochemistry* 24: 6586-6591.

Sanchez ER (1990): Hsp56: a novel heat shock protein associated with untransformed steroid receptor complexes. *J Biol Chem* 265: 22067-22070.

Sanchez ER, Toft DO, Schlesinger MJ and Pratt WB (1985): Evidence that the 90-kDa phosphoprotein associated with the untransformed L-cell glucocorticoid receptor is a murine heat shock protein. *J Biol Chem* 260: 12398-12401.

Sanchez ER and Pratt WB (1986): Phosphorylation of L-cell glucocorticoid receptors in immune complexes: Evidence that the receptor is not a protein kinase. *Biochemistry* 25: 1378-1382.

Sanchez ER, Redmond T, Scherrer LC, Bresnick EH, Welsh MJ and Pratt WB (1988): Evidence that the 90-kilodalton heat shock protein is associated with tubulin-containing complexes in L cell cytosol and in intact PtK cells. *Mol Endocrinol* 2: 756-760.

Sando JJ, Hammond ND, Stratford CA and Pratt WB (1979): Activation of thymocyte glucocorticoid receptors to the steroid binding form. The roles of reducing agents, ATP, and heat-stable factors. *J Biol Chem* 254: 4779-4789.

Savouret JF, Misrahi M, Loosfelt H, Atger M, Bailly A, Perrot-Applant M, Vu Hai MT, Guiochon-Mentel A, Jolivet A, Lorenzo F, Logeat F, Pichon MF, Bouchard P and Milgrom E (1989): Molecular and cellular biology of mammalian progesterone receptors. *Rec Prog Horm Res* 45: 65-120.

Scheidereit C and Beato M (1984): Contacts between hormone receptor and DNA double helix within a glucocorticoid regulating element of mouse mammary tumor virus. *Proc Natl Acad Sci USA* 81: 3029-3033.

Scherrer LC, Dalman FC, Massa E, Meshinchi S and Pratt WB (1990): Structural and functional reconstitution of the glucocorticoid receptor-Hsp90 complex. *J Biol Chem* 265: 21397-21400.

Schrader WT and O'Malley BW (1972):Progesterone binding proteins of chick oviduct. IV. Characterization of purified subunit. *J Biol Chem* 247: 51-59.

Schrader WT, Birnbaumer ME, Hughes MR, Weigel NL, Grody WW and O'Malley BW (1981): Studies on the structure and function of the chicken progesterone receptor. *Rec Prog Horm Res* 37: 583-633.

Schuh SS, Yonemoto W, Brugge J, Bauer VJ, Riehl RM, Sullivan WP and Toft DO 1985: A 90,000-dalton binding protein common to both steroid receptors and the Rouse Sarcoma Virus transforming protein pp60[v-src]. *J Biol Chem* 26: 14292-14296.

Schule R, Muller M, Kaltschmidt C and Renkawitz R (1988): Many transcription factors interact synergistically with steroid receptors. *Science* 242: 1418-1420.

Sheridan PJ, Buchanan JM, Anselmo VC and Martin PM (1979): Equilibrium: the intracellular distribution of steroid receptors. *Nature* 282: 579-582.

Sheridan PL, Krett NL, Gordon JA and Horwitz KB (1988): Human progesterone receptor transformation and nuclear down-regulation are independent of phosphorylation. *Mol Endocrinol* 2: 1329-1342.

Sheridan PL, Evans RM and Horwitz KB (1989): Phosphotryptic peptideanalysis of human progesterone receptors. New phosphorylated sites formed in nuclei after hormone treatment. *J Biol Chem* 264: 6520-6528.

Singh VB and Moudgil VK (1984): Protein kinase activity of purified rat liver glucocorticoid receptor. *Biochem Biophys Res Commun* 125: 1067-1073.

Singh VB and Moudgil VK (1985): Phosphorylation of rat liver glucocorticoid receptor. *J Biol Chem* 260: 3684-3690.

Skafar DF and Notides AC (1985): Modulation of the estrogen receptor's affinity for DNA by estradiol. *J Biol Chem* 22: 12208-12213.

Smith DF, Faber LE and Toft DO (1990a): Purification of unactivated progesterone receptor and identification of novel receptor-associated proteins. *J Biol Chem* 265: 3996-4003.

Smith DF, Schowalter DB, Kost SL and Toft DO (1990b): Reconstitution of progesterone receptor with heat shock proteins. *Mol Endocrinol* 4: 1704-1711.

Sullivan WP, Smith DF, Beito TG, Krco CJ and Toft DO (1988a): Hormone-dependent processing of the avian progesterone receptor. *J Cell Biochem* 36: 103-119.

Sullivan WP, Madden BJ, McCormick DJ and Toft DO (1988b): Hormone-dependent phosphorylation of the avian progesterone receptor. *J Biol Chem* 263: 14717-14723.

Tai P-KK, Maeda Y, Nakao K, Wakim NG, Duhring JL and Faber L (1986): A 59-kilodalton protein associated with progestin, estrogen, androgen and glucocorticoid receptors. *Biochemistry* 25: 5269-5275.

Tai P-KK, Albers MW, Chang H, Faber L and Schreiber SL (1992): Association of a 59 kilodalton immunophilin with the glucocorticoid receptor. *Science* 256: 1315-1318.

Tilley WD, Marcelli M, Wilson JD and McPhaul MJ (1989): Characterization and expression of a cDNA encoding the human androgen receptor. *Proc Natl Acad Sci USA* 86: 327-331.

Toft D and Gorski J (1966): A receptor molecule for estrogens: Isolation from the rat uterus and preliminary characterization. *Proc Natl Acad Sci USA* 55: 1574-1581.

Toft DO and Nishigori H 1979: Stabilization of the avian progesterone receptor by inhibitors. *J Steroid Biochem* 11: 413-416.

Tora L, Gronemeyer H, Turcotte B, Gaub M-P and Chambon P (1988): The N-terminal region of the chicken progesterone receptor specifies target gene activation. *Nature* 333: 185-188.

44

Ulmann A, Teutsch G and Philibert D (1990): RU486. *Sci Amer* 262: 42-48.

Van Laar JH, Berrevoets CA, Trapman J, Zegers ND and Brinkmann AO (1991): Hormone-dependent androgen receptor phosphorylation is accompanied by receptor transformation in human lymph node carcinoma of the prostate cells. *J Biol Chem* 266: 3734-3738.

Vegeto E, Allan GF, Schrader WT, Tsai M-J, McDonnell DP and O'Malley BW (1992): The molecular mechanism of RU486 antagonism is dependent on the conformation of the carboxy-terminal tail of the human progesterone receptor. *Cell* 69: 703-713.

Wei LL, Sheridan PL, Krett NL, Fransis MD, Toft DO, Edwards DP and Horwitz KB (1987): Immunological analysis of human breast cancer progesterone receptors. II, Structure, phosphorylation and processing. *Biochemistry* 26: 6262-6272.

Weigel NL, Carter TH, Schrader WT and O'Malley BW (1992): Chicken progesterone receptor is phosphorylated by a DNA-dependent protein kinase during in vitro transcription assays. *Mol Endocrinol* 6: 8-14.

Weinberger C, Hollenberg SM, Rosenfeld MG and Evans RM (1985): Domain structure of human glucocorticoid receptor and its relationship to the v-erb-A oncogene. *Nature* 318: 670-672.

Welshons WV, Lieberman ME and Gorski J (1984): Nuclear localization of unoccupied oestrogen receptors. *Nature* 307: 747-749.

Welshons WV, Krummel M and Gorski J (1985): Nuclear localization of unoccupied receptors for glucocorticoids, estrogens, and progesterone in GH_3 cells. *Endocrinology* 117: 2140-2147.

Wikstrom AC, Bakke O, Okret S, Bronnegard M and Gustafsson JA (1987): Intracellular localization of the glucocorticoid receptor: Evidence for cytoplasmic and nuclear localization. *Endocrinology* 120: 1232-1242.

Willmann T and Beato M (1986): Steroid-free glucocorticoid receptor binds specifically to mouse mammary tumor virus DNA. *Nature* 324: 688-691.

Wrange O, Carlstedt-Duke J and Gustafsson JA (1986): Stoichiometric analysis of the specific interaction of the glucocorticoid receptor with DNA. *J Biol Chem* 261: 11770-11778.

Yamamoto KR (1985): Steroid receptor regulated transcription of specific genes and gene networks. *Ann Rev Genet* 19: 209-252.

Yoshinaga SK, Peterson CL, Herskowitz I and Yamamoto KR (1992) Roles of SWI1, SWI2, and SWI3 proteins for transcriptional enhancement by steroid receptors. *Science* 258: 1598-1604.

STEROID RECEPTOR INTERACTIONS
WITH DNA AND NUCLEAR COMPONENTS

STRUCTURAL AND FUNCTIONAL STUDIES OF SELECTIVE DNA BINDING BY STEROID/NUCLEAR RECEPTORS

Leonard P. Freedman
Cell Biology & Genetics Program
Sloan-Kettering Institute

Ben F. Luisi
MRC Virology Unit
Glasgow, UK

Iris Alroy
Terri L. Towers
Cell Biology & Genetics Program
Sloan-Kettering Institute

INTRODUCTION

Nuclear hormone receptors directly transduce signals presented by levels of hormones and other small molecules into effects on gene expression. Upon binding these ligands, the receptors can associate with specific DNA sequences and modulate the transcription of target genes. Unified by functional analogy and a characteristic, punctuated sequence homology, these molecules have been grouped as the "nuclear receptor superfamily", and members have been identified and isolated from species as diverse as mammals and arthropods. Vertebrate receptors have been

STEROID HORMONE RECEPTORS
V. K. Moudgil, Editor
© 1993 Birkhäuser Boston

characterized that specifically bind hormonal forms of vitamins A and D, thyroid hormone, peroxisomal activators, and steroid hormones, such as glucocorticoid, progesterone, estrogen, androgen, and aldosterone (reviewed in Parker, 1991 and references therein). A number of "orphan receptors" whose ligands and function are unknown, have also been discovered. Several putative nuclear receptors have been isolated from *Drosophila* that may play important regulatory roles in embryogenesis, including a receptor for the insect steroid, ecdysone (Seagraves, 1991).

Like many eukaryotic transcription factors, nuclear hormone receptors are composed of domains that correspond to discrete functions. The domains of the receptors encompass the functions of ligand-binding (carboxyl terminal region), DNA-binding (preceding the ligand-binding domain), nuclear localization (coinciding with DNA-and ligand-binding domains), and transcriptional modulation (localized to more variable regions, including the amino terminus). The amino termini of nuclear receptors vary considerably in length and composition, but the sequences of the DNA and ligand binding domains have been well conserved (Amero et al, 1992; Laudet et al, 1992). The sequence homology is greatest in the DNA-binding domain, and indeed, many receptor-encoding genes were cloned by screening cDNA libraries with probes against this region. The ligand and DNA binding domains, as well as transcriptional activation regions, all can confer specific function when linked to unrelated, nonreceptor proteins (for example, see Picard et al, 1988 and Godowski et al, 1988). The DNA binding specificity of the receptors has been shown to be encoded entirely by the DNA binding domain and can be switched by swapping the corresponding segment between different receptors (Green et al, 1988).

The DNA-binding segment of the glucocorticoid receptor (GR) has been shown to be a true domain in that if folds stably and retains function in isolation from the remainder of the receptor (Freedman et al, 1988). This has been a great convenience for experimental characterization. Several amino acids in the DNA-binding domain are invariant throughout the family, including eight cysteines that were shown in the case of GR to tetrahedrally coordinate two Zn ions (Freedman et al, 1988). Mutagenesis experiments have indicated that specific residues within the Zn module region of GR are critical for DNA binding specificity (Mader et al, 1989; Danielson et al, 1989; Umesono and Evans, 1989), DNA-dependent dimerization (Umesono and Evans, 1989; Dahlman-Wright et al, 1991) and positive control of transcription

(Schena et al, 1989). These results have recently been confirmed and expanded by three-dimensional structural analysis (Hard et al, 1990; Schwabe et al, 1990; Luisi et al, 1991).

Here, we review and correlate the various genetic and biochemical experiments that have been described primarily for the DNA binding domains of the glucocorticoid and estrogen receptors with the structural information. We describe how the structure and function have permitted us to predict properties of the DNA binding domains of the other nuclear receptors, such as the vitamin D_3 receptor. In particular, we postulate how receptors discriminate closely related response elements through sequence specific contacts and distinguish symmetry of target sites through protein-protein interactions. This mechanism explains in part how the receptors regulate diverse sets of genes from a limited repertoire of core response elements.

Structure of the DNA Binding Domain

The three dimensional structures of the DNA binding domains of the glucocorticoid and estrogen receptors have been established by nuclear resonance spectroscopy (Hard et al, 1990; Schwabe et al, 1990) and the crystal structure of the glucocorticoid receptor DNA binding domain (GRdbd) in complex with DNA has been determined (Luisi et al, 1991). These structures agree well with each other, but there is a difference in secondary structure in a small segment that may be nucleated by DNA binding, as we will describe.

The salient feature of the DNA binding domain is its Zn coordination sites. In agreement with stoichiometry determinations and spectroscopic evaluation (Freedman et al, 1988), two Zn ions are each tetrahedrally coordinated by four cysteines to stabilize two peptide loops and cap amino termini of two amphipathic α-helices (Fig. 1). The metal binding sites differ structurally and functionally from those found in the other eukaryotic transcription factors bearing "Zn fingers", such as TFIIIA, ADR1, and Xfin, where the metal is coordinated by two histidines and two cysteines (reviewed in Berg, 1990). It also differs from the Zn coordination site of GAL4 (Marmomstein et al, 1992), where two metal ions share cysteine ligands. We therefore refer to the nuclear receptor loop/helix subdomains as "Zn modules" to distinguish them from the other Zn-bearing structures. Residues coordinating the Zn and supporting the fold of the domain are conserved throughout the family, which strongly suggests that the Zn module structures are also conserved among the receptors.

50

Although they appear to be structurally similar, the Zn modules of the GRdbd serve different functions. The amino terminal module exposes an α-helix to the major groove and directs contacts with the bases of the target site. The carboxy terminal module forms a dimerization interface, which is mediated principally through contacts made by residues 476-482 (Fig. 1), in a region referred to as the "D box" for its postulated role in dimerization (Umesono and Evans, 1989). The loops of both modules make phosphate backbone contacts. Although structurally distinct, the Zn

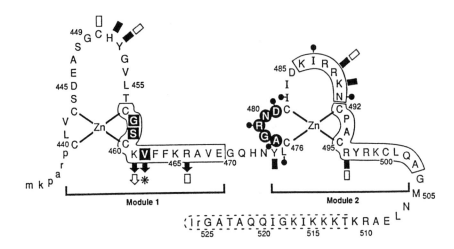

FIGURE 1. Zinc coordination scheme of the rat glucocorticoid receptor DNA binding domain. Numbering scheme is based on the full-length receptor. Indicated residues and regions are based on the crystal structure of the protein bound to a GRE (Luisi et al, 1991). Residues making specific phosphate backbone contacts are indicated by solid rectangles; those making nonspecific contacts are indicated by open rectangles. Residues making direct contacts with bases are depicted by solid arrows. Amino acids involved in dimer interface interactions are indicated by a solid dot. Three amino acids (residues 458, 459, 462) that confer specificity in mutagenesis experiments (see text) are shown in solid boxes; those demonstrated to confer half-site spacing requirements (residues 477-481) are shown in solid circles. α-helical regions are enclosed by solid lines. A disordered section at the C-terminus is enclosed by dashed lines. Amino acids as lower-case letters derive from vector sequence of the overexpression plasmid. Reprinted by permission from NATURE vol. 352 pp. 497-505. Copyright (C) 1991 Macmillan Magazines Ltd.

fingers, GAL4 Zn center, and the nuclear receptor Zn modules do share the general feature that they all serve to stabilize and orient an α-helix for interaction with the DNA target through major groove contacts (Luisi, 1992). In the case of the Zn fingers and the GAL4 Zn center, the recognition helix lies within the loop region, while it begins at the carboxy terminal end of the loop in the nuclear receptor modules.

The apparent structural similarity of the two modules gives the protein the appearance of having an approximate structural repeat. Moreover, the modules are encoded by separate exons (Ponglikimong et al, 1988), which suggests that they may have arisen by gene duplication and then evolved with different functions. However, the modules are actually topologically non-equivalent, so their relationship may not be so simple. The topology is defined here as the chirality of Zn coordinating residues, which can be R or S. The amino terminal Zn module, like the Zn fingers, (Berg, 1988) has the S configuration. The carboxy terminal module, however, has the mirror image R configuration about the metal, and it cannot be changed to that found for the amino terminal module without breaking a Zn-S bond. Despite their structural similarities, the two Zn modules may not have arisen by gene duplication unless one refolded catastrophically about the Zn ion in the very early stages of its evolution.

The two amphipathic helices of GRdbd pack together to form a hydrophobic core which is important for maintaining the globular fold of the domain (Luisi et al, 1991). The residues in and supporting this core are very strongly conserved in the superfamily. Functionally non-conservative changes here result in loss of function of the glucocorticoid receptor *in vivo* (Schena et al, 1989) and in the vitamin D and androgen receptors, a substitution here is associated with two clinical disorders (Freedman, 1992).

The NMR structure of the protein in the absence of DNA and the crystal structure of the protein/DNA complex agree quite well. There is a difference, however, in the secondary structure of the carboxyl terminal module. In the crystal structure, there is a short α-helix in the loop region of this module that makes phosphate backbone and dimer interface contacts. This helix is not present in the NMR structures, and it may be nucleated by DNA binding. There is precedent for this effect, for interaction with nucleic acid may nucleate secondary structure in the DNA-binding domains of GAL4 and leucine zipper transcription factors (Marmomstein et al, 1992; Patel et al, 1990). It is not clear if folding of the short α-helix occurs in the context of the full length glucocorticoid receptor, which may already exist as a dimer before it binds DNA.

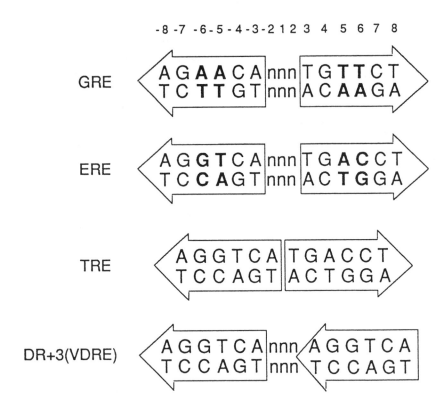

FIGURE 2. Idealized hormone response elements are organized as two half-sites. Arrows indicate the directionality of the half sites.

Response Elements of Steroid Receptors

The glucocorticoid response element (GRE) was first identified in DNA binding studies with the mouse mammary tumor virus long-terminal repeat (MMTV LTR) (Payvar et al, 1983). Glucocorticoid-inducibility was found to be mediated through a 15 base-pair, partially palindromic sequence that consists of two hexameric half-sites separated by three bases. A similar sequence was discovered in the controlling region of the genes for human metallothionein IIA, chicken lysozyme, and rat tyrosine aminotransferase (reviewed in Beato, 1989). Based on these and other subsequently characterized GREs, a functional consensus was proposed: 5'-GGTACAnnnTGTTCT-3' (Beato, 1989), where n can be any nucleotide. In nature, the sequences of the half-sites of GREs may vary considerably; nonetheless, the spacing between half-sites is always three bases. A GRE can be constructed that has a perfect palindrome and which will impart glucocorticoid-inducible gene expression *in vivo* (Strahle et al, 1987): 5'-AGAACAnnnTGTTCT-3' (Fig. 2). LaBaer and Yamamoto (1989) and Hard and colleagues (1990) have shown that the GRdbd association constant for a specific site relative to a non-specific site differs by only two orders of magnitude (10^8 M^{-1} versus 10^6 M^{-1}, respectively). This is a much weaker interaction than found for the prokaryotic gene regulatory proteins, and the crystal structure shows that there are relatively fewer protein/DNA contacts. The relatively weak DNA binding is fairly typical of eukaryotic transcription factors.

The hormone response elements of other steroid receptors are also palindromic with a fixed, intervening spacing. These receptors also bind their targets with relatively low affinity. The estrogen response element consensus sequence has been identified and has similarity to the GRE: 5'-AGGTCAnnnTGACCT-3' (Klock et al, 1987).

Receptor Dimerization: Grading of Targets

The palindromic nature of GREs suggests that the receptor binds its targets as dimers. Furthermore, the center-to-center separation of half-sites in the palindrome is nine bases, which is nearly a helical repeat, suggesting that the two subunits lie on nearly the same face and could contact each other. Indeed, crystallographic analysis (Luisi et al, 1991) and electrophoretic mobility studies (LaBaer and Yamamoto, 1989; Hard et al, 1990; Dahlman-Wright et al, 1990; Alroy and Freedman, 1992) show this to be the case for the GRdbd. Two monomers of GRdbd bind to the target site and lie on one surface of the DNA, making extensive protein-protein contacts. The subunits expose recognition α-helices into adjacent major grooves. The GRdbd/DNA complex resembles that of prokaryotic proteins having the helix-turn-helix motif. These proteins also bind to palindromic targets as dimers on one surface of the DNA and interact in the major groove via a recognition α-helix.

The GRdbd is monomeric in solution (Freedman et al, 1988b; Hard et al, 1990), but two molecules bind a GRE cooperatively (LaBaer and Yamamoto, 1989; Hard et al, 1990; Dahlman-Wright et al, 1990; Alroy and Freedman, 1992). The binding of the first monomer increases the affinity of the second by two orders of magnitude (LaBaer and Yamamoto, 1989; Hard et al, 1990). Cooperativity results from favorable protein-protein contacts made through an interface which is aligned by DNA binding. If the spacing between half-sites is increased or decreased by a single base, the contacts should be disrupted and indeed, cooperativity is lost *in vitro* (Dahlman-Wright et al, 1990; Freedman and Towers, 1991) and transactivation is abolished *in vivo* (Dahlman-Wright et al, 1990; Nordeen et al, 1990). As already mentioned, crystallographic analysis shows that the protein-protein contacts are mediated through residues in the carboxy-terminal module, including a region referred to as the D box. Freedman and Towers (1991) found that substitution of the D-box residues of the GRdbd with those of the vitamin D_3 receptor (VDR) abolishes cooperativity. The swap substitutes residues that cannot provide all the favorable interactions and would weaken association of subunits. Interestingly, VDR binds non-palindromic targets, as we will discuss below.

The crystal structure of the GRdbd has been studied in two complexes with DNA (Luisi et al, 1991). In one, the consensus target was employed; in the other, the spacing between half-sites was increased from three to four bases. In both cases, the contacts of the dimer interface are the same. One subunit of the dimer is forced out of alignment with the recognition sequence, and it does not make specific base contacts with DNA. We proposed that this interaction represents a non-specific complex (Luisi et al, 1991). In nature, GREs may not be perfectly palindromic, and deviations from the consensus may be exploited for grading of affinity and activity of the response elements. The structural results show that the receptor could still form a dimer on the surface of the most extreme deviation, but one subunit would be forced to make a non-specific complex. The affinity of such a complex would clearly not be as great as for that with a target conforming to the consensus. Thus, dimerization may serve as a means of modulating target affinities, and possibly, transcriptional responsiveness. The hormone response elements of other steroid receptors are also palindromic and presumably bind to these sites in an analogous fashion to the GR/GRE complex. For instance, the ERdbd is also monomeric in solution and binds its target cooperatively (J. Schwabe, Ph. D. thesis, MRC Cambridge, 1992). As we describe later, dimerization is also mediated by the ligand binding domain present in the full length receptors.

DNA Binding Specificity and Discrimination of Response Elements

The glucocorticoid and estrogen receptors discriminate each other's response elements *in vivo,* and the consensus elements differ at two bases in each half-site: 5'-AGGTCAnnnTGACCT-3' of the ERE versus 5'-AGAACAnnnTGTTCT of the GRE (Fig. 2). Studies using *in vivo* transactivating assays have identified three amino acids in the DNA-binding domain of each receptor that direct this discrimination (Mader et al, 1989; Danielson et al, 1989; Umesono and Evans, 1989). These residues are located in the amino terminus of the α-helical region of the first Zn module, in a segment referred to as the "P box" (see Fig. 4). Mader et al (1989) demonstrated that changing three amino acids, Glu-Gly-cys-lys-Ala, as it occurs in ER, to Gly-Ser-cys-lys-Val, as it occurs in GR, completely changed specificity so that this mutant ER transactivated strongly from a GRE-driven reporter and not at all from an ERE. Substitution of two of these amino acids in GR to the corresponding ER residues (Gly-Ser-> Glu-Gly) partially switches the receptor's specificity from a GRE to an ERE; the third substitution, Val -> Ala is required for the full switch (Danielson et al, 1989; Umesono and Evans, 1989). The binding affinities of these mutants in the context of the DNA-binding domain have been investigated *in vitro* and correlate with corresponding *in vivo* affects (Alroy and Freedman, 1992).

The crystal structure of the GRdbd/DNA complexes show that only one residue of the discriminatory triplet (Gly-Ser-cys-lys-Val) makes a base contact (Luisi et al, 1991). The Val makes a favorable van der Waals contact with 5-methyl group of the T of the GRE half-site TGTTCT (Fig. 3a). The ERE has an A at the corresponding position (see Fig 2). Corroborating the importance of the Val/methyl contact, Alroy and Freedman (1992) found that substitution of Val with Ala, as it occurs in the ER, decreases binding affinity *in vitro* by a factor of ten. Surprisingly, they found that substitution of the Val with Ala does not increase GRdbd's affinity for an ERE. This residue may not be making an attractive interaction in the ER/ERE complex, but may instead have been selected in evolution for the effective repulsion it presents to a GRE. We propose that only one residue of the ER triplet makes an attractive interaction with the ERE: Glu of the ER recognition helix (of Glu-Gly-cys-lys-Ala) accepts a hydrogen bond from the C of TGACCT (Fig. 3b). Indeed, *in vitro* DNA binding of the GRdbd carrying the Glu at the first position supports the importance of this residue in directing ERE binding specificity (Alroy and Freedman, 1992).

The GR and ER probably share two base contacts in common, and these do not contribute to ERE/GRE discrimination, although they do contribute to recognition of the common features of the hormone response element motif: TGxxCT, where x is a discriminating base. Lys461 and Arg466 in the recognition α-helix of GR contact G of AGAACT and its complement TGTTCT, respectively. Lys and Arg are also found at the corresponding positions in the recognition helix of ER (and in fact, in all

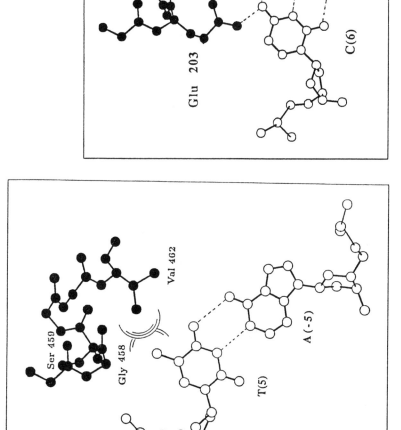

FIGURE 3. Specific side-chain contacts with bases in the GRE or ERE. (a, *left panel*) The methyl group of GR Val 462 makes a van der Waals contact with the 5-methyl group of T(5) of GRE; (b, *right panel*) proposed interaction of ER Glu203 with 4-amino group of C(6) of ERE.

nuclear receptors reported to date) and probably make corresponding base contacts in the ER/ERE complex: T<u>GA</u>CCT and A<u>GG</u>TCA (or T<u>G</u>xxCT and A<u>G</u>xxCA of the general response element).

Response Elements Fall into Classes

Nuclear receptors tend to fall into subgroups which recognize the same response element "core", and the amino acid sequence of the P box can be used to predict core preference. As summarized in Fig. 4, receptors carrying the GS--V motif at the amino terminus of the recognition α-helix (i.e., glucocorticoid, progesterone, mineralocorticoid, androgen receptors) all recognize a GRE with high affinity. Receptors carrying the EG--A/G/S motif at the corresponding positions (i.e., estrogen, vitamin D_3, thyroid hormone, retinoic acid, ecdysone receptors and many orphan receptors) all appear to be able to bind to the ERE core (5'-AGGTCA-3') with high affinity, and probably in all these cases, the Glu of the P boxes accepts a hydrogen bond from one of the two <u>C</u>'s of the complement (5'-TGA<u>CC</u>T-3') half-site (see Fig. 3). Within the latter group, specificity may be conferred by the spacing and relative orientation of the two half-sites (see below). It is interesting to note that GREs can function virtually as efficiently as response elements for progesterone, androgen, and mineralocorticoid hormones although these hormones elicit distinct physiological effects in organisms. The distinguishing effects of these hormones may arise from tissue-specific expression of the corresponding receptors or the role of auxiliary factors in directing specific gene activation.

Wilson et al (1991) have found that the mammalian orphan receptor NGFI-B recognizes A-T rich sequences flanking the 5' end of the ERE hexameric core. A peptide segment at the carboxy terminal end of the DNA-binding domain, which Wilson et al (1992) term the "A box", was shown to direct interactions with these sequences . The crystal structure of GRdbd/DNA complex suggests that the corresponding residues of the orphan receptors could interact with the flanking sequence in the minor groove (Luisi et al, 1991). This segment does exhibit sequence variation in nuclear receptors, and may impart different DNA-binding properties. For instance, the ERE may be tuned by flanking sequences (Alroy and Freedman, 1992), and it is possible that the A box of this receptor makes analogous minor groove contacts.

Half-site	P-Box	Receptor
AGAACA TCT TGT	G--SV	glucocorticoid, progesterone, androgen, mineralocorticoid
AGGTCA TCCAGT	E--GA	estrogen
AGGTCA TCCAGT	E--GG	vitamin D_3, thyroid hormones, retinoids, NGFI-B, ecdysone, ultraspiracle, E75
AGGTCA TCCAGT	E--GS	v-erbA, COUP-TF, knirps, knirps-related, seven-up

FIGURE 4. The "P box" determines nuclear hormone receptor DNA sequence selectivity. For simplicity, consensus half-site sequences are shown.

Discrimination of DNA Targets by Protein-Protein Contacts

The subgroup of non-steroid receptors that recognize the ERE core half site, AGGTCA, may distinguish targets by recognizing the relative orientation and spacing of two such sites. This group includes the thyroid hormone, retinoic acid and vitamin D_3 receptors (TR, RAR, and VDR, respectively) and certain orphan receptors. In some cases, the response elements are arranged as direct repeats (DRs), which suggests that the proteins may bind as asymmetrical dimers (i.e., in a head-to-tail orientation). It is intriguing to note that when the D-box of GRdbd is changed to that of VDR, cooperative binding to the palindromic GRE is abolished (Freedman and Towers, 1991). This supports the idea that VDR (and the related receptors) may not form symmetrical dimers on the DNA target. Instead, VDR and other receptors may form asymmetrical dimers stabilized by protein-protein interactions, as we shall describe.

TR can activate genes from elements with two ERE half-sites arranged in an inverted repeat with no spacing between half sites (see Fig. 1) (Glass et al, 1988) or as a direct repeat with a spacer of four bases (Umesono et al, 1991; Naar et al 1991). For the former element, the center-to-center spacing of the two half-sites is six bases, which is a little more than half a structural repeat of the DNA helix. If two monomers bind to this target, they would lie on opposite faces of the DNA and could not make protein-protein contacts through the DNA-binding domain. However, they could contact through other uncharacterized parts of the receptor, such as the ligand binding domain. It is formally possible that two dimers bind this element, in which a tetramer would cover both faces of the DNA. In the TRE arranged as a direct repeat with a four base spacer, the center-to-center separation of half-sites is 10 bases. Although this distance corresponds to roughly a helical repeat, modeling suggests that the DNA-binding domains would be spaced too far to contact, but would be too close for two pairs of dimers to bind this element side-by-side.

RAR preferentially induces reporter genes under the control of a target with ERE half-sites arranged as direct repeats with a spacing of five bases (Umesono et al, 1991; Naar et al 1991). In this case, the center-to-center spacing would be 11 bases, but two monomers would also be too far to contact through the DNA binding domain, and two pairs of dimers could not bind as they would clash. Again, it is possible that contacts could be mediated through another part of the full-length receptor.

The human vitamin D_3 receptor (hVDR) appears to bind to and activate transcription from a direct repeat element with a three base pair spacer (Umesono et al, 1991). In order to characterize the nature of target selection by hVDR, we over expressed hVDR in *Escherichia coli* and purified the protein to homogeneity. The purified hVDR binds with high affinity to the vitamin D response element of the mouse osteopontin (Spp-1) gene (Noda et al, 1990). This target is composed of a direct repeat of the half-site 5'-GGTTCA-3' with a three base pair separation. We designed a synthetic binding site that had equally high affinity for hVDR carrying a similar half site repeat 5'-AGTTCA-3' (DR+3'; see Fig. 2). Two half-sites appear to be required for high affinity hVDR binding, since the protein will not bind well to an element composed of a single half-site. hVDR has a roughly 30-fold lower affinity for sequences when the separation between the half-sites is increased by one or two bases (DR+4' and DR+5'. In solution and when bound to the DNA target, the purified hVDR is a homodimer. Mixing hVDR with a larger glutathione-S-transferase (GST)-hVDR fusion protein results in three shifted species

60

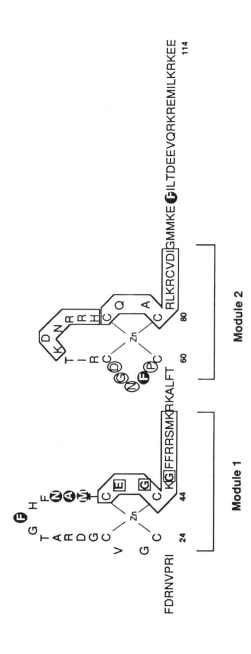

VDR DNA BINDING DOMAIN (VDRF)

FIGURE 5. The primary sequence and secondary structures of the DNA-binding domain of the human vitamin D_3 receptor (Baker et al, 1988). Structural assignments are based on the crystal structure of the GR_{dbd}/GRE complex (Luisi et al, 1991). The helical regions are enboxed. Residues that in the GRdbd direct self-complementary dimerization are encircled; those residues shown to be important in target-site discrimination are boxed. Residues that are proposed here to form an asymmetric dimerization interface (see text) are enboxed and darkened.

when bound to the DR+3' target, suggesting that hVDR binds DNA as a dimer. A similar result is seen when hVDR is mixed with a GST-RXRα fusion and bound to the target probe, confirming that hVDR also forms heterodimers with RXR; RXR appears to be a functional partner of several nuclear receptors (see below). Like homodimeric hVDR, the heterodimer preferentially associates with the DR+3' element over DR+4' and DR+5'.

The spacing preferences of hVDR were further examined using the isolated DNA-binding domain (VDRF; Fig. 5 [Freedman and Towers, 1991]). At low concentrations, a monomeric species will bind with equal affinity to the DR+3', DR+4', and DR+5' sites, suggesting that VDRF is a monomer in solution and consistent with previous findings that the DNA-binding domains of both the glucocorticoid and estrogen receptors are also monomers when not bound to DNA (Hard et al, 1990; Schwabe et al, 1990). As the protein concentration is increased, a second VDRF monomer binds to the DR+3' element; much higher levels are required for the second monomer to bind DR+4' and DR+5'. Two VDRF monomers bind the DR+3' cooperatively (*i.e.*, binding of the first monomer favors binding of a second; see below) but bind DR+4' and DR+5' non-cooperatively. This suggested to us that the binding of the domain to a DR+3' target could occur through a stereospecific alignment of a dimerization interface.

The interactions of two vitamin D_3 receptor DNA-binding domain molecules on a direct repeat element were modelled using the crystal structure of the glucocorticoid receptor DNA-binding domain (GRdbd)-GRE complex (Luisi et al, 1991). This was accomplished by simply rotating one of the GRdbd monomers by 180° in its binding site. Although this type of modelling cannot give accurate details of the interaction, it can reveal regions where contacts between monomers on a DR+3 element could be made. The modelling suggests that intermolecular contacts form between residues in the vicinity of the carboxy-terminus of one monomer with residues in Module 2 of the neighboring monomer. This interface may be supported by intramolecular contacts between residues in the tip of Module 1 and residues in the vicinity of the carboxy-terminus of Module 2 (summarized in Fig. 6).

We used site-directed mutagenesis to test the role of certain residues from the indicated regions in target site selection. Substitutions in the tip of Module 1, the beginning of Module 2, and the boundary of Module 2 all relax spacing specificity. The mutations replace hVDR residues with corresponding amino acids from the thyroid hormone or retinoic acid receptors, which preferentially bind to DR+4 and DR+5 sites, respectively. For example, if the substitutions F_{34}->Y, in the tip of the first finger (Module 1; Fig. 5), and F_{62}->Y, in the D-box of Module 2 (Fig. 5), are introduced into the full-length hVDR, the mutant protein bound equally well as a homodimer to DR+3', DR+4', and DR+5' sites. F_{34} appears to be a critical residue, since a single mutation here also is sufficient to relax specificity, although not quite to the same extent as found for the F_{34}, F_{62} double mutant. Introducing the same double mutation into VDRF also relaxes its spacing selectivity: this mutant protein now also exhibits the same titration behavior irrespective of half-site spacing. Our results further indicate that while the wild-type VDRF occupies the two half-sites of a DR+3' cooperatively (Hill's coefficient = 1.4), this cooperativity is abolished in the F_{34}, F_{62} double mutant (Hill's coefficient = 0.9), and the mutations, do not affect the DNA-binding affinity of the monomer. The mutations must disrupt favorable protein-protein interactions occurring when the monomers are aligned on the DR+3' target.

An additional key region involved in the putative asymmetric interface is located just outside the carboxy terminus of Module 2. We found that a substitution of F_{93} (Fig. 5) to alanine, the corresponding residue of the retinoic acid receptor γ (RARγ), together with additional mutations near the tip of Module 1 ($N_{37}A_{38}M_{39}$ to G, V, S, respectively, as the sequence occurs in RARs) and F_{34}->Y, also lead to a relaxed selectivity for hVDR. A bound heterodimer of this hVDR mutant and RXRα (as a GST-RXR fusion) did not change the relaxed phenotype of this mutant. Cumulatively, these results implicate three distinct segments of the hVDR DNA-binding domain as encoding the asymmetrical dimerization function: i) the tip of the loop region in Module 1; ii) segments of Module 2, including the previously defined "D-box", a region that plays an important role in the formation of the symmetrical dimer interface of steroid receptors; and iii) a region about five residues beyond the carboxy terminus of Module 2.

Using a yeast genetic screen, Wilson et al (1992) have identified a region of the RXRβ receptor that confers transcriptional activation from an element composed of direct repeats of the estrogen response element half-site (AGGTCA) with a single base spacer. Amino acids mediating this effect lie in a region C-terminal to Module 2 which they term the "T box". Wilson et al have suggested that residues in the T box direct protein-protein interactions for RXRβ. We find that specific residues in the corresponding region of hVDR (i.e., F_{93}) play an analogous role and define in part hVDR's preference for a DR+3 element.

We propose that the residues in the T box region of hVDR make an intramolecular packing against residues in the tip of Module 1 (Fig. 6a). For instance, the C-terminal residues F_{93}, I_{94}, and L_{95} may make hydrophobic interactions with F_{34} and with the residues N_{37} A_{38} M_{39} in Module 1. This interaction does not occur in the crystal structure of the GRdbd/DNA complex (Luisi et al, 1991), where the corresponding residues are overall less hydrophobic. The proposed interaction would orient and stabilize the T box of one hVDR monomer to make contacts with the D-box region of Module 2 of an adjacently bound monomer on the DR+3 element (Fig. 6b). The hydrophobic interactions might be disrupted or diminished by the substitutions introduced here, and so affect DNA-binding selectivity through affects on the asymmetrical dimerization interface. The details of the interface may differ from receptor to receptor, but it is possible that other receptors might fold similarly to hVDR so that they too bind selectively to their respective targets through asymmetrical interactions of DNA binding domains.

If full-length nuclear receptors were to associate with DNA in a fashion that mirrors the direct repeat, then these proteins would form virtually endless polymers [Fig. 7, Model (i)] or alternatively, closed rings. Instead, hVDR and other nuclear receptors form stable dimers: either homodimers, or in complex with RXR, heterodimers. This association probably occurs through a well-conserved, self-complementary interface made by the ligand binding domain (Forman and Samuels, 1990). The linkage between the interfaces made by the ligand-binding and DNA binding domains must be sufficiently flexible to permit the latter domain to match the orientation of the sequence repeat of the DNA target site [Fig. 7, Models (ii) and (iii)]. The ligand-binding domain's dimer interface provides the principal organizational effect of bringing two DNA reading heads within proximity of each other, and would reduce the entropy penalty of association of the DNA-binding domain to its target. This interaction would in turn allow the proper interface between two DNA-binding domains to form and selectively match a direct or inverted repeat, depending on the interface. An analogous effect probably also occurs in the trimeric heat-shock factor (Sorger, 1991). An extreme example of nuclear receptor function is provided by the case of COUP-TF, which has been shown to bind as a homodimer and with nearly equal affinity to elements having either direct or inverted repeats with a variety of spacings (Cooney et al, 1992). The DNA-binding and dimerization (i.e., ligand-binding) domains of the COUP-TF monomer may be connected through a linker with high flexibility that permits a

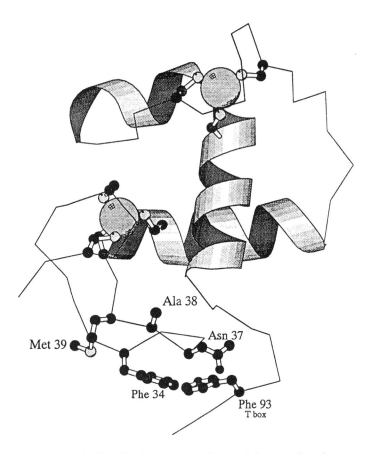

FIGURE 6. Models for the intramolecular and intermolecular contacts involved in asymmetric target-site binding.

(A) A three-dimensional model for the intramolecular interaction of the T box of hVDR with the tip of Module 1. Helices are depicted by ribbons, and the Cα backbone trace by lines. The side-chains of selected residues of the T box and Module 1 tip are shown in ball-and-stick representation, as are the metal coordinating cysteines. The two zinc ions are represented by spheres. The recognition α-helix of Module 1, which lies in the major groove of the DNA, is oriented such that its axis is on the horizon. The principal helix of Module 2 lies vertically. Phe 93 of the T box may pack against Asn 37 and Phe 34, which is supported by Ala 38, and possibly, Met 39. Ile 94 and Leu 95 may also pack against the tip (not shown for clarity). The hydrophobic interactions may orient the T box to make intermolecular contacts, as suggested in (B).

great degree of freedom, and the reading heads themselves probably have little interaction. An analogous effect might occur to a certain extent with the thyroid hormone and retinoic acid receptors, since they both can bind and activate from an inverted repeat element with no spacing (Glass et al, 1988) as well as from direct repeats.

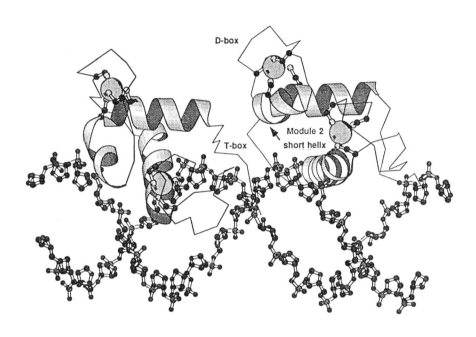

FIGURE 6. Models for the intramolecular and intermolecular contacts involved in asymmetric target-site binding.

(**B**) The orientation of monomers on the DNA. Intermolecular contacts could form between residues of the T box of one monomer and residues of the D box and short α-helix in Module 2 (such as His 75) of the neighboring monomer. For clarity, only the phosphate backbone of the DNA is shown. (A) & (B) were prepared using MOLSCRIPT.

Dimerization through the Ligand Binding Domain

The preference of the VDR, TR and RAR to activate genes from artificial controlling elements that are direct repeats of ERE half sites with a spacing of three, four and five bases, respectively, has been termed the "3-4-5 rule" (Umesono et al, 1991). Binding preferences *in vivo* may not be so simple, because these receptors may form heterodimers with the receptor for 9 cis-retinoic acid, RXR (Yu et al, 1991; Leid et al 1992; Zhang et al, 1992; Kliewer et al, 1992; Bugge et al, 1992; Marks et al, 1992). This interaction potentiates receptor DNA-binding and transactivation, presumably by forming a more stable dimer than the homodimeric species. Whether the "3-4-5 rule" represents a biologically relevant means to confer specific hormone receptor responsiveness to genuine target genes is not presently clear. However, these observations do suggest that binding site selectivity of this group of receptors may be influenced more by organization of the response elements than by particular nucleotide differences *per se,* as appears to be true for the

FIGURE 7. Schematic model showing the relationship between target-site symmetry and protein-protein interactions. In model (i), receptors may form a continuous polymer if the asymmetrical association were sufficiently strong to yield a spacing preference. [Note that this complex superficially resembles the binding of multiple TFIIIA-type zinc fingers to a target, but in that case the individual fingers are physically linked with an appropriate spacing and need not associate by protein-protein interactions]. Model (ii) shows that symmetrical dimerization through the ligand-binding domain (shaded rectangles) maintains the DNA reading heads (blocked arrows) in proximity for association. The two domains are connected by a flexible peptide linker. Spacing selectivity could then be achieved by comparatively weaker asymmetrical association of the DNA binding domains than that required for model (i). Polymers could occur by interactions of adjacently bound dimers (in particular, homodimers), analogous to model (i), and might result in cooperative binding. For heterodimers, such as hVDR/RXR, the asymmetrical dimerization interfaces of the monomers must be equivalent to favor higher order binding, but this may not necessarily be the case. Model (iii) shows the representative case of the steroid receptors, where dyad-symmetric dimers bind to elements composed of inverted repeats. Here, the interfaces are both self-complementary.

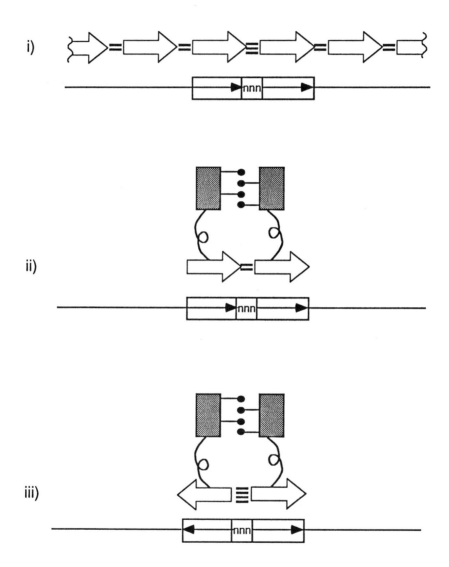

FIGURE 7. Schematic model showing the relationship between target-site symmetry and protein-protein interactions.

discrimination of GRE and EREs. We suggest that spacing and orientation may be influenced by protein-protein contacts in homo- and hetero-dimers in VDR, TR, RAR and related receptors.

In the absence of DNA, the full length GR and ER are predominantly homodimers, and bind exclusively as dimers to their targets. Kumar and Chambon (1988) showed that ER dimerization occurs principally through residues within the ligand-binding domain. This dimerization function is stronger than that of the DNA-binding domain. By deletion and site-directed mutagenesis, this region has been sub-localized to ER residues between 500-540 (Farwell et al, 1990). Interestingly, this region, when aligned with other members of the nuclear receptor superfamily, contains a conserved heptad repeat of hydrophobic amino acids that could form a dimerization interface analogous to a coiled-coil interface of leucine zippers (Landschultz et al, 1988). Conceivably, coiled-coil interactions could take place within this region. Forman et al (1989) have proposed that this region can direct both homo- and hetero-dimerization. Indeed, this region appears to mediate heterodimerization between several nuclear receptors and RXR (see above). Recent work also suggests that GR can functionally trimerize with jun-fos and jun-jun complexes, possibly through this domain. This interaction will in turn determine whether transcription of a given gene will be enhanced or represented by the receptor (reviewed in Miner and Yamamoto, 1991). The ligand binding domain may also mediate interactions with the heat-shock protein hsp90, which forms a complex with the receptor in the cytoplasm (Pratt et al, 1988, and this volume). The regulatory consequences of such baroque interactions of glucocorticoid and, possibly, other nuclear receptors with different classes of transcription factors are clearly profound.

CONCLUSIONS

The crystal structure of the GRdbd/DNA complex has provided stereochemical details of protein-DNA and protein-protein interactions, and the importance of these contacts have been corroborated by mutational analyses. The marriage of structural and functional methods has permitted a better understanding of not only GR DNA binding, but also binding by its highly homologous relatives comprising the nuclear receptor superfamily. The structural and functional data from GR and ER, which allowed us to explain how a few amino acids direct discrimination between closely related GREs and EREs, have been consolidated and extrapolated to describe the DNA binding modes of other nuclear receptors, such as the vitamin D_3, thyroid hormone, and retinoic acid receptors. These latter receptors appear to bind DNA specifically through discrimination of the spacing and orientation of homologous half-sites.

The results described here represent an early stage of an expanding investigation of an intriguing and important superfamily. In our work, only one aspect of nuclear receptor action has been addressed, leaving many questions concerning other functions unanswered. How, for example, are ligands recognized and discriminated? How does ligand binding activate nuclear localization and/or transcriptional activation? How do interactions with other proteins, such as hsp90 and AP-1, affect function, and where is the site of interaction? What converts a nuclear receptor from a transcriptional activator to a repressor? These and other questions are fundamental to a detailed understanding of nuclear receptor action, and will be greatly enhanced by further studies combining structural and functional approaches.

REFERENCES

Alroy I and Freedman LP (1992): DNA binding analysis of glucocorticoid receptor specificity mutants. *Nucl Acids Res* 20: 1045-1052.

Amero SA, Kretsinger RH, Moncrief ND, Yamamoto KR and Pearson WR (1992): The origin of nuclear receptor proteins: a single precursor distinct other transcription factors. *Mol Endocrinol* 6: 3-7.

Beato M (1989): Gene regulation by steroid hormones. *Cell* 56: 335-344.

Berg JM (1988): Proposed structure for the zinc-binding domains from transcription factor IIIA and related proteins. *Proc Natl Acad Sci USA* 85: 99-102.

Berg J (1990): Zinc fingers and other metal-binding domains. *J Biol Chem* 265: 6513-6516.

Bugge TH, Pohl J, Lonnoy O and Stunnenberg H (1992): RXRα, a promiscuous partner of retinoic acid and thyroid hormone receptors. *EMBO J* 11: 1409-1417.

Conney AJ, Tsai SY, O'Malley BW and Tsai M-J (1992): *Mol Cell Biol* 12: 4153-4160.

Dahlman-Wright K, Siltata-Roos H, Carstedt-Duke J and Gustafsson J-A (1990): Protein-protein interactions facilitate DNA binding by the glucocorticoid receptor DNA binding domain. *J Biol Chem* 265: 14030-14035.

Dahlman-Wright K, Wright A, Gustafsson J-A and Carstedt-Duke J (1991): Interaction of the glucocorticoid receptor DNA-binding domain with DNA as a dimer is mediated by a short segment of five amino acids. *J Biol Chem* 266: 3107-3112.

Danielson M, Hinck L and Ringold GM (1989): Two amino acids within the knuckle of the first zinc finger specify response element activation by the glucocorticoid receptor. *Cell* 57: 1131-1138.

Farwell SE, Lees JA, White R and Parker MG (1990): Characterization and localization of steroid binding and dimerization activities in the mouse estrogen receptor. *Cell* 60: 953-962.

Forman BM, Yang C-R, Au M, Cassanova J, Ghysdael J and Samuels HH (1989): A domain containing leucine zipper-like motifs mediate novel *in vivo* interactions between the thyroid hormone and retinoic acid receptors. *Mol Endocrinol* 3: 1610-1626.

Freedman LP, Luisi BF, Korszun ZR, Basavappa R, Sigler PJ and Yamamoto KR (1988a): The function and structure of the metal coordination sites within the glucocorticoid receptor DNA binding domain. *Nature* 334: 543-546.

Freedman LP, Yamamoto KR, Luisi BF and Sigler PJ (1988b): More fingers in hand. *Cell* 54: 444.

Freedman LP and Towers T (1991): DNA binding properties of the vitamin D3 receptor zinc finger region. *Mol Endocrinol* 5: 1815-1826.

Glass CK, Holloway JM, Devary OV and Rosenfeld MG (1988): The thyroid hormone receptor binds with opposite transcriptional effects to common sequence motif in thyroid hormone and estrogen response elements. *Cell* 54: 313-323.

Godowski PJ, Picard D and Yamamoto KR (1988): Signal transduction and transcriptional regulation by glucocorticoid receptor-LexA fusion proteins. *Science* 241: 812-816.

Green S, Kumar V, Theulaz I, Whali W and Chambon P (1988): The N-terminal DNA binding zinc finger of the oestrogen and glucocorticoid receptors determines target gene specificity. *EMBO J* 7: 3037-3044.

Hard T, Kellenbach E, Boelens R, Maler BA, Dahlman K, Freedman LP, Carlstedt-Duke J, Yamamoto KR, Gustafsson J-A and Kaptein R (1990): Solution structure of the glucocorticoid receptor DNA binding domain. *Science* 249: 157-160.

Hard T, Dahlman K, Carstedt-Duke J, Gustafsson J-A and Rigler R (1990): Cooperativity and specificity in the interactions between DNA and the glucocorticoid receptor DBA-binding domain. *Biochemistry* 29: 5358-5364.

Hughes MR, Malloy PJ, Kieback DG, Kesterson RA, Pike JW, Feldman D and O'Malley BW (1989): Point mutations in the human vitamin D receptor gene associated with hypocalcemic rickets. *Science* 242: 1702-1705.

Kliewer SA, Umesono K, Mangelsdorf DJ and Evans RM (1992): Retinoid X receptor interacts with nuclear receptors in retinoic acid, thyroid hormone, and vitamin D_3 signalling. *Nature* 355: 446-449.

Klock G, Strahle U and Schutz G (1987): Oestrogen and glucocorticoid responsive elements are closely related but distinct. *Nature* 329: 734-736.

Kumar V and Chambon P (1988): The estrogen receptor binds tightly to its responsive element as a ligand-induced homodimer. *Cell* 55: 145-156.

LaBaer J (1989): Ph.D. thesis, University of California, San Francisco, USA.

Landschulz WH, Johnson PF and McKnight SL (1988): The leucine zipper: a hypothetical structure common to a new class of DNA binding proteins. *Science* 240: 1759-1764.

Laudet V, Hanni C, Coll J, Catzflis F and Stehelin D (1992): Evolution of the nuclear receptor gene superfamily. *EMBO J* 11: 1003-1013.

Leid M, Kastner P, Lyons R, Nakshatri H, Saunders M, Zacharewski T, Chen J-A, Staub A, Garnier J-M, Mader S and Chambon P (1992): Purification, cloning, and RXR identity of the HeLa cell factor with which RAR or TR heterodimerizes to bind target sequences. *Cell* 68: 377-395.

Luisi BF, Xu WX, Otwinowski Z, Freedman LP, Yamamoto KR and Sigler PB (1991): Crystallographic analysis of the interaction of the glucocorticoid receptor with DNA. *Nature* 352: 497-505.

Luisi BL (1992): Zinc standard for economy. *Nature* 356: 379-380.

Mader S, Kumar V, deVereneuil H and Chambon P (1989): Three amino acids of the oestrogen receptor are essential to its ability to distinguish an oestrogen from a glucocorticoid-responsive receptor. *Nature* 338: 271-274.

Marks MS, Hallenbeck PL, Nagata T, Segars JH, Appella E, Nikodem VM and Ozato K (1992): H-2RIIBP (RXRβ) heterodimerization provides a mechanism for combinatorial diversity in the regulation of retinoic acid and thyroid hormone responsive genes. *EMBO J* 11: 1419-1435.

Marmomstein R, Carey M, Ptashne M and Harrison SC (1992): DNA recognition by GAL4: structure of a protein-DNA complex. *Nature* 356: 408-414.

Miner JN and Yamamoto KR (1991): Regulatory cross-talk at composite response elements. *TIBS* 16: 423-426.

Naar AM, Boutin J-M, Lipkin SM, Yu VC, Holloway JM, Glass CK and Rosenfeld MG (1991): The orientation and spacing of core DNA-binding motifs dictate selective transcriptional responses to three nuclear receptors. *Cell* 65: 1267-1279.

Nordeen SK, Suh BJ, Kuhnel B and Hutchison CA (1990): Structural determinants of a glucocorticoid receptor recognition element. *Mol Endocrinol* 4: 1866-1873.

Parker MG (1991): "Nuclear Hormone Receptors". London: Academic Press.

Patel L, Abate, C and Curran T (1990): Altered protein conformation on DNA binding by Fos and Jun. *Nature* 347: 572-575.

Payvar F, DeFranco D, Firestone GL, Edgar B, Wrange O, Okret S, Gustafsson JA and Yamamoto KR (1983): Sequence-specific binding of the glucocorticoid receptor to MTV-DNA at sites within and upstream of the transcribed region. *Cell* 35: 381-392.

Picard D, Salser SJ and Yamamoto KR (1988): A movable and regulable inactivation function within the steroid binding domain of the glucocorticoid receptor. *Cell* 54: 1073-1080.

Ponglikimongkol M, Green S and Chambon P (1988): Genomic organization of the human oestrogen receptor gene. *EMBO J* 7: 3385-3388.

Pratt WB, Jolly DJ, Pratt DV, Hollenberg SM, Gigure V, Cadepond FM, Schweizer-Groyer G, Catelli M-G, Evans RM and Baulieu E-E (1988): A region of in the steroid binding domain determines formation of the non-DNA-binding 9S glucocorticoid receptor complex. *J Biol Chem* 263: 267-273.

Sally A, Kretsinger RH, Moncrief ND, Yamamoto KR and Pearson WR (1992): Minireview: The origin of nuclear receptor proteins: A single precursor distinct from other transcription factors. *Mol Endocrinol* 6: 3-7.

Schena M, Freedman LP and Yamamoto KR (1989): Mutations in the glucocorticoid receptor zinc finger region that distinguish interdigitated DNA binding and transcriptional enhancement activities. *Genes Develop* 3: 1590-1601.

Schwabe JWR, Neuhaus D and Rhodes D (1990): Solution structure of the DNA binding domain of the oestrogen receptor. *Nature* 348: 458-461.

Seagraves WA (1991): Something old, some things new: the steroid receptor superfamily in Drosophila. *Cell* 67: 225-228.

Sorger PK (1991): Heat shock factor and the heat shock response. *Cell* 65: 363-366.

Strahle U, Klock G and Schutz G (1987): A DNA sequence of 15 base pairs is sufficient to mediate both glucocorticoid and progesterone induction of gene expression. *Proc Natl Acad Sci USA* 84: 7871-7875.

Umesono K and Evans RM (1989): Determinants of target gene specificity for steroid/thyroid hormone receptors. *Cell* 57: 1139-1146.

Umesono K, Murakami KK, Thompson CC and Evans RM (1991): Direct repeats as selective response elements for the thyroid hormone, retinoic acid, and vitamin D3 receptors. *Cell* 65: 1255-1266.

Wilson TE, Fahrner TJ, Johnston M and Milbrandt J (1991): Identification of the DNA binding site for NGFI-B by genetic selection in yeast. *Science* 252: 1296-1300.

Wilson TE, Paulsen RE, Padgett KA and Milbrandt J (1992): Participation of non-zinc finger residues in DNA binding by two orphan receptors. *Science* 256: 107-110.

Yu VC, Delsert C, Andersen B, Holloway JM, Devary OV, Naar AM, Kim SY, Boutin J-M, Glass CK and Rosenfeld MG (1991): RXRβ: a coregulator that enhances binding of retinoic acid thyroid hormone, and vitamin D receptors to their cognate response elements. *Cell* 67: 1251-1266.

Zhang X-k, Hoffmann B, Tran PB-V, Graupner G and Pfahl M (1992): Retinoid X receptor is an auxiliary protein for thyroid hormone and retinoic acid receptors. *Nature* 355: 441-446.

Note added in proof: The description of direct repeat binding by the vitamin D3 receptor appearing on pages 59-63 has since been published. The reference is:

Towers, T.L., Luisi, B.F., Asianov, A., and Freedman, L.P. (1993) DNA target selectivity by the vitamin D3 receptor: mechanism of dimer binding to an asymmetric repeat element. *Proc Natl Acad Sci USA 90: 6310-6314.*

ESTROGEN RECEPTOR INDUCED DNA BENDING

David J. Shapiro
Department of Biochemistry
University of Illinois

Geoffrey L. Greene
The Ben May Institute
The University of Chicago

Ann M. Nardulli
Department of Physiology
University of Illinois

ABSTRACT

We have examined the ability of the full-length estrogen receptor (ER) and the purified estrogen receptor DNA binding domain to bend DNA on binding to estrogen response element (ERE). Purified ER DNA binding domain (DBD) expressed in *E. coli*, crude human ER expressed in MCF-7 cells, and partially purified human ER expressed in yeast, bound specifically to a series of circularly permuted ERE-containing DNA fragments. Binding of ER to a single ERE induced a reproducible DNA bend of 56°. This was 1.65 fold greater than the 34° bending angle induced by binding of the ER DNA binding domain. The DNA bending angle induced was the same whether the salt-extracted receptor was unoccupied, occupied by 17β-estradiol, or occupied by trans-hydroxytamoxifen. To determine if proteins associated with ER in

STEROID HORMONE RECEPTORS
V. K. Moudgil, Editor
© 1993 Birkhäuser Boston

MCF-7 cells affect the degree of bending, we examined the ability of partially purified human ER, expressed in yeast, to bend DNA. The degree of bending induced by the partially purified yeast ER, and by the crude MCF-7 cell ER, was the same. When two EREs were present in the DNA fragment, 1.2 and 1.55 fold increases in DNA bending were observed for the full-length ER, and for the ER DNA binding domain, respectively. Our demonstration that binding of human estrogen receptor to the ERE induces DNA bending suggests a role for DNA bending in ER-induced transcription activation.

INTRODUCTION

Because many prokaryotic and eukaryotic transcription factors bend DNA upon binding to their cognate recognition sequences (Giese et al, 1992; Nardulli et al, 1993 and references therein), it has been proposed that DNA bending is involved in transcription activation. A role for DNA bending has been clearly demonstrated in studies of the role of integration host factor in site-specific integration of bacteriophage lambda DNA into the bacterial chromosome (reviewed in Landy, 1989; Goodman et al, 1990; Nash, 1990). The fact that intrinsically bent DNA can replace protein binding sites and mediate both repression and activation of prokaryotic transcription provides additional evidence that DNA bending plays a role in the transcription (Bracco et al, 1989; Rojo and Salas, 1991). There is less data directly supporting a role for DNA bending in transcription regulation by eukaryotic proteins. The lymphoid enhancer binding factor, LEF-1, and the related testis determining factor, SRY, are transcription regulatory proteins which bind to DNA and induce sharp DNA bends. The ability of LEF-1 to functionally replace the prokaryotic DNA bending protein, integration host factor, in facilitating communication between distant protein binding sites in a recombination assay (Giese et al, 1992), strongly supports the view that LEF-1 and SRY regulate transcription through their ability to induce DNA bending. The observations that the transcriptionally active fos-jun heterodimer bends DNA in the opposite direction as the transcriptionally inactive jun-jun homodimer (Kerppola and Curran, 1991), and that several yeast promoters contain T-rich regions, which are essential for efficient transcription (Lue et al, 1989), also suggests a role for DNA bending in eukaryotic transcription activation.

The estrogen receptor is a member of the steroid/nuclear receptor gene superfamily of ligand regulated transcription factors (Reviewed in Beato, 1989; Evans 1988). Despite extensive study, a detailed mechanism for transcription activation by estrogen receptor (ER), and by other nuclear receptors, has not been described. It has been hypothesized that steroid hormone receptors activate transcription either directly by protein-protein contacts, which facilitate interaction between components of the basal transcription apparatus and the promoter, or indirectly by stabilizing binding of other transcription factors to their recognition sequences (Schule et al, 1988; Strahle et al, 1988; Tasset et al, 1990). It has also been suggested that steroid receptors may act by changing the organization of nucleosomes which contain transcription factor binding sites (Pina et al, 1990; Hager and Archer, 1991), and that DNA supercoiling may influence the interaction of steroid receptors with hormone response elements (Pina et al, 1990). The possibility that steroid receptors activate transcription by inducing changes in DNA structure had not been extensively studied. We therefore decided to examine in more detail the ability of ER to induce DNA bending.

Because it is difficult to prepare substantial quantities of pure, biologically active, estrogen receptor, we employed the ER DNA binding domain in our initial studies. This region of the receptor is responsible for specific interaction with the ERE (reviewed in Nardulli et al, 1991; Chang et al, 1992). Detailed information about the structure of the DNA binding domain and its interaction with hormone response elements has been obtained by NMR and x-ray crystallography of the purified glucocorticoid and estrogen receptor DNA binding domains (Hard et al, 1990; Schwabe et al, 1990; Luisi et al, 1991). In addition, the estrogen (Nardulli et al, 1991; Waterman et al, 1988), progesterone (Guiochon-Mantel et al, 1988; Klein-Hitpass et al, 1990), and glucocorticoid (Miesfeld et al, 1987; Freedman et al, 1989; Debs et al, 1990) receptor DNA binding domains have been shown to retain at least some ability to activate transcription of responsive genes.

In order to extend these studies to crude extracts containing full-length ER, it was necessary to develop conditions for studying the specific interaction of estrogen receptor with the ERE on the relatively long (400-500 nucleotide) DNA fragments used in DNA bending studies. These modified conditions enabled us to obtain quantitative measurements of DNA bending induced by a memeber of the steroid/nuclear receptor gene superfamily, and to compare the DNA bending angles induced by crude MCF-7 cell ER, by partially purified yeast ER, and by the purified ER DNA binding expressed in *E. coli.*

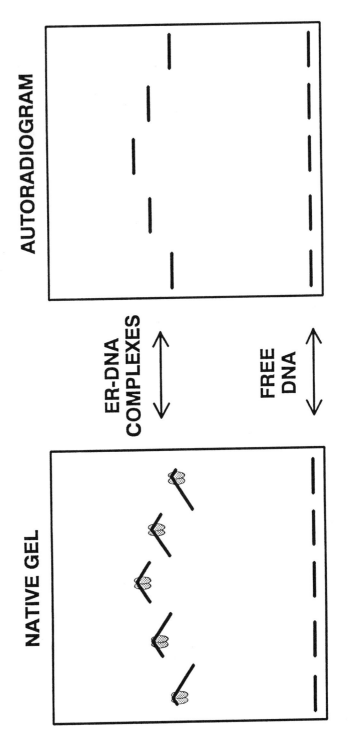

FIGURE 1. A gel electrophoresis assay for DNA bending.

RESULTS

A gel electrophoresis assay for DNA bending. Altered electrophoretic mobility has been used in several prokaryotic systems to demonstrate that transcription activators and repressors bind to target DNA and cause it to bend (reviewed in Crothers et al, 1991; Dieckman, 1992). These assays are based on the observation that DNA fragments that are bent near the middle migrate more slowly than DNA fragments that are bent near the end. In order to study DNA bending induced by binding of a protein to the DNA fragment, a plasmid is constructed which allows the isolation of DNA fragments containing the protein binding site of interest in a variety of locations, ranging from near the end of the DNA fragment, to the middle of the fragment. If binding of the protein induces DNA bending, the fragments with the binding site near the end will migrate more rapidly than those with the binding site near the middle. This type of DNA bending assay is illustrated in Fig. 1. The resulting bow shaped pattern, with the slowest migrating fragment corresponding to the fragment with the binding site in the middle, and the most rapidly migrating fragments corresponding to the fragments with binding site near the end, indicates DNA bending. We applied this type of DNA bending assay in our studies of the binding of estrogen receptor to the estrogen response element.

Because the ERE is located in different positions in each DNA fragment, it was important to eliminate the possibility that the flanking DNA sequence surrounding the ERE could induce a DNA bend of its own, and complicate our analysis of ER-induced DNA bending. We therefore constructed a vector containing a single ERE with identical 5' and 3' fragments on either side of the ERE. This plasmid is illustrated schematically in Fig. 2. By digesting this plasmid, which we designated ERE Bend I, with different restriction enzymes we generated a series of six 430 base pair fragments containing an ERE at various positions, from the middle of the fragment, to near the ends of the fragment (Fig. 2). These fragments were identical in size and base composition and differed only in the location of the ERE in the fragment.

FIGURE 1. A gel electrophoresis assay for DNA bending. The left panel illustrates the protein-DNA complexes formed by binding of a dimeric protein which induces DNA bending at its binding site. The right panel illustrates the appearance of the autoradiogram corresponding to this type of protein-DNA complex.

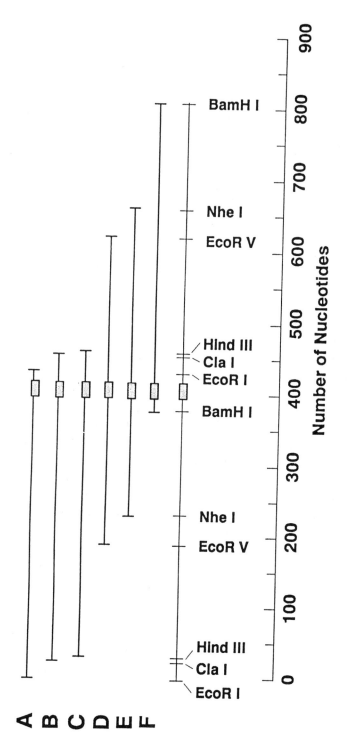

FIGURE 2. DNA fragments used in bending analysis.

Because we had previously carried out extensive studies on the interaction of the ER DNA binding domain with various EREs (Nardulli et al, 1991; Chang et al, 1992), we began our studies of DNA bending by examining the binding of the purified *Xenopus laevis* estrogen receptor (XER) DNA binding domain to the ERE. In earlier work we had expressed the DNA binding domain at high levels in *E. coli*, purified the XER DNA binding domain (DBD) to near homogeneity and examined its properties (Nardulli et al, 1991). Although the isolated XER DNA binding domain was shown to be a monomer in solution, it appeared to occupy both halves of the ERE palindrome. The DBD activated transcription of an estrogen-responsive vitellogenin-derived promoter (Nardulli et al, 1991), but bound less tightly to the ERE than the full-length estrogen receptor (Ponglikitmongkol et al, 1990; Nardulli et al, 1991; Chang et al, 1992). Studies of the interaction of the purified XER DNA binding domain with the ERE therefore provide useful information about the properties of the full-length ER.

Binding of the XER DNA Binding Domain to the ERE Induces DNA Bending. We prepared labeled DNA fragments containing a single ERE in various positions ranging from the middle to the end of the fragment (Fig. 2), and examined their electrophoretic mobility in gel mobility shift assays. In the absence of the purified *Xenopus* estrogen receptor DBD, the migration of all the ^{32}P-labeled DNA fragments was similar, irrespective of the position of the ERE in the fragment (Fig. 3, panel A, -DBD). After incubation with purified DBD, the migration of the ^{32}P labeled fragments was affected by the position of the ERE in the fragment (Fig. 3, panel A, + DBD). The mobility of the DBD-DNA complex was greatest when the ERE was present near the end of the fragment (Fig. 2, + DBD, lanes A and F), lowest when the ERE was located in the middle of the fragment (Fig. 3 + DBD, lane D) and intermediate when the ERE was located at internal positions (Fig. 3, + DBD, lanes B, C, and E); (Nardulli and Shapiro, 1992).

FIGURE 2. DNA fragments used in bending analysis. The DNA bending vector ERE Bend I was constructed (Nardulli and Shapiro, 1992) and digested with either Eco RI, Cla I, Hind III, Eco RV, Nhe I, or Bam HI to produce 430 base pair DNA fragments with a single consensus ERE at the indicated positions. The consensus ERE sequence (GATAGGTCACTGTGACCTATC) is designated by the shaded box. The restriction sites in each fragment are noted. (From Nardulli and Shapiro, 1992, reprinted with permission).

82

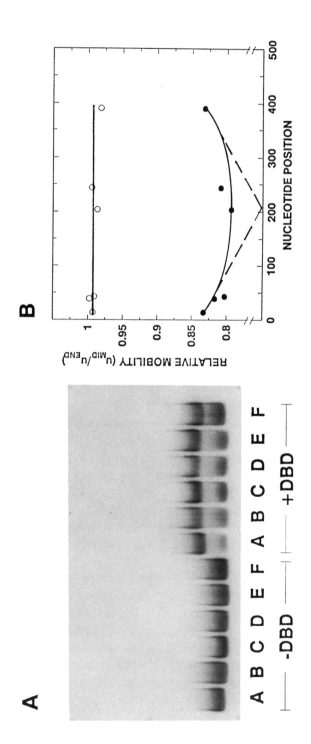

FIGURE 3. Induction of DNA binding.

The data from five independent experiments is summarized diagrammatically in Figure 3, panel B, and supports the idea that the electrophoretic mobility of DBD-DNA complexes is dependent on the position of the ERE within the DNA fragment. In order to be sure that the gel conditions used were not a factor in the apparent difference in mobility of the DBD-DNA complexes, these experiments were carried out with 3.5, 6, 8, and 10% gels for varying periods of time. In all cases the DBD-DNA complex formed with fragment D in which the ERE was located in the center of the DNA fragment had the lowest mobility, complexes with fragments B, C, and E had intermediate mobility, and complexes with fragments A and F had the highest mobility (Nardulli and Shapiro, 1992). This pattern indicated that binding of the XER DBD to the ERE induced DNA bending. These data also supported the view that the site of DNA bending was located in or very near the ERE (Fig. 3, Panel B).

Determination of the degree of DBD-induced DNA bending. We next quantitated the degree of DNA bending. Although the differences in the mobility of the DNA fragments was small, it was possible to use the method of Thompson and Landy (Thompson and Landy, 1988) to determine the degree of DNA bending. In this method, DNA fragments containing $(dA)_6$ tracts, which contain intrinsic bends $(18°/(dA)_6$ tract), were used as bending standards. When 2-7 $(dA)_6$ tracts were present in the center of the fragments (Fig. 4, Panel A upper bands), migration was slower than when equivalent numbers of tracts were present near the end of the fragments (Fig. 4 panel, A lower bands). To obtain a standard curve for DNA bending, we calculated the relative mobility of these DNA bending standards (Nardulli and Shapiro, 1992). The degree of bending

FIGURE 3. DNA bending is induced by binding of the XER DBD to the ERE. In Panel A labeled ERE BEND I DNA fragments were incubated in the absence (-DBD) or presence (+DBD) of purified estrogen receptor DBD and fractionated on an acrylamide gel. The gel was dried and subjected to autoradiography. In panel B the relative mobility (migration of a DNA fragment relative to the most rapidly migrating fragment) of fragments A-F is plotted as a function of the distance of the ERE from the 3' end of the DNA (average of 5 independent experiments). The migration minimum (dashed lines) was determined by extrapolating from the linear ends of the curve, and is near the position of the ERE. Data for uncomplexed fragments (O) and for the DBD-DNA complexes (●) are shown. (From Nardulli and Shapiro, 1992, reprinted with permission).

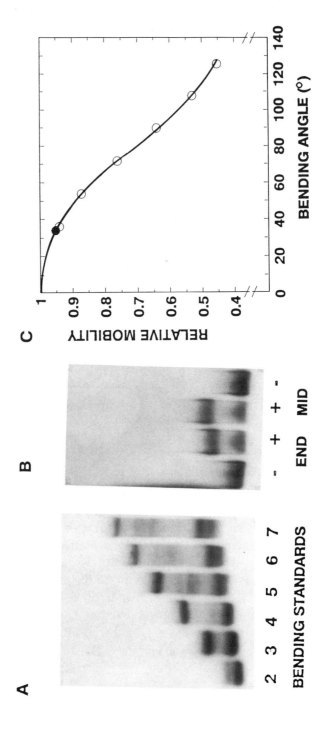

FIGURE 4. Determination of the bending angle of a DBD-ERE complex.

of an unknown DNA fragment could then be determined by comparing its relative mobility with the known standards on the curve.

To calculate the bending angle induced by binding of the XER DBD to the ERE, the relative mobilities of DBD-DNA complexes with EREs in the middle or at the end of the DNA fragments were compared (Fig. 4, Panel B, MID and END). The relative mobility of the DBD-DNA complexes from 14 separate determinations was 0.95 ± 0.004 (S.E.M.). This was very similar to the relative mobility of the bending standard containing 2 $(dA)_6$ tracts, and corresponds to a bending angle of 34° (Fig. 4, Panel C); (Nardulli and Shapiro, 1992).

Binding of the DBD to two EREs increased DNA bending. Many estrogen responsive genes contain multiple EREs (Walker et al, 1984; Seiler-Tuyns et al, 1986; Chang et al, 1992), not the single ERE used in our initial bending studies. We therefore examined whether DBD binding to two EREs would result in increased bending. We constructed a different series of bending vectors, containing one and two EREs, respectively. The center-to-center spacing between the two EREs was 21 nucleotides, which separated the EREs by two turns of the DNA helix. This is similar to the 20 nucleotide spacing which separates the two EREs in the vitellogenin B1 promoter (Chang and Shapiro, 1990; Chang et al, 1992). We found that the degree of DNA bending induced by binding of the XER DBD to 2 EREs was 1.55 fold greater than the degree of bending seen on binding of the DBD to 1 ERE (Nardulli and Shapiro, 1992).

Although these studies clearly demonstrated that binding of the XER DNA binding domain to the ERE induces DNA bending, we still wished to determine if the full-length estrogen receptor induced DNA bending. In addition, the actual degree of DNA bending induced by binding to a hormone response element had not been determined for any member of the steroid/nuclear receptor gene superfamily. We therefore

FIGURE 4. Determination of the bending angle of a DBD-ERE complex. In panel A ^{32}P-labeled DNA bending standards containing 2-7 $(dA)_6$ tracts in the middle (upper bands), or at the end (lower bands) of the DNA fragments were fractionated on an acrylamide gel and visualized by autoradiography. In panel B fragments A and D, which contained a single ERE at either the end (END) or the middle (MID) of the DNA fragment, were incubated in the absence (-) or presence (+) of XER DBD. In panel c the relative mobility of the DNA bending standards (O) and of the XER DBD (●) are plotted as a function of the bending angle. (From Nardulli and Shapiro, 1992, reprinted with permission).

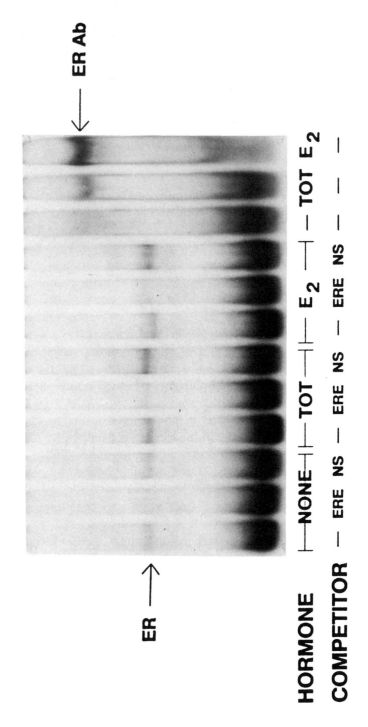

FIGURE 5. Binding of ER in crude extracts from MCF-7 cells to an ERE-containing DNA fragment.

examined DNA bending induced by binding of crude and partially purified human estrogen receptor (hER) to the ERE.

Human ER from MCF-7 cells binds specifically to ERE-containing DNA fragments. We elected to use crude extracts from MCF-7, human breast cancer cells. These cells contain significant levels of functional ER and any cellular proteins which may be required for ER action. The moderate level of hER present in crude MCF-7 cell extracts, and the relatively long 430 nucleotide DNA fragments used in the DNA bending assay required the development of modified conditions for gel mobility shift assays. We found that preincubation of the crude MCF-7 cell extracts with unlabeled sheared salmon sperm DNA prior to addition of the labeled ERE-containing DNA fragments was required in order to reduce non-specific binding of extract proteins to the DNA fragments. Other unlabeled competitor DNAs including poly dI/dC, pBR322, and pUC were unable to effectively titrate out the non-specific DNA binding proteins (Nardulli et al, 1993). These studies appear to be the first to use crude extracts of naturally occurring ER to achieve sequence-specific binding to large DNA fragments. Using these modified gel shift conditions we examined the binding of crude MCF-7 cell hER to the ERE-containing DNA fragments.

hER from MCF-7 cells interacts specifically with ERE-containing DNA fragments. Since we wished to examine the effect of estrogen and antiestrogen binding on DNA bending, we prepared whole cell extracts from MCF-7 cells that had been exposed to ethanol vehicle, trans-hydroxytamoxifen, or 17β-estradiol. The extracts were incubated with a labeled DNA fragment containing a single ERE and fractionated on a nondenaturing polyacrylamide gel. ER present in each of these extracts

FIGURE 5. Binding of ER in crude extracts from MCF-7 cells to an ERE-containing DNA fragment. Whole cell extracts were prepared from MCF-7 cells that had been exposed to ethanol vehicle (NONE), tans-hydroxytamoxifen (TOT), or 17β-estradiol (E_2) for 40 minutes prior to harvest. Extracts were incubated with a 430 base pair ^{32}P-labeled ERE-containing DNA fragment (Fragment C, see Fig. 2) using the modified gel shift conditions we developed (Nardulli et al, 1993). The data indicates ER-specific binding (denoted by ER→) in the presence of no additional competitor DNA (-), a 100 fold excess of an unlabeled non-specific DNA fragment (NS), or a 100 fold excess of unlabeled ERE-containing DNA fragment (ERE). An ER-specific polyclonal antibody bound to the ER-DNA complex and reduced its mobility (←ER Ab). (From Nardulli et al, 1993, reprinted with permission).

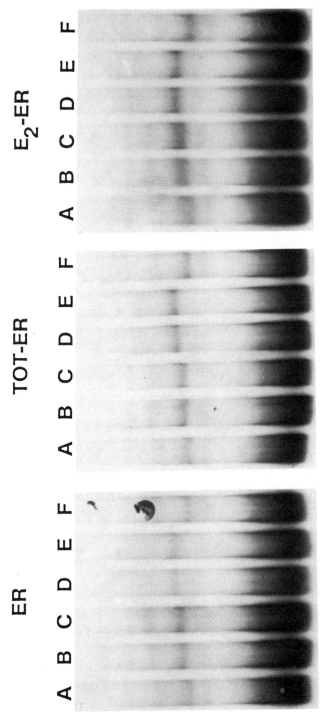

FIGURE 6. Human estrogen receptor bound to an ERE induces DNA bending.

yielded a single major protein-DNA complex (Fig. 5 ER→). Complexes formed with extracts from cells which had been exposed to 17β-estradiol (Fig. 5, E_2), or to trans-hydroxytamoxifen (Fig. 5, TOT) migrated slightly faster or slower, respectively, than complexes from cells that had not been exposed to hormone (Fig. 5, NONE). Other laboratories have also noted these differences in mobility of ER-ERE complexes, which are usually attributed to conformational changes in the receptor induced by ligand binding (Kumar and Chambon, 1988; Brown and Sharp, 1990; Tzukerman et al, 1990; Curtis and Korach, 1990; Fawell et al, 1990).

The large size (430 nucleotides) of the ERE-containing DNA fragment, and the use of crude MCF-7 cell extracts, made it important to demonstrate that the gel shifted band represented ER bound to the ERE. The interaction appeared to be specific for the ERE since the gel shifted protein-DNA complex disappeared when a 100 fold excess of an unlabeled 51 base pair fragment containing a single consensus ERE was present in the incubations (Fig. 5, ERE), and was unaffected by the presence of a 100 fold excess of a non-specific 51 base pair DNA fragment (Fig. 5, NS). To demonstrate that it was ER in the MCF-7 cell extracts which was bound to the ERE, we showed that the addition of ER specific polyclonal antibodies to the incubations greatly reduced the mobility of the protein-DNA complexes (Fig. 5 ←ER Ab).

The intact estrogen receptor bends an ERE-containing DNA fragment. We analyzed the interaction of hER with an ERE, using our ERE Bend I plasmid (Fig. 2). To determine the degree of DNA bending induced by unoccupied ER, by ER occupied with antiestrogen, and by ER occupied with estrogen, whole cell extracts were prepared from MCF-7 cells that had been treated with ethanol vehicle, trans-hydroxytamoxifen, or 17β-estradiol, respectively. Extracts were incubated with the same set of ^{32}P labeled DNA fragments used to study DNA bending induced by the XER DNA binding domain (Fig. 2), and separated on nondenaturing acrylamide gels. Unoccupied receptor (Fig. 6, ER), receptor occupied by antiestrogen (Fig. 6, TOT-ER), and receptor occupied by estrogen (Fig. 6, E_2-ER), formed complexes with very similar electrophoretic mobility. Complexes formed with fragment A or fragment F, which contained an

FIGURE 6. Human estrogen receptor bound to an ERE induces DNA bending. Whole cell extracts from MCF-7 cells that had been exposed to no hormone (ER), transhydroxytamoxifen (TOT-ER), or 17β-estradiol (E_2-ER) were incubated with labeled DNA fragments A-F (see Fig. 2), and analyzed by polyacrylamide gel electrophoresis. (From Nardulli et al, 1993, reprinted with permission).

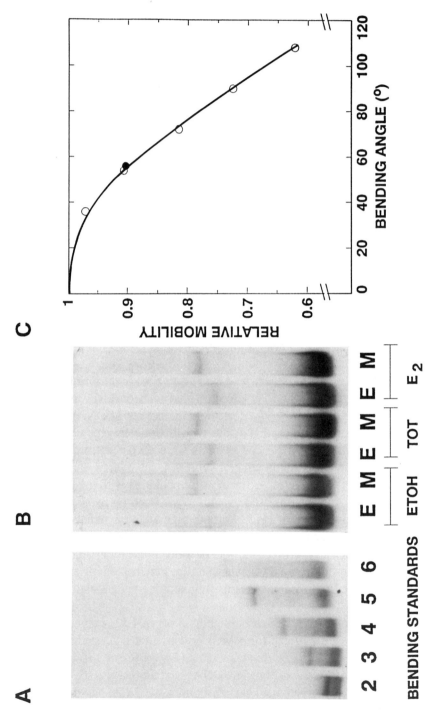

FIGURE 7. Determination of the DNA bending angle induced by MCF-7 cell ER.

ERE at the 5' or 3' end of the DNA fragment, respectively, migrated most rapidly. The complex formed with fragment D, which contained an ERE in the middle of the DNA fragment, migrated most slowly. The bow-shaped pattern of migration observed with each of these ER-containing extracts (Fig. 6) was indicative of DNA bending (Nardulli et al, 1993).

Determination of the degree of DNA bending induced by binding of ER to the ERE. Quantitation of the degree of DNA bending was carried out by using the same series of DNA bending standards used to determine the degree of DNA bending induced by the XER DBD. By comparing the electrophoretic mobilities of DNA fragments containing 2-6 $(dA)_6$ tracts in the middle or near the end of the DNA fragments, we were able to measure bending angles of 36°-108° (Fig. 7, panel A). MCF-7 cell extracts were prepared from cells that had been exposed to ethanol vehicle, trans-hydroxytamoxifen or 17β-estradiol, and incubated with DNA fragments which contain an ERE at the end (Fig. 7, Panel B, E) or in the middle (Fig. 7, Panel B, M) of the DNA fragments, respectively. The protein-DNA mixture was then fractionated on the same gel as the DNA bending standards (Fig. 7, panel B). Using a standard curve derived from the relative migration of the DNA bending standards (Fig. 7, panel C), we determined that the degree of bending brought about by binding of each of the three ER preparations to the ERE-containing DNA fragment was 56°.

FIGURE 7. Determination of the DNA bending angle induced by MCF-7 cell ER. In panel A labeled DNA bending standards containing two to six $(dA)_6$ tracts in the middle (upper bands) or at the end (lower bands) of the DNA fragments were fractionated on a polyacrylamide gel. In panel B labeled fragments A and D (see Fig. 2), which contained an ERE near the end (E) or Middle (M) of the DNA fragment, were incubated with extracts from MCF-7 cells that had been exposed to ethanol vehicle (ETOH), trans-hydroxytamoxifen (TOT), or 17β-estradiol (E_2) for 40 min prior to harvest. The protein-DNA complexes were fractionated on the same polyacrylamide gel as the DNA bending standards. In panel C the relative mobilities of the DNA bending standards (O), are plotted as a function of the bending angle. The relative mobility of the ER-DNA complex in the presence and absence of hormone is shown (●), and corresponds closely to the location of the DNA bending standard containing 3 dA_6 tracts. (From Nardulli et al, 1993, reprinted with permission).

A second independent method was also used to calculate the bending angle induced by ER binding. Gels were scanned with a laser densitometer to precisely locate the top of the gel, the complexes formed, and the uncomplexed DNA fragments. The bending angle was then calculated using the expression $\mu_M/\mu_E = \cos(\alpha/2)$ (Thompson and Landy, 1988), where μ is the migration of the ER-DNA complex divided by the migration of the corresponding uncomplexed DNA fragment when the ERE is near the middle (M) or the end (E) of the DNA fragment. Calculations of μ_M/μ_E were $0.903 \pm .004$ (S.E.M., n=8), $0.906 \pm .004$ (S.E.M., n=10) and 0.899 ± 0.009 (S.E.M., n = 10) for complexes formed with extracts from ethanol, trans-hydroxytamoxifen, and 17β-estradiol treated cells, respectively. These corresponded to bending angles (α) of $56.5° \pm 1.2°$, $55.6° \pm 1.2°$, and $57.7° \pm 2.6°$, respectively, and were in excellent agreement with the 56° bending angles observed using the DNA bending standards. Thus, the bending angle induced by the unoccupied ER was not statistically different from the bending angle induced by the receptor occupied by 17β-estradiol or receptor occupied by trans-hydroxytamoxifen.

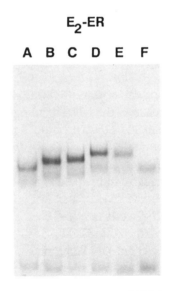

E$_2$-ER

A B C D E F

FIGURE 8. Partially purified, yeast-expressed ER induces DNA bending. The labeled DNA fragments A-F (See Fig. 2) were combined with 385 fmoles of partially purified yeast-expressed hER. The protein-DNA mixture was fractionated on a nondenaturing polyacrylamide gel and the bands visualized by autoradiography (From Nardulli et al, 1993, reprinted with permission).

We do not believe that the absence of hormone-dependence for ER-induced DNA bending supports the view that DNA bending is unrelated to the process of transcription activation. In intact cells, ER activation is strongly dependent on estrogen exposure and is usually inhibited by antiestrogens. However, to date, it has not been possible to reproduce the process of transcription activation using either naturally occurring or recombinant ER extracted from cells. Neither binding of the ER to ERE-containing DNA fragments (Kumar and Chambon, 1988; Lees et al, 1989; Brown and Sharp, 1990; Tzukerman et al, 1990; Curtis and Korach, 1990; Fawell et al, 1990; Murdoch et al, 1990), nor activation of transcription by ER in cell-free extracts, shows any hormone dependence (Elliston et al, 1990). It seems probable that salt extraction or some other step used in preparing crude ER-containing extracts results in receptor activation. It was therefore not surprising that unoccupied receptor and occupied hER induced an equivalent 56° bend in the DNA.

Partially purified human ER expressed in yeast induced the same degree of DNA bending as crude MCF-7 cell ER. To test the possibility that tissue or species-specific proteins found in MCF-7 cells influence the extent to which the ER bends DNA, we examined the ability of partially purified human ER expressed in yeast to induce DNA bending. Partially purified yeast-expressed ER exhibited specific interaction with the ERE as shown by competition with an unlabeled ERE, but not with random DNA, and by the ability of ER-specific antibodies to reduce the mobility of the protein-DNA complex (Nardulli, et al, 1993).

Incubation of the partially purified yeast-expressed ER with ERE-containing DNA fragments A-F (Fig. 2) revealed the same ER induced DNA bending pattern seen with crude MCF-7 cell ER preparations (Compare Fig. 8 with Fig. 6). In addition to the major complex formed by binding of yeast-expressed ER to each of the labeled DNA fragments, a less prominent more rapidly migrating complex was observed, and probably represents a complex between the ERE and a proteolytic fragment of ER.

Using the same DNA bending standards that were used to determine the bending angle induced by crude MCF-7 cell ER (Fig. 7), we calculated that the degree of bending induced by the hER expressed in yeast was 56°. This was identical to the degree of bending seen with the crude MCF-7 ER. Since the yeast ER preparations contained the 17β-estradiol used to elute the ER from the estradiol affinity column, the hER was probably still complexed by hormone. To confirm that hormone would not alter the degree of DNA bending observed, we added 17β-estradiol to the receptor DNA mixture. Addition of 17β-estradiol did not affect the migration of ER-DNA complexes, or alter the DNA bending angle (Nardulli, et al, 1993). These data provided a quantitative demonstration that human ER expressed in yeast possesses the posttranslational modifications required for interaction with the ERE.

ER binding to an ERE may require additional accessory proteins. To more directly evaluate the possibility that additional proteins are required for ER-ERE interaction, we examined bending brought about by more extensively purified yeast-expressed hER. As described above, after fractionation of the yeast extract on an estradiol-affinity column, the partially purified ER bound effectively to the ERE (Fig. 8) and induced the same 56° bend as the crude MCF-7 cell hER. Further purification of the yeast-expressed hER using CM-Sepharose chromatography yielded an ER preparation that bound hormone and was of the appropriate molecular weight, but was incapable of binding to the ERE-containing DNA fragments, even when a 2,300 fold excess of the more highly purified yeast ER was used (Nardulli et al, 1993).

Since the majority of the ER in these preparations shows no evidence of proteolytic degradation and retains normal capacity to bind hormone, the inability of this more highly purified ER to bind to the ERE may be due to the removal of required accessory proteins. These findings are in agreement with those of others (Mukherjee and Chambon, 1990), who reported that highly purified ER expressed in either yeast or HeLa cells lost the ability to bind to an ERE. They found that when the purified ER was supplemented with either whole cell extracts from yeast or HeLa cells, or with a 45 kd protein purified from yeast, the ability of purified ER to bind to an ERE-containing DNA fragment was restored. Since the yeast and MCF-7 cell ER-ERE complexes induce similar DNA bending angles and exhibit similar migration patterns on polyacrylamide gels, the yeast and human proteins involved in facilitating interaction with the ERE may be quite similar. Because the yeast ER is eluted from the affinity column using urea, it is also probable that a fraction of the partially purified ER does not renature correctly and regain activity. This is consistent with the observation that the partially purified ER is less active than the ER in crude extracts from yeast.

DNA fragments containing two EREs exhibit increased DNA bending. Since we had previously shown that the binding of the XER DNA binding domain to a DNA fragment containing two EREs induced a 1.55 fold increase in DNA bending when compared to a DNA fragment containing a single ERE, we carried out a similar study with the full length partially purified yeast hER. Although we used the same DNA fragments, in which the EREs were separated by two turns of the DNA helix, that we used for our study of the XER DNA binding domain, we observed only a small, but reproducible, 1.2 fold increase in DNA bending with two EREs.

Summary of DNA bending induced by the XER DBD and by full-length hER. Our findings on DNA bending induced by the XER DNA binding domain and by the full-length hER can be summarized as follows:

1. Binding of hER, or of the XER DBD, to the ERE induces DNA bending.
2. Full-length hER binds to the ERE with higher affinity than the XER DBD, and induces a 1.55 fold greater bend in the DNA.
3. The DNA bending angle is the same when salt-extracted ER is unliganded, complexed to 17β-estradiol, or complexed to trans-hydroxytamoxifen.
4. Partially purified hER, expressed in yeast, and crude MCF-7 cell hER, induce the same degree of DNA bending.
5. Increased DNA bending is observed when 2 EREs (on the same side of the DNA helix) are present in the DNA fragment.

Table 1. DNA Bending Angles Induced by hER and by XER DBD

	Bending Angle[*]
XER DNA Binding Domain	34°
hER/E_2 (MCF-7 Cells)	56°
hER/E_2 (Expressed in Yeast	56°
Salt Extracted MCF-7 Cell hER	
hER/Unliganded	56°
hER/E_2	56°
hER/TOT	56°
Increase in Bending with 2 EREs	
XER DBD	1.55 fold
hER (Yeast)	1.20 fold

[*]DNA bending angles are derived from the data presented in Nardulli and Shapiro, 1992 and Nardulli et al, 1993.

A major feature of this work was the quantitative determination of DNA bending angles. The data summarized in Table 1 represent the quantitative measurements of the DNA bending angle induced by binding of a member of the steroid/nuclear receptor gene superfamily to its hormone response element.

DISCUSSION

The full-length hER induced a larger DNA bend than the DNA binding domain. The 56° bending angle induced by full-length ER is 1.65 fold greater than the 34° bending angle induced by the purified ER DNA binding domain (Nardulli and Shapiro, 1992; Nardulli et al, 1993). Similar findings have been reported for the transcription factor AP1, in which the leucine zipper DNA binding domain induces a 57° bend in the DNA, and the full-length protein induces a 90° bend (Kerppola and Curran, 1991). In contrast, the HMG DNA binding domain of the lymphoid enhancer factor 1 and the intact protein both induce similar 130° bends in the DNA (Giese et al, 1992).

Our quantitative determination of the DNA bending angles induced by the full-length ER and by the ER DNA binding domain allows direct comparison with bending angles reported for other DNA binding proteins. Bending angles reported for the prokaryotic Cro (30°), Int (38-51°), P_4 (40°), Eco RI (50°), Xis (44-90°), cyclic AMP binding protein (140°), and integration host factor (> 140°)proteins (Liu-Johnson et al, 1986; Thompson and Landy, 1988; Kim et al, 1989; Rojo and Salas, 1991; Crothers et al, 1991) and the eukaryotic POU (37°), NF-κB (110°), Jun-Jun (79°), Fos-Jun (94°), testis determining factor (85°), and lymphoid enhancer factor 1 (130°) proteins vary greatly (Schreck et al, 1990; Kerppola and Curran, 1991; Verrijzer et al, 1991; Giese et al, 1992). However, the 34° bending angle induced by the purified ER DNA binding domain is smaller than the DNA bending angles reported for most prokaryotic and eukaryotic proteins that bend DNA. Because the bending angle induced by the ER DNA binding domain was so small, it was uncertain whether the observed bend would play a role in ER-induced gene transcription. The considerably larger bending angle induced by binding of the full-length ER to the ERE is in the range of bending angles observed for other proteins, and therefore lends credence to the hypothesis that DNA bending is an integral part of the process of ER-induced gene transcription.

Although the precise relationship between DNA bending and gene transcription rates is not understood, it has been suggested that a larger bending angle is associated with a higher rate of gene transcription. Thyroid receptor bound to a thyroid response element that is a strong activator of transcription induces a larger DNA bend than thyroid receptor bound to a weak thyroid response element (Leidig et al, 1992). The inability of the ER DNA binding domain to induce the same level of DNA bending as the full-length ER, may be partially responsible for its reduced ability to activate transcription.

Enhanced bending by the full-length ER may be due to increased affinity for the ERE. Determination of the degree of bending observed for the prokaryotic cyclic AMP binding protein and for the eukaryotic lymphoid enhancer factor 1 bound to strong and weak binding sites suggests that the more tightly a protein binds to a DNA sequence, the greater the degree of bending it can induce (Verrijzer et al, 1991; Giese et al, 1992; Liu-Johnson et al, 1986). Our data and that of others (Fawell et al, 1990; Ponglikitmongkol et al, 1990) indicate that full-length ER has a higher affinity for the ERE than does the ER DNA binding domain. We find that a 300 fold molar excess of yeast-expressed ER is required to fully occupy both EREs and completely shift the mobility of a DNA fragment containing 2 EREs (Nardulli et al, 1993). In contrast, a 77,000 fold excess of the purified estrogen receptor DNA binding domain is required to completely shift the mobility of the 2 ERE DNA fragment (Nardulli and Shapiro, 1992). We therefore believe that the increased DNA bending observed with the full-length ER may result from its increased affinity for the ERE.

It is possible that ER domains which lie outside of the classical DNA binding domain may be responsible for the increased affinity of the full-length receptor for the ERE. In addition, these domains may interact with accessory proteins to enhance the binding of ER to the ERE. Co-activators, which modulate DNA binding and transcription activation, have been reported for several members of the nuclear receptor family including thyroid, vitamin D, and retinoic acid receptors (Beebe et al, 1991; Yu et al, 1991), and have also been implicated in the action of other steroid receptors, including ER (Spelsberg et al, 1983; Nelson et al, 1986; Schule et al, 1988; Mukherjee and Chambon, 1990).

Although the *Xenopus* ER DNA binding domain and the human ER are from different species, it is unlikely that species differences are responsible for the increase in DNA bending seen with full-length ER.

The *Xenopus* and human ERs contain highly conserved DNA binding domains (Green et al, 1986; Greene et al, 1986; Weiler et al, 1987) and bind similarly to ERE-containing DNA fragments *in vitro* (Martinez and Wahli, 1989).

DNA bending is increased when 2 EREs are present. Several estrogen responsive genes contain multiple EREs. We were, therefore, interested in determining if the full-length ER bends a DNA fragment containing 2 EREs to a greater extent than a DNA fragment containing a single ERE. In these studies we used a 21 nucleotide center-to-center spacing between the two EREs, which is very similar to the 20 nucleotide center-to-center spacing in the Xenopus vitellogenin B1 gene and the 21 nucleotide spacing used in our standard 2 ERE reporter plasmid (Walker et al, 1984; Chang et al, 1992). This represents two turns of the DNA helix. We therefore expect that ER dimers bound to these EREs will be on the same side of the DNA helix. In our initial study we demonstrated that the ER DNA binding domain induces a 1.55 fold increased in bending of a DNA fragment containing two EREs relative to a DNA fragment containing a single ERE (Nardulli and Shapiro, 1992). The ER DNA binding domain and the partially purified yeast-expressed ER induce bending angles of 34° and 56°, respectively, when one ERE is present, and 53° and 67°, respectively, when two EREs are present. The modest increase in DNA bending we observe when the full-length ER binds to 2 EREs may reflect constraints on the total degree of DNA bending. Since we readily observe ER binding to the second ERE (Nardulli et al, 1993), it seems unlikely that the 1.2 fold increase in bending observed with two EREs is due to failure of the second ERE to function.

Role of DNA bending in transcription activation. In this work we describe the beginning of the process of associating the changes in DNA topology we refer to as DNA bending to the action of estrogen receptor. We showed that intact ER and the ER DNA binding domain bend DNA on binding to the ERE. The degree of bending induced by full-length hER was substantially greater than that observed with the purified, bacterially-expressed ER DNA binding domain. The ER DNA binding domain, which induces a smaller DNA bend, was also a weaker transcription activator than the full-length ER (Kumar et al, 1987; Kumar and Chambon, 1988; Lees et al, 1989). Although these data provided a general correlation between the ability of the DNA binding domain and the full-length ER to bend DNA and to activate transcription, a direct determination of the role that DNA bending plays in transcription activation of estrogen responsive genes awaits further investigation.

The most probable role of DNA bending in steroid receptor action cannot be readily deduced by analogy to the effect of DNA bending in other systems. It is true that many, but not all, prokaryotic and eukaryotic transcription factors bend DNA on binding to their cognate recognition sequences *in vitro* (Wu and Crothers, 1984; Shuey and Parker, 1986; Crothers and Crothers, 1988; Thompson and Landy, 1988; Vignais and Sentenac, 1989; Yang and Nash, 1989; Kim et al, 1989; Bracco et al, 1989; Verrijzer et al, 1991). Although these observations have led to the hypothesis that DNA bending plays a significant role in transcription regulation, a direct role for DNA bending in transcription regulation has only been demonstrated in a few prokaryotic (Bracco et al, 1989; Rojo and Salas, 1991; Landy 1989; Goodman et al, 1990; Nash, 1990) and eukaryotic (Giese et al, 1992) systems. However, there is considerable indirect data suggesting a role for DNA bending. Sequences, which might be expected to bend DNA, are contained in the transcription regulatory regions of several yeast genes (Lue et al, 1989). Change in the direction of DNA bending correlate with the ability of jun-fos and jun-jun dimers to regulate transcription (Kerpolla and Curran, 1991).

There are several potential roles for DNA bending in estrogen receptor action:

1. By curving the DNA, DNA bending could increase the contact area between the estrogen receptor protein and the DNA.
2. By bending the DNA at the ERE, contact between other transcription regulatory proteins which bind at separated sites could be facilitated. This appears to be the mechanism through which DNA bending mediated by IHF-1 and LEF-1 function (Landy 1989; Goodman et al, 1990; Nash, 1990; Giese et al, 1992). However, those proteins lack the ability to act as independent activators of transcription and require the presence of other transcription factor binding sites in the promoters they regulate (Giese et al, 1992). In contrast, the estrogen receptor, and other steroid receptors, effectively activate transcription from simple synthetic promoters containing only estrogen response elements and a TATA box (Ponglikitmongkol et al, 1990; Chang et al, 1992).
3. Bends or kinks in the DNA could provide a recognition motif for DNA binding proteins or transcription factors.

4. By altering the interaction of histones with the DNA, bends or kinks could change the location, or phasing, of nucleosomes on the DNA, and expose or sequester recognition sequences for transcription factors.

Future experiments will doubtless center on both the role of DNA bending in steroid/nuclear receptor action, and on determining the extent to which any, or all, of these processes contribute to steroid receptor regulation of gene expression.

ACKNOWLEDGMENTS

This research was supported by National Institutes of Health Grants HD-16720 (to D.J.S.) and CA-02897 (to G.L.G.). We are grateful to Ms. Stephanie Teeple for art work.

REFERENCES

Beato, M (1989): Gene regulation by steroid hormones. *Cell* 56: 335-344.

Beebe JS, Darling DS and Chin WW (1991): 3,5,3'-Triiodothyronine receptor auxiliary protein (TRAP) enhances receptor binding by interactions within the thyroid hormone response element. *Mol Endocrinol* 5: 85-93.

Bracco L, Kotlarz D, Kolb A, Diekmann S and Buc H (1989): Synthetic curved DNA sequences can act as transcriptional activators in *Escherichia coli*. *EMBO J* 8: 4289-4296.

Brown M and Sharp PA (1990): Human estrogen receptor forms multiple protein DNA complexes. *J Biol Chem* 265: 11238-11243.

Chang T-C, Nardulli AM, Lew D and Shapiro DJ (1992): The role of estrogen response elements in expression of the *Xenopus laevis* vitellogenin B1 gene. *Mol Endocrinol* 6: 346-354.

Chang T-C and DJ Shapiro (1990): An Nf1-related vitellogenin activator element mediates transcription from the estrogen regulated *Xenopus laevis* vitellogenin promoter. *J Biol Chem* 265: 8176-8182.

Crothers MR and Crothers DM (1988): DNA sequence determinants of CAP-induced bending and protein binding affinity. *Nature* 333:824-831.

Crothers DM, Crothers MR and Shrader TS (1991): DNA bending in protein-DNA complexes. *Methods Enzymol* 208: 118-146.

Curtis SW and Korach KS (1990): Uterine estrogen receptor interaction with estrogen-responsive DNA sequences *in vitro*: effects of ligand binding on receptor-DNA complexes. *Mol Endocrinol* 4: 276-286.

Debs RJ, Freedman LP, Edmunds S, Gaensler KL, Dzgnes N and Yamamoto KR (1990): Regulation of gene expression *in vivo* by liposome-mediated delivery of a purified transcription factor. *J Biol Chem* 265: 10189-10192.

Dieckman S (1992): Analyzing DNA curvature in polyacrylamide gels. *Methods Enzymol* 212: 30-46.

Elliston JF, Falwell SE, Klein-Hitpass L, Tsai SY, Tsai M-J, Parker MG and O'Malley BW (1990): Mechanism of estrogen receptor-dependent transcription in cell-free system. *Mol Cell Biol* 12: 6607-6612.

Evans RM (1988): The steroid and thyroid hormone receptor superfamily. *Science* 240: 889-895.

Falwell SE, Lees JA, White R and Parker MG (1990): Characterization and colocalization of steroid binding and dimerization activities in the mouse estrogen receptor. *Cell* 60: 953-962.

Freedman LP, Yoshinaga SK, Vanderbilt JN and Yamamoto KR (1989): *In vitro* transcription enhancement by purified derivatives of the glucocorticoid receptor. *Science* 245: 298-301.

Giese K, Cox J and Grosschedl R (1992): The HMG domain of lymphoid enhancer factor 1 bends DNA and facilitates assembly of functional nucleoprotein structures. *Cell* 69: 185-195.

Goodman S, Young C-C, Nash H, Savai A and Jernigan R (1990): Bending of DNA by IHF Protein. *In Structure and Methods, Volume 2, DNA Protein Complexes and Proteins*, RH Sarma and MH Sarma, eds (New York: Adenine Press), pp. 51-62.

Green S, Walter P, Kumar V, Krust A, Bornert J-M, Argos P and P Chambon (1986): Human oestrogen receptor cDNA: sequence, expression and homology to v-*erb*-A. *Nature* 320: 134-139.

Greene GL, Gilna P, Waterfield M, Baker A, Hort Y and Shine J (1986): Sequence and expression of human estrogen receptor complementary DNA. *Science* 231: 1150-1154.

Guiochon-Mantel A, Loosfelt H, Ragot T, Bailly A, Atger M, Misrahi M, Perricaudet M and Milgrom E (1988): Receptors bound to antiprogestin form abortive complexes with hormone responsive elements. *Nature* 336: 695-698.

Hager GL and Archer TK (1991): The interaction of steroid receptors with chromatin, p. 217-234. In: MG Parker (ed), Nuclear hormone receptors: molecular mechanism, cellular functions and clinical abnormalities. Academic Press, San Diego, CA.

Hard T, Kellenbach E, Boelens R, Maler BA, Dahlman K, Freedman LP, Carlstedt-Duke J, Yamamoto KR, Gustafsson J-A and Kaptein R (1990): Solution structure of the glucocorticoid receptor DNA-binding domain. *Science* 249: 157-160.

Kerppola TK and Curran T (1991): Fos-jun heterodimers and jun homodimers bend DNA in opposite orientations: implications for transcription factor cooperativity. *Cell* 66: 317-326.

Kim J, Zwieb C, Wu C and Adhya S (1989): Bending of DNA by gene-regulatory proteins: construction and use of a DNA bending vector. *Gene* 85: 15-23.

Klein-Hitpass L, Tsai SY, Weigel NL, Allan GF, Riley D, Rodriguez R, Schrader WT, Tsai M-J and O'Malley BW (1990): The progesterone receptor stimulates cell-free transcription by enhancing the formation of a stable preinitiation complex. *Cell* 60: 247-257.

Kumar V, Green S, Stack G, Berry M, Jin J-R and Chambon P (1987): Functional domains of the human estrogen receptor. *Cell* 51: 941-951.

Kumar V and Chambon P (1988): The estrogen receptor binds tightly to its responsive element as a ligand-induced homodimer. *Cell* 55: 145-156.

Landy A (1989): Dynamic, structural and regulatory aspects of site-specific recombination. *Annu Rev Biochem* 58: 913-949.

Lees JA, Falwell SE and Parker MG (1989): Identification of two transactivation domains in the mouse estrogen receptor. *Nucleic Acids Res* 17: 5477-5488.

Leidig F, Shepard AR, Zhang W, Stelter A, Cattini PA, Baxter JD and Eberhardt NL (1992): Thyroid hormone responsiveness in human growth hormone-related genes: possible correlations with receptor-induced DNA conformational changes. *J Biol Chem* 267: 913-921.

Liu-Johnson H-N, Crothers MR and Crothers DM (1986): The DNA binding domain and bending angle of *E. coli* CAP protein. *Cell* 47: 995-1005.

Lue NF, Buchman AR and Kornberg RD (1989): Activation of yeast RNA polymerase II transcription by a thymidine-rich upstream element *in vitro*. *Proc Natl Acad Sci USA* 86:486-490.

Luisi BF, Xu WX, Otwinowski Z, Freedman LP, Yamamoto KR and Sigler PB (1991): Crystallographic analysis of the interaction of the glucocorticoid receptor with DNA. *Nature* 352: 497-505.

Martinez E and Wahli W (1989): Cooperative binding of estrogen receptor to imperfect estrogen-responsive DNA elements correlates with their synergistic hormone-dependent enhancer activity. *EMBO J* 8: 3781-3791.

Miesfeld R, Godowski PJ, Maler BA and Yamamoto KR (1987): Glucocorticoid receptor mutants that define a small region sufficient for enhancer activation. *Science* 236: 423-427.

Mukherjee R and Chambon P (1990): A single-stranded DNA-binding protein promotes the binding of the purified oestrogen receptor to its responsive element. *Nucl Acids Res* 18: 5713-5716.

Murdoch, FE, Meier DA, Furlow JD, Grunwald K and Gorski J (1990): Estrogen receptor binding to a DNA response element *in vitro* is not dependent on estradiol. *Biochem* 29: 8377-8385.

Nardulli AM, Lew D, Erijman L and Shapiro DJ (1991): Purified estrogen receptor DNA binding domain expressed in *Escherichia coli* activates transcription of an estrogen-responsive promoter in cultured cells. *J Biol Chem* 266: 24070-24076.

Nardulli AM and Shapiro DJ (1992): Binding of the estrogen receptor DNA-binding domain to the estrogen response element induces DNA bending. *Mol Cell Biol* 12: 2037-2042.

Nardulli AM, Greene GL and Shapiro DJ (1993): Human estrogen receptor bound to an estrogen response element bends DNA. *Mol Endocrinol* 7: 331-340.

Nash H (1990): Bending and supercoiling of DNA at the attachment site of bacteriophage lambda. *Trends Biochem Sci* 15: 222-227.

Nelson WG, Pienta KJ, Barrack ER and Coffey DS (1986): The role of the nuclear matrix in the organization and function of DNA. *Annu Rev Biophys Chem* 15: 457-475.

Pina B, Hach RJG, Arnemann J, Chalepakis G, Slater EP and Beato M (1990): Hormonal induction of transfected genes depends on DNA topology. *Mol Cell Biol* 10: 625-633.

Pina B, Bruggemeier U and Beato M (1990): Nucleosome positioning modulates accessibility of regulatory proteins to the Mouse Mammary Tumor Virus promoter. *Cell* 60: 719-731.

Ponglikitmongkol M, White JH and Chambon P (1990): Synergistic activation of transcription by the human estrogen receptor bound to tandem responsive elements. *EMBO J* 9: 2221-2231.

Rojo F and Salas M (1991): A DNA curvature can substitute phage d29 regulatory protein p4 when acting as a transcriptional repressor. *EMBO J* 10:3429-3438.

Schreck R, Zorbas H, Winnacker E-L and Baeuerle PA (1980): The NF-κB transcription factor induces DNA bending which is modulated by its 65-kD subunit. *Nucl Acids Res* 18: 6497-6502.

Schule R, Muller M, Kattschmidt C and Renkawitz R (1988): Many transcription factors interact synergistically with steroid receptors. *Science* 242: 1418-1420.

Schwabe JWR, Neuhaus D and Rhodes D (1990): Solution structure of the DNA-binding domain of the oestrogen receptor. *Nature* 348: 458-461.

Seiler-Tuyns A, Walker P, Martinez E, Merillat AM, Givel F and Wahli W (1986): Identification of estrogen-responsive DNA sequences by transient expression experiments in a human breast cancer cell line. *Nucl Acids Res* 14: 8755-8770.

Shuey DJ and Parker CS (1986): Bending of promoter DNA on binding of heat shock transcription factor. *Nature* (London) 323: 459-561.

Spelsberg TC, Littlefield BA, Seelke R, Dani GM, Toyoda H, Boyd-Leinen P, Thrall CT and Kon OL (1983): Role of specific chromosomal proteins and DNA sequences in the nuclear binding sites for steroid receptors. *Recent Prog Horm Res* 39: 463-513.

Strahle U, Schmid W and Schutz G (1988): Synergistic action of the glucocorticoid receptor with transcription factors. *EMBO J* 7: 3389-3395.

Tasset D, Tora L, Fromental C, Scheer E and Chambon P (1990): Distinct classes of transcriptional activating domains function by different mechanisms. *Cell* 62: 1177-1187.

Thompson JF and Landy A (1988): Empirical estimation of protein-induced DNA bending angles: applications to G site-specific recombination complexes. *Nucl Acids Res* 16: 9687-9705.

Tzukerman M, Zhang X-K, Hermann T, Wills KN, Graupner G and Pfahl M (1990): The human estrogen receptor has transcriptional activator and repressor functions in the absence of ligand. *New Biol* 2: 613-620.

Verrijzer CP, van Oosterhout JAWM, van Weperen WW and van de Vliet PC (1991): POU proteins bend DNA via the POU-specific domain. *EMBO J* 10: 3007-3014.

Vignais M-L and Sentenac A (1989): Asymmetric DNA bending induced by the yeast multifunctional factor TUF. *J Biol Chem* 264: 8463-8466.

Walker P, Germond JE, Brown-Luedi M, Givel F and Wahli W (1984): Sequence homologies in the region preceding the transcription initiation site of the liver estrogen-responsive vitellogenin and apo-VLDL II genes. *Nucl Acids Res* 12: 8611-8625.

Waterman ML, Adler S, Nelson C, Greene GL, Evans RM and Rosenfeld MG (1988): A single domain of the estrogen receptor confers deoxyribonucleic acid binding and transcriptional activation of the rat prolactin gene. *Mol Endocrinol* 2: 14-21.

Weiler IJ, Lew D and Shapiro DJ (1987): The *Xenopus laevis* estrogen receptor: sequence homology with human and avian receptors and identification of multiple messenger ribonucleic acids. *Mol Endocrinol* 10: 355-362.

Wu H-M and Crothers DM (1984): The locus of sequence-directed and protein-induced DNA bending. *Nature* (London) 308: 509-513.

Yang C-C and Nash HA (1989): The interaction of *E coli* IHF-protein with its specific binding sites. *Cell* 57: 869-880.

Yu VC, Delsert C, Andersen B, Holloway JM, Devary OV, Nr AM, Kim SY, Boutin J-M, Glass CK and Rosenfeld MG (1991): RXR: a coregulator that enhances binding of retinoic acid, thyroid hormone, and vitamin D receptors to their cognate response elements. *Cell* 67: 1251-1266.

Zwieb C, Kim J and Adhya S (1989): DNA bending by negative regulatory proteins: gal and lac repressors. *Genes and Develop* 3: 606-611.

THE NUCLEAR ENVIRONMENT AND ESTROGEN ACTION

Jerry R. Malayer
Jack Gorski
Department of Biochemistry
University of Wisconsin-Madison

INTRODUCTION

It is generally agreed that the actions of estrogens and other steroid hormones are mediated by specific cellular receptor proteins that, certainly in the case of the estrogen receptor (ER), and probably all members of the steroid receptor group, are localized in the nucleus of the cell (for reviews, see Gorski, 1986; Evans, 1988; Beato, 1989; O'Malley, 1990). These receptors are members of a large family of receptor proteins, including steroid, thyroid hormone, retinoic acid, and vitamin D_3 receptors, which act as ligand-induced transcriptional activators or repressors (Evans, 1988; Diamond et al, 1990). In general, several essential steps in this signal transduction pathway of steroid action have been identified: 1) interaction of the receptor protein with a specific regulatory sequence in the DNA called a hormone response element (HRE) (Yamamoto, 1985); 2) specific, high-affinity ligand binding, which results in as yet undefined conformational changes in the receptor protein (Hansen and Gorski, 1985; Hansen et al, 1988; Fritsch et al, 1992); and 3) recruitment of transcription factors and RNA polymerase to initiate transcription of target genes (Cordingley et al, 1987; Corthesy

STEROID HORMONE RECEPTORS
V. K. Moudgil, Editor
© 1993 Birkhäuser Boston

et al, 1989). However, a number of questions remain concerning both the details of these events and other, undiscovered and perhaps even more important events.

Understanding the complexities of ER action has become increasingly entangled with understanding the complexities of the eukaryotic cell nucleus and the DNA-protein and protein-protein interactions that play critical roles in the biological processes of DNA replication and transcription. Great effort has been expended in identifying various regulatory proteins, their DNA binding sites, and the factors that modulate their interactions; for review, see Ptashne and Gann (1990). Among the various factors that might modulate steroid receptor action are the nuclear matrix or nucleoskeleton, chromatin organization, and the DNA topology; for review, see Landers and Spelsberg (1992). In discussing some of the recent findings from our laboratory concerning some of these aspects of ER action in transcriptional activation of the rat prolactin gene (rPRL), we will also briefly review related findings of others in the larger context of a more general mechanism of ER action. We suggest that the influence of ER on the local topology of the DNA at a flanking enhancer site plays a critical role in the estrogen-induced activation of the rPRL gene. The transient formation of non-B DNA structures under the influence of negative supercoiling may be trapped by the ER, perhaps acting in concert with other proteins, and this thermodynamic energy is used to regulate the local topology and chromatin structure. The general idea of proteins acting to trap non-B DNA structure to regulate topology and influence gene transcription and DNA replication has been suggested previously (Frank-Kamenetskii, 1989).

The nuclear matrix

The nucleus is a highly organized structure consisting of a surrounding membrane complex, the DNA, histone and nonhistone chromatin proteins, and a scaffolding system called the nuclear matrix. The nucleus must be highly organized to account for the 50,000-fold compaction of the DNA known to be required to fit within the nucleus; for review, see Getzenberg et al (1990). The nuclear matrix accounts for about 10% of nuclear protein and consists of the peripheral pore complex lamina, an internal fibrogranular network containing both protein and RNA, and a residual nucleolus. The nuclear matrix is believed to be the dynamic structural subcomponent of the nucleus that directs the functional organization of the DNA into supercoiled loop domains and provides sites for the specific control of nucleic acids (Nelson et al, 1986). It has been suggested that the nuclear matrix is intimately involved in the regulation of several key cellular processes, including DNA organization, DNA replication, mRNA synthesis and processing, DNA loop attachment sites, DNA polymerase α activity, and topoisomerase activity; all of these are closely coupled to the transcriptional regulation of specific genes. There is much evidence that these events occur not in solution but rather in association with relatively insoluble structural components firmly bound to the nuclear matrix (for reviews, see Nelson et al, 1986; Cook, 1989). As reviewed by Cook (1989), the model of a soluble transcription complex moving along a fixed DNA template is based in large part upon *in vitro* experiments with naked DNA and semi-purified reaction components in hypotonic salt. The conditions used for those experiments were chosen because of the difficulty in handling reaction components such as RNA polymerase and chromatin under more physiological conditions; a state in which they are relatively insoluble. Transcriptionally active areas of the chromatin or enhancer elements exhibit specific attachment sites to the nuclear matrix when extracted under isotonic salt (Jackson and Cook, 1985; Cockerill and Garrard, 1986; Gasser and Laemmli, 1986; Cockerill et al, 1987). These matrix-associated regions (MARs) may provide the means to localize specific DNA-binding proteins with their binding sites in the chromatin.

Nuclear DNA is nonrandomly organized into supercoiled loop domains (Udvardy et al, 1985; Bonifer et al, 1991). This linear chromosomal DNA must be anchored in the nucleus to impede free rotation, allowing torsional stress to be introduced into the chromatin. It has been suggested that transcriptionally poised chromatin is under torsional stress (Weintraub et al, 1986; Pina et al, 1990) and that this torsional stress is introduced and regulated by topoisomerase II acting at sites near the base of the DNA loop domain (Cockerill and Garrard, 1986; Bode et al, 1992). Topoisomerase II binding sites occur near MARs (Cockerill and Garrard, 1986; Razin et al, 1991). More recently, it has been proposed that stable unwinding of the DNA in the region of MARs may act to relieve negative superhelicity within the loop domain (Bode et al, 1992). This stable unwinding could provide the means to store and release thermodynamic energy to regulate superhelicity within the transcriptionally active chromatin, relax positively supercoiled DNA, and prevent superhelical stress from being transmitted to neighboring loop domains.

An alternate model for transcription, a mobile template passing over a fixed transcription complex (Cook, 1989), explains several observations regarding ER action; these include the presence of matrix attachment sites for ER (Ruh and Ruh, 1988), the estradiol-binding character of immobilized ER, and ER binding to specific estrogen response elements (EREs) in the absence of hormone (Murdoch and Gorski, 1991). When it was recognized that ER was localized in the nucleus of the cell, it became apparent that some interaction with nuclear components was necessary to hold it there (Gorski, 1986). The nature of this interaction remains unclear, although the nuclear matrix has been suggested to be the major site of steroid hormone binding in the nucleus (Barrack, 1987). ER binding to nuclear components such as acceptor sites on the chromatin (Ruh and Ruh, 1988) and to structural elements such as the nuclear matrix or nucleoskeleton (Metzger and Korach, 1990) have been demonstrated. The ER, progesterone receptor (PR), and glucocorticoid receptor (GR) bind in a saturable manner to the chromatin. Equilibrium dissociation constants have been determined and show a high affinity of receptor for these chromatin acceptor sites. These sites are specific; i.e., the ER, PR, and GR do not compete with each other for their individual sites. These acceptor sites exhibit target tissue specificity and DNA sequence specificity (reviewed by Ruh and Ruh, 1988). Specific proteins that act as the chromatin acceptor site for the PR have been identified (Horton et al, 1991; Schuchard et al, 1991).

Protein-DNA interactions and the chromatin

Alterations in chromatin structure have been associated with transcriptionally active genes (Weisbrod, 1983; Weintraub, 1984). These changes include depletion of histone protein H1 (Weintraub, 1984; Tazi and Bird, 1990) and the appearance of DNaseI-hypersensitive sites (Burch and Weintraub, 1983; Zaret and Yamamoto, 1984; Durrin and Gorski, 1985). Modification of chromatin proteins by acetylation, methylation, or phosphorylation has also been associated with transcriptionally active regions (Murdoch et al, 1983; Tazi and Bird, 1990; Krajewski and Luchnik, 1991).

In intact nuclei isolated from prolactin-expressing rat pituitary cells, the 5' flanking region of the rPRL gene contains two tissue-specific DNaseI-hypersensitive sites (Durrin et al, 1984; Durrin and Gorski, 1985). Each of these sites in the rPRL gene contains several cis-acting DNA elements that mediate transcriptional responses to several regulatory proteins, including the homeodomain Pit-1 protein (Nelson et al, 1986;

Day and Maurer, 1989; Mangalam et al, 1989; Ingraham et al, 1990). These areas, as well as part of the coding sequence, were found to be undermethylated in the rPRL expressing rat pituitary cells (Figure 1) (Durrin et al, 1984). Interestingly, estradiol does not seem to affect the time course of the appearance of these sites *in vivo*, because they appear prior to sexual maturity. Estradiol administration does not alter the time course (Durrin and Gorski, 1985), but does exert quantitative effects on these sites (Seyfred and Gorski, 1990). It remains unclear exactly how these sites are generated. Using indirect end-labeling of nuclease digested DNA from intact nuclei, these DNaseI-hypersensitive sites were mapped: from: a) -100 to -280 ± 10 (HS I) base pairs in the 5' direction from the transcription start site, a region consisting of about 200 base pairs and corresponding to the area covered by one nucleosome; and b) from -1545 to -1985 ± 20 (HS II) base pairs in the 5' direction from the transcription start site, corresponding to the area covered by two to three nucleosomes. It was originally hypothesized that these regions of open DNA structure corresponded to areas where nucleosomes had been dissociated from the DNA, or where movement of nucleosomes along the DNA was less restricted. The presence of nucleosomes in these regions during estradiol-mediated transcriptional activation was confirmed by micrococcal-nuclease digestion (Seyfred and Gorski, 1990).

It is not clear how the positions of nucleosomes are established along the length of the DNA and what role the ER plays in this positioning. It has been suggested that the inherent bendability of a DNA sequence may determine the affinity for binding the nucleosome octamer to a specific site (Pina et al, 1990), a phenomenon referred to by Travers (Travers, 1989) as indirect readout; i.e., the DNA sequence encodes a structural configuration that promotes protein-DNA interactions. Alternatively, it was found that the acidic activation domains of the yeast GAL4 transcription factor and the herpes simplex virus-VP16 transcriptional activator compete with nucleosome cores during nucleosome assembly *in vitro* for sites along the DNA (Workman et al, 1991). This competition is independent of DNA binding, specifically requiring the presence of the acidic activation domain; thus, specific protein binding in a region may act to exclude nucleosomes. Nucleosome positioning involves two important aspects: translational control, in which DNA sequences are wrapped about the nucleosome and are in the linker region; and rotational control, the specific folding of the DNA around the nucleosome, which fixes the path of the double helix across the protein core. The site of glucocorticoid binding in the mouse mammary tumor virus (MMTV) promotor is a DNaseI-hypersensitive site (Zaret and

112

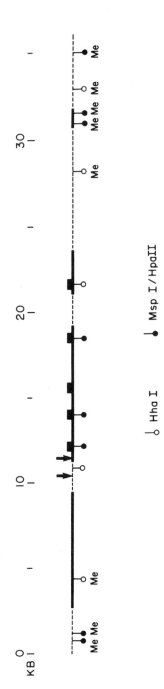

FIGURE 1. Summary of the methylation pattern, location of repetitive DNA sequences, and DNaseI-hypersensitive sites associated with the prolactin gene domain in rat pituitary tumors. Translated regions of the prolactin gene are represented by filled boxes. Unique DNA sequences within and around the prolactin gene are represented by the solid line; regions of DNA containing repetitive sequences are represented by the dotted line. MspI/HpaII and HhaI restriction sites are shown as solid and open circles, respectively. Sites that are methylated in pituitary tumors or control pituitaries of rats are indicated (Me). The two arrows 5' to the first exon indicate the location of the hypersensitive sites observed in pituitary tumors. From Durrin and Gorski (1985). Reprinted with permission.

Yamamoto, 1984). Nucleosome positioning across this promotor is very precise, both *in vivo* and *in vitro* and the GR interacts in a specific fashion with the chromatin. This interaction results in a repositioning of a nucleosome octamer and the unmasking of a binding site for the transcription factor NF1, required for transcription from the promotor (Pina et al, 1990). Nucleosomes remain in place throughout this interaction, suggesting a rotational control effect. Cordingley et al (1987) demonstrated the recruitment of NF1 *in vivo* following glucocorticoid treatment using an exonuclease protection assay. A similar modification of nucleosome positioning by GR appears in the rat tyrosine aminotransferase gene promotor (Carr and Richard-Foy, 1990; Adom et al, 1991; Reik et al, 1991). Recently, Gilbert et al (1992) examined the nuclease digestion patterns of the chromatin structure around a chromosomally integrated estradiol-responsive promotor transfected into *Saccharomyces cerevisiae*. The chromatin around this promotor is significantly altered in the presence of ligand-bound ER; whereas similar changes were not evident in the presence of the DNA-binding domain alone. DNA binding is believed to be ligand-independent (Murdoch and Gorski, 1991). This suggests that the role of ligand may be to induce a conformational change in DNA-bound ER that results in specific modification of chromatin at the target promotor. The role of nucleosomes in the regulation of transcription and chromatin remodeling has been reviewed (see Kornberg and Lorch, 1991; Felsenfeld, 1992).

Seyfred and Gorski (1990) found estradiol modulation of the DNaseI-hypersensitive sites in the 5' flanking region of the rPRL gene in a rPRL-expressing rat pituitary tumor-derived cell line. The quantitative increase in appearance of the DNaseI-hypersensitive sites differed between estradiol and the antiestrogen monohydroxytamoxifen (MHT). Estradiol induced an increase in the appearance of DNaseI-hypersensitivity at both HS I and HS II and caused increased transcription from the rPRL promotor. However, MHT increased the hypersensitivity at the more distal HS II only and did not increase transcription. These data indicate a critical interaction between HS I and HS II in the induction of transcription from the rPRL promotor, possibly mediated through a looping out of the intervening DNA to bring HS I and HS II into close proximity, resulting in protein-protein contact at the promotor. The single-strand endonuclease S1 was used to probe the HS I and HS II regions of rPRL (Seyfred and Gorski, 1990). Two areas of hypersensitivity were located in HS II, adjacent to or within locations believed to contain ERE sequences, -1721 and -1581 base pairs from the transcription start site (Figure 2). Both sites showed increased S1

114

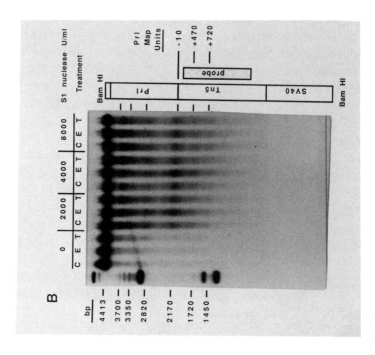

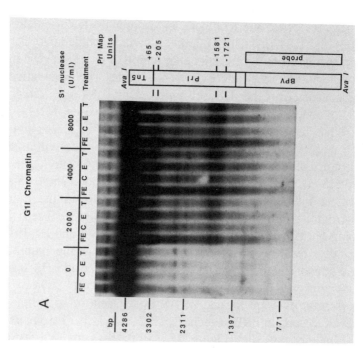

FIGURE 2. Effects of estradiol and monohydroxytamoxifen on the PRL gene.

hypersensitivity following estradiol treatment. A region of S1 hypersensitivity was found in HS I and surrounding the transcriptional start site as well (Seyfred and Gorski, 1990). Changes in the nuclease hypersensitive sites closely correlated with increases in the rates of prolactin gene transcription induced by estradiol (Seyfred and Gorski, 1990). Open complex formation at the promotor may be a common feature of transcription by RNA polymerase II (Wang et al, 1992).

Protein-DNA interactions and DNA structure

In order to understand these protein-DNA interactions further, it is necessary to recognize that the double-stranded B-DNA helix is not a static or even rigid docking lattice, but is a dynamic molecule that exhibits considerable structural polymorphism. Under appropriate conditions for certain types of sequences the B-DNA helix can assume conformations that are left handed (Z-DNA), triple stranded (H-DNA), stem looped (hairpins), bent, and stably unwound or single stranded. The potential exists that some or all of these altered DNA conformations could affect DNA-protein interactions. It has been suggested that transcribing RNA polymerase positively supercoils the DNA in front and negatively supercoils the DNA behind (Liu and Wang, 1987). *In vitro* and *in vivo* evidence for the generation of such supercoils has been presented (Brill and Sternglanz, 1988; Giaever and Wang, 1988; Wu et al, 1988; Frank-Kamenetskii, 1989; Tsao et al, 1989). As discussed earlier, it has been suggested that transcriptionally poised chromatin is under torsional stress that is introduced and regulated by topoisomerase II acting at sites near the base of the DNA loop domain (Cockerill and Garrard, 1986; Bode et al, 1992). It then appears likely that considerable thermodynamic energy is sequestered in the DNA helix, and this energy

FIGURE 2. Effects of estradiol and monohydroxytamoxifen on the S1 nuclease-hypersensitive sites present in the 5' flanking region of the PRL gene in PRL-minichromosomes. (A) Nuclei were isolated from PRL-minichromosome-containing cells treated with fetal bovine serum plus E_2 (FE), ethanol vehicle (C), estradiol (E), or monohydroxytamoxifen (T). Nuclei were then digested with increasing concentrations of S1 nuclease for 30 min at 37°C. The DNA was isolated, digested to completion with *Ava*I, and subjected to end-labeling analysis. (B) Similar to (A), except the isolated DNA was digested with *Bam*HI and then probed with a 923 base pair ^{32}P-labeled Tn5 DNA fragment from *Pst*I to *Pst*I. From Seyfred and Gorski (1990). Reprinted with permission.

may drive conformational alterations in the DNA, critical for protein-DNA interactions. Such a dynamic flux between positively and negatively supercoiled DNA has been suggested as a model for nucleosome repositioning during transcription (Clark and Felsenfeld, 1991).

Each of the DNaseI-hypersensitive sites in the rPRL gene contains several *cis*-acting DNA elements that mediate transcriptional responses to several regulatory proteins on the rPRL promotor, including binding sites for the homeodomain Pit-1 protein (Nelson et al, 1986; Mangalam et al, 1989; Ingraham et al, 1990). Additionally, there are two alternating purine-pyrimidine sequences, a 58 base pair (dA-dC)-(dG-dT) tract and 178 base pair (dT-dG)-(dA-dC) tract that flank the distal regulatory region of the rPRL gene. A region of high A+T (70%) content lies between these alternating purine-pyrimidine sequences (Kladde et al, 1993; L. Durrin, unpublished data).

Kladde et al (1993) obtained detailed structural information about sequences within the rPRL HS II via a chemical-probing primer-extension assay *in vitro*. The two (dA-dC)-(dG-dT) rich repeats undergo a series of stepwise transitions to establish a full-length, left-handed helix under the stabilizing influence of negative superhelicity. As more transitions occur, the length of the altered helix increases in the direction of the rPRL distal regulatory region. At high levels of superhelicity, the region of the high A+T content undergoes an additional transition to a stably unwound or single-stranded structure that propagates toward the binding sites for the *trans*-acting regulatory factors. The ER is among these *cis*-acting regulatory factors and has been reported to bind preferentially to a putative single-stranded region in the coding sequence of the ERE (Lannigan and Notides, 1989). This binding may be assisted by the presence of a single-stranded binding protein other than the ER, which localizes to the same region of DNA (Mukherjee and Chambon, 1990). However, ER binding to single-stranded DNA is a controversial concept not supported by studies of ER binding to the *Xenopus* vitellogenin A_2 ERE. It remains to be seen whether this putative single-stranded region is in reality a stem-loop or some other type of structure. This putative single-stranded region exhibits several characteristics of MARs, including the high A+T content of the region, its enhancer-like characteristics, and the capacity to attain a stably unwound structure (Cockerill and Garrard, 1986). Additionally, there are (dA-dT) sequences present that may encode preferential topoisomerase II sites. It has yet to be established that this particular region is itself a MAR, though it is clear that the structural characteristics of the DNA in this region occur within a loop domain constrained at the ends by MARs.

The ER has been shown to directly modify the structure of DNA upon binding. Fragments of DNA containing EREs underwent specific bending when the DNA-binding domain of the ER was bound as measured by perturbation of migration on an electrophoretic gel in a protein-DNA gel shift assay (Nardulli and Shapiro, 1992). The linking number of closed-circular plasmids containing glucocorticoid response elements was altered upon GR binding, as demonstrated by altered migration in a two-dimensional electrophoretic matrix in the presence of chloroquine, which unwinds the DNA and reveals its superhelical properties (Carballo and Beato, 1990). This effect was limited to intact, 97 kilodalton GR and was shown to be dependent upon the so-called "modulator domain," not the DNA-binding or ligand-binding domains present in the 40 kilodalton form of the receptor (Carballo and Beato, 1990). A similar finding for the ability of ER to alter the linking number of closed-circular plasmids containing EREs *in vitro* has been reported (Ishibe et al, 1991). A number of other examples of DNA-binding proteins that induce bending or twisting of the DNA exist (Gustafson et al, 1989; Schreck et al, 1990; Kerppola and Curran, 1991; Giese et al, 1992).

Protein-DNA interactions and protein structure

In addition to structural polymorphism of the DNA helix and its potential modulation by protein binding, a number of regulatory proteins also undergo allosteric modification upon binding to DNA. Using dissociation kinetics, Fritsch et al (1992) demonstrated that the ligand-binding capacity of the ER from rat uterine cytosols was modified by binding to DNA containing the *Xenopus* vitellogenin A_2 ERE (Figure 3). When DNA was bound *in vitro* to the ER, the rate of estrogen dissociation increased twofold. This change was dependent upon the DNA-binding domain, since the steroid-binding domain alone exhibited no change in estrogen dissociation in the presence of DNA. The effect appeared limited to the ligand-binding domain of the ER because proteolytic cleavage mapping revealed no change in the amino-terminal end of the protein. The change in the dissociation rate of estradiol was correlated with the equilibrium dissociation constants of the ER for multiple DNA sequences as determined previously (Murdoch et al, 1990). Specific binding of ER to DNA sequences with lower binding affinities

118

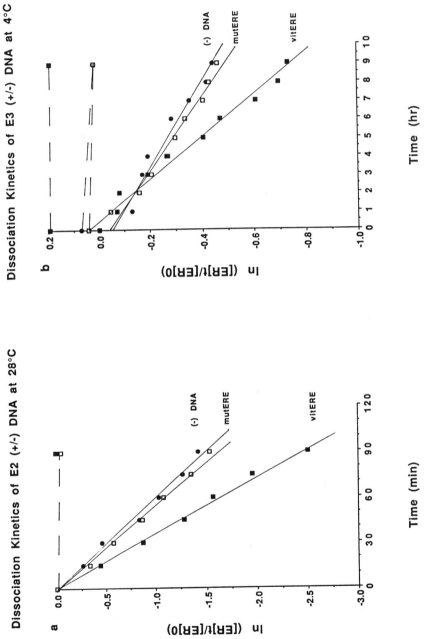

FIGURE 3. Dissociation kinetics of estrogen from the ER in the absence and presence of DNA.

FIGURE 3. Dissociation kinetics of estrogen from the ER in the absence and presence of DNA. (A) Rat uterine cytosolic ER occupied with [^{3}H]estradiol was placed at 28°C for varying amounts of time either in the presence of 3 μM diethylstilbestrol (solid lines), or in the absence of diethylstilbestrol (dashed lines) with no DNA (closed circles), 20 nM *Xenopus* vitellogenin A$_2$ ERE (vitERE; closed squares), or 20 nM *Xenopus* vitellogenin A$_2$ ERE with a one base pair substitution in each arm of the palindrome (mutERE; open squares). The natural logarithm of the ratio of the amount of specific [^{3}H]estradiol bound at each time ([ER]$_t$) to the amount of specific [^{3}H]estradiol bound at time 0 ([ER]$_0$) was plotted against time at 28°C. The data shown are representative of four independent experiments. (B) The experiment was performed as in (A) except ER was occupied with [^{3}H]estriol and dissociation performed at 4°C for the indicated times. From Fritsch et al (1992). Reprinted with permission.

for the ER had a lesser effect upon the dissociation rate of [^{3}H]estradiol-ER complexes than binding to the so-called consensus ERE of *Xenopus* vitellogenin A$_2$, which has the greatest affinity for the ER. Estrogen binding, however, has no effect upon the binding of ER to DNA in this system (Murdoch and Gorski, 1991). *In vivo*, the ER of MCF-7 cells exhibited a 7-fold lower rate of estradiol dissociation in a whole-cell exchange assay than was observed *in vitro*, indicating that nuclear factors in addition to DNA binding act to modulate the steroid-binding domain of the ER. This allosteric modulation of the ER may serve to restrict other nuclear proteins from interacting with the ER unless it is already bound to DNA (Fritsch et al, 1992). Thus, DNA and estrogen may act together to transform the receptor to the proper conformation for transcriptional activation, allowing the receptor to make contacts with other mediating factors. Among the important regulatory proteins that undergo allosteric modulation upon binding to DNA, as demonstrated by circular dichroism, fluorescence spectroscopy and nuclear magnetic resonance studies, is the *fos* and *jun* heterodimer (Patel et al, 1990), which undergoes specific conformational changes when bound to its specific AP-1 site, and the yeast transcription factor GCN4 (Weiss et al, 1990), which is stabilized in a particular folded conformation when bound to DNA.

120

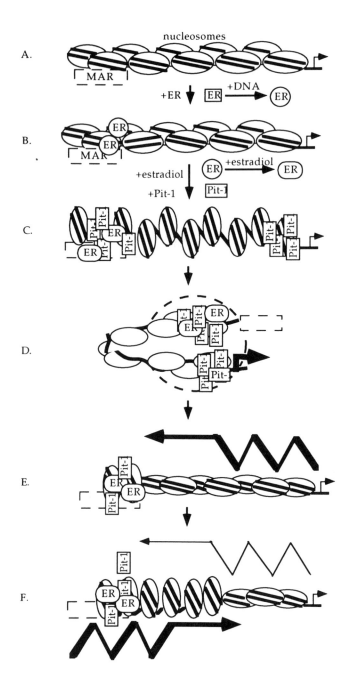

FIGURE 4. ER modulation of torsional stress within the regulatory domain of the rat prolactin gene.

SUMMARY

Several important components of the nuclear environment are manifested in estrogen regulation of the rPRL gene. The ER itself may interact with a variety of ligands, including DNA, estrogens, and other nuclear proteins in a dynamic process involving multiple conformations of ER. This complex of equilibrium states of the ER domains may serve to regulate interactions with nuclear components to maintain specificity of sites of action. Developmental (Durrin et al, 1984; Durrin and Gorski, 1985) and hormonal (Seyfred and Gorski, 1990) modification of chromatin structure is evident in the appearance of DNaseI-hypersensitive regions in the 5' flanking region of the rPRL gene; these occur at sites corresponding to *cis*-regulatory elements. There is evidence that hormone-independent ER binding results in bending of the DNA (Nardulli and Shapiro, 1992) and that ER is directly responsible for alterations in the chromatin in a hormone-dependent interaction (Gilbert et al, 1992) either with other regulatory factors or with the nucleosomes directly. The role of the nuclear matrix is evident in the binding of ER

FIGURE 4. ER modulation of torsional stress within the regulatory domain of the rat prolactin gene. (A) Inactive chromatin structure adjacent to a putative MAR (hatched box) and the site of rPRL transcription initiation (arrow). The intervening 1800 to 1900 base pairs provide sites for 9 or 10 nucleosomes. (B) The action of developmental factor(s) or perhaps a ligand-independent action of the ER supports an alteration of the local DNA structure mediated through regulation of the local torsional stress as a result of ligand-independent bending of the DNA. The resulting increase in thermodynamic energy could shift nucleosome positioning, forming a more open chromatin structure and thus uncover *cis*-regulatory sites or otherwise modify the DNA structure to modulate interactions with regulatory molecules. (C) In the presence of Pit-1 and estradiol, protein-protein and protein-DNA interactions facilitate and stabilize (D) the formation of the transcription complex (dashed circle). (E) The negative supercoiling generated behind the transcription complex is propagated to the distal regulatory region where it can be absorbed and stabilized by the non-B DNA structure generated there. (F) ER may act to trap some of this thermodynamic energy and reverse the propagation of supercoils once the transcription complex has dissociated from the DNA. In the continued presence of estrogen, the ER could again generate torsional stress in the 3' direction to modify the local topology and reintroduce the open chromatin configuration.

to matrix proteins (Ruh and Ruh, 1988; Metzger and Korach, 1990). A region with characteristics of an MAR is present within the 5' flanking region of the gene, as well as sequences having homology to topoisomerase II sites. The ER has been demonstrated to be capable of bending DNA and of modifying the linking number of the DNA helix, possibly by altering the recognition by topoisomerase II of the superhelical state of the loop domain, thereby modulating topoisomerase activity. Structural polymorphism of the DNA is apparent in the hormonal induction of S1 nuclease-hypersensitivity and in the superhelicity-dependent appearance of stably unwound DNA in the HS II region (Kladde et al, 1993). This unwinding occurs at sites believed to be responsible for ER binding to the DNA. The appearance of a single-stranded region or some form of stem-and-loop structure has been suggested as a preferred binding site for the ER in rPRL (Lannigan and Notides, 1989). Even though it is unclear exactly what the structure of the binding site might be, a dynamic equilibrium between multiple states of DNA structure is likely. Such a dynamic state of flux between B-DNA, single-stranded DNA, a hairpin structure, or some other form of the DNA might constitute a region from which torsional stress might be controlled by a stabilizing protein. Such a protein might exhibit the ability to influence the topology of the local region of DNA, such as the ER (Gilbert et al, 1992). As previously mentioned, it has been suggested that transcriptionally poised chromatin is under torsional stress introduced and regulated by topoisomerase II acting at sites near the base of the DNA loop domain. Further, the stable unwinding of the DNA in the region of MARs may act to relieve negative superhelicity within the loop domain. This stable unwinding could provide the means to store and release thermodynamic energy to regulate superhelicity within the transcriptionally active chromatin, and might possibly be controlled, in part, by the ER.

Taken together, these data suggest a model in which the ER modulates the torsional stress within the loop domain to modify other protein-DNA interactions (Figure 4). The action of some unknown developmental factor(s) *in vivo*, or perhaps a ligand-independent action of the ER supports an alteration of the local DNA structure mediated through regulation of the local torsional stress (Figure 4B and 4C). This would likely result from a ligand-induced conformational change in the ER already bound to the DNA. The resulting increase in thermodynamic energy could either shift nucleosome positioning to uncover *cis*-regulatory sites or otherwise modify the DNA structure to modulate interactions with regulatory molecules. Having established an active chromatin configuration, other regulatory molecules such as the Pit-1 protein may interact with the transcription complex to initiate transcription (Figure 4C and 4D); this is similar to the GR modulation of NF-1 binding in the MMTV promotor (Pina et al, 1990). Both Pit-1 and ER are necessary to initiate rPRL transcription *in vivo*. The negative supercoiling generated behind the transcription complex (Liu and Wang, 1987) is propagated to the distal regulatory region where it can be absorbed and stabilized by the non-B DNA structure generated there (Figure 4D and 4E). ER may act to trap some of this thermodynamic energy and reverse the propagation of supercoils once the transcription complex has dissociated from the DNA (Figure 4E and 4F). Alternatively, the ER may simply prevent topoisomerase from modifying the superhelical state of the loop domain in response to the increase in negative supercoiling, allowing the process to reverse itself without additional input of energy. In the continued presence of estrogen, the ER could reinitiate the process by again generating torsional stress in the 3' direction to modify the local topology and reintroduce the open chromatin configuration. Analysis of the chromatin state and of the DNA topology in transcriptionally active and inactive situations generated in the presence of estrogens, anti-estrogens, and other transcriptional activators and inhibitors will determine whether such a model, or any part of it, has relevance to the *in vivo* regulation of rPRL gene transcription.

ACKNOWLEDGMENTS

The authors would like to thank Drs. S. Lundeen, D. Gregg, and C. Ying for their helpful discussions and critical reviews of the manuscript, and K. Holtgraver for editorial assistance.

This work was sponsored in part by the University of Wisconsin College of Agricultural and Life Sciences, by NRSA Fellowship HD07478 awarded to J. M., and by NIH Grants HD08192, HD07259, and CA18110 awarded to J. G.

REFERENCES

Adom J, Carr KD, Gouilleux F, Marsaud V and Richard-Foy H (1991): Chromatin structure of hormono-dependent promoters. *J Steroid Biochem Mol Biol* 40: 325-332.

Barrack ER (1987): Steroid hormone receptor localization in the nuclear matrix: interaction with acceptor sites. *J Steroid Biochem* 27: 115-121.

Beato, M (1989): Gene regulation by steroid hormones. *Cell* 56: 335-344.

Bode J, Kohwi Y, Dickinson L, Joh T, Klehr D, Mielke C and Kohwi-Shigematsu T (1992): Biological significance of unwinding capability of nuclear matrix-associating DNAs. *Science* 255: 195-197.

Bonifer C, Hecht A, Saueressig H, Winter DM and Sippel AE (1991): Dynamic chromatin: the regulatory domain organization of eukaryotic gene loci. *J Cell Biochem* 47: 99-108.

Brill SJ and Sternglanz R (1988): Transcription-dependent DNA supercoiling in yeast topoisomerase mutants. *Cell* 54: 403-410.

Burch JBE and Weintraub H (1983): Temporal order of chromatin structural changes associated with activation of the major chicken vitellogenin gene. *Cell* 33: 65-76.

Carballo M and Beato M (1990): Binding of the glucocorticoid receptor induces a topological change in plasmids containing the hormone-responsive element of mouse mammary tumor virus. *DNA Cell Biol* 9: 519-525.

Carr KD and Richard-Foy H (1990): Glucocorticoids locally disrupt an array of positioned nucleosomes on the rat tyrosine aminotransferase promotor in hepatoma cells. *Proc Natl Acad Sci* USA 87: 9300-9304.

Clark DJ and Felsenfeld G (1991): Formation of nucleosomes on positively supercoiled DNA. *EMBO J* 10: 387-395.

Cockerill PN and Garrard WT (1986): Chromosomal loop anchorage of the kappa immunoglobulin gene occurs next to the enhancer in a region containing topoisomerase II sites. *Cell* 44: 273-282.

Cockerill PN, Yuen M-H and Garrard WT (1987): The enhancer of the immunoglobulin heavy chain locus is flanked by presumptive chromosomal loop anchorage elements. *J Biol Chem* 262: 5394-5397.

Cook PR (1989): The nucleoskeleton and the topology of transcription. *Eur J Biochem* 185: 487-501.

Cordingley MG, Riegel AT and Hager GL (1987): Steroid-dependent interaction of transcription factors with the inducible promoter of the mouse mammary tumor virus *in vivo*. *Cell* 48: 261-270.

Corthesy B, Cardinaux J-R, Claret F-X and Wahli W (1989): A nuclear factor I-like activity and a liver-specific repressor govern estrogen-regulated in vitro transcription from the Xenopus laevis vitellogenin B1 promoter. *Mol Cell Biol* 9: 5548-5562.

Day RN and Maurer RA (1989): The distal enhancer region of the rat prolactin gene contains elements conferring response to multiple hormones. *Mol Endocrinol* 3: 3-9.

Diamond MI, Miner JN, Yoshinaga SK and Yamamoto KR (1990): Transcription factor interactions: selectors of positive or negative regulation from a single DNA element. *Science* 249: 1266-1272.

Durrin LK and Gorski J (1985): The prolactin gene hypersensitive sites are present early in development and are not induced by estrogen administration. *Endocrinology* 117: 2098-2105.

Durrin LK, Weber JL and Gorski J (1984): Chromatin structure, transcription, and methylation of the prolactin gene domain in pituitary tumors of Fischer 344 rats. *J Biol Chem* 259: 7086-7093.

Evans RM (1988): The steroid and thyroid hormone receptor superfamily. *Science* 240: 889-895.

Felsenfeld G (1992): Chromatin as an essential part of the transcriptional mechanism. *Nature* 355: 219-224.

Frank-Kamenetskii M (1989): Waves of DNA supercoiling. *Nature* 337: 206.

Fritsch M, Leary C, Furlow JD, Ahrens H, Schuh T, Mueller G and Gorski J (1992): A ligand-induced conformational change in the estrogen receptor is localized in the steroid binding domain. *Biochemistry* 31: 5303-5311.

Fritsch M, Welch RD, Murdoch FE, Anderson I and Gorski J (1992): DNA allosterically modulates the steroid binding domain of the estrogen receptor. *J Biol Chem* 267: 1823-1828.

Gasser SM and Laemmli UK (1986): Cohabitation of scaffold binding regions with upstream enhancer elements of three developmentally regulated genes of *D. melanogaster. Cell* 46: 521-530.

Getzenberg RH, Pienta KJ and Coffey DS (1990): The tissue matrix: cell dynamics and hormone action. *Endocr Rev* 11: 399-417.

Giaever GN and Wang JC (1988): Supercoiling of intracellular DNA can occur in eukaryotic cells. *Cell* 55:849-856.

Giese K, Cox J and Grosschedl R (1992): The HMG domain of lymphocyte enhancer factor 1 bends DNA and facilitates assembly of functional nucleoprotein structures. *Cell* 69: 185-195.

Gilbert DM, Losson R and Chambon P (1992): Ligand dependence of estrogen receptor induced changes in chromatin structure. *Nucleic Acids Res* 20: 4525-4531.

Gorski J (1986): The nature and development of steroid hormone receptors. *Experientia* 42:744-750.

Gustafson TA, Taylor A and Kedes L (1989): DNA bending is induced by a transcription factor that interacts with the human c-fos and α-actin promotors. *Proc Natl Acad Sci* USA 86: 2162-2166.

Hansen JC and Gorski J (1985): Conformational and electrostatic properties of unoccupied and liganded estrogen receptors determined by aqueous two-phase partitioning. *Biochemistry* 24: 6078-6085.

Hansen JC, Welshons WV and Gorski J (1988): Toward a more unified model of steroid hormone receptor structure and function. In: *Steroid Receptors and Disease: Cancer, Autoimmune, Bone and Circulatory Disorders.* Sheridan PJ, Blum K, Trachtenberg, MC eds. New York: Marcel Dekker, Inc.

Horton M, Landers JP, Subramaniam M, Goldberger A, Toyoda H, Gosse B and Spelsberg TC (1991): Enrichment of a second class of native acceptor sites for the avian oviduct progesterone receptor as intact chromatin fragments. *Biochemistry* 30: 9523-9530.

Ingraham HA, Flynn SE, Voss JW, Albert VR, Kapiloff MS, Wilson L and Rosenfeld MG (1990): The POU-specific domain of Pit-1 is essential for sequence-specific, high affinity DNA binding and DNA-dependent Pit-1-Pit-1 interactions. *Cell* 61: 1021-1033.

Ishibe Y, Klinge CM, Hilf R and Bambara RA (1991): Estrogen receptor alters the topology of plasmid DNA containing estrogen responsive elements. *Biochem Biophys Res Commun* 176: 486-491.

Jackson DA and Cook PR (1985): Transcription occurs at a nucleoskeleton. *EMBO J* 4: 919-925.

Kerppola TK and Curran T (1991): Fos-jun heterodimers and jun homodimers bend DNA in opposite orientations-implications for transcription factor cooperativity. *Cell* 66: 317-326.

Kladde MP, D'Cunha J and Gorski J (1993): Multiple transitions to non-B-DNA structures occur in the distal regulatory region of the rat prolactin gene. *J Mol Biol* 229: 344-367.

Kornberg RD and Lorch Y (1991): Irresistible force meets immovable object: transcription and the nucleosome. *Cell* 67: 833-836.

Krajewski WA and Luchnik AN (1991): High rotational mobility of DNA in animal cells and its modulation by histone acetylation. *Mol Gen Genet* 231: 17-21.

Landers JP and Spelsberg TC (1992): New concepts in steroid hormone action: transcription factors, proto-oncogenes, and the cascade model for steroid regulation of gene expression. *Crit Rev Eukaryotic Gene Expression* 2: 19-63.

Lannigan DA and Notides AC (1989): Estrogen receptor selectively binds the "coding strand" of an estrogen responsive element. *Proc Natl Acad Sci* USA 86: 863-867.

Liu LF and Wang JC (1987): Supercoiling of the DNA template during transcription. *Proc Natl Acad Sci* USA 84: 7024-7027.

Mangalam HJ, Albert VR, Ingraham HA, Kapiloff M, Wilson L, Nelson C, Elsholtz H and Rosenfeld MG (1989): A pituitary POU domain protein, Pit-1, activates both growth hormone and prolactin promotors transcriptionally. *Genes Develop* 3: 946-958.

Metzger DA and Korach KS (1990): Cell-free interaction of the estrogen receptor with mouse uterine nuclear matrix: evidence of saturability, specificity, and resistance to KCl extraction. *Endocrinology* 126: 2190-2195.

Mukherjee R and Chambon P (1990): A single-stranded DNA-binding protein promotes the binding of the purified oestrogen receptor to its responsive element. *Nucleic Acids Res* 18: 5713-5716.

Murdoch FE and Gorski J (1991): The role of ligand in estrogen receptor regulation of gene expression. *Mol Cell Endocrinol* 78: C103-C108.

Murdoch FE, Meier DA, Furlow JD, Grunwald KAA and Gorski J (1990): Estrogen receptor binding to a DNA response element in vitro is not dependent upon estradiol. *Biochemistry* 29: 8377-8385.

Murdoch GH, France R, Evans RM and Rosenfeld MG (1983): Polypeptide hormone regulation of gene expression. *J Biol Chem* 258: 15329-15335.

Nardulli AM and Shapiro DJ (1992): Binding of the estrogen receptor DNA-binding domain to the estrogen response element induces DNA bending. *Mol Cell Biol* 12: 2037-2042.

Nelson C, Crenshaw EB, Franco R, Lira SA, Albert VR, Evans RM and Rosenfeld MG (1986): Discrete cis-active genomic sequences dictate the pituitary cell type-specific expression of rat prolactin and growth hormone genes. *Nature* 322: 557-562.

Nelson WG, Pienta KJ, Barrack ER and Coffey DS (1986): The role of the nuclear matrix in the organization and function of DNA. *Annu Rev Biophys Biophys Chem* 15: 457-475.

O'Malley B (1990): The steroid receptor superfamily: more excitement predicted for the future. *Mol Endocrinol* 4: 363-369.

Patel L, Abate C and Curran T (1990): Altered protein conformation on DNA binding by fos and jun. *Nature* 347: 572-578.

Pina B, Barettino D, Truss M and Beato M (1990): Structural features of a regulatory nucleosome. *J Mol Biol* 216: 975-990.

Pina B, Bruggemeier U and Beato M (1990): Nucleosome positioning modulates accessibility of regulatory proteins to the mouse mammary tumor virus promotor. *Cell* 60: 719-731.

Pina B, Hache RJG, Arnemann J, Chalepakis G, Slater EP and Beato M (1990): Hormonal induction of transfected genes depends on DNA topology. *Mol Cell Biol* 10: 625-633.

Ptashne M and Gann AAF (1990): Activators and targets. *Nature* 346: 329-331.

Razin SV, Petrov P and Hancock R (1991): Precise localization of the α-globin gene cluster within one of the 20- to 300-kilobase DNA fragments released by cleavage of chicken chromosomal DNA at topoisomerase II sites *in vivo*: evidence that the fragments are DNA loops or domains. *Proc Natl Acad Sci* USA 88: 8515-8519.

Reik A, Schutz G and Stewart AF (1991): Glucocorticoids are required for establishment and maintenance of an alteration in chromatin structure: induction leads to a reversible disruption of nucleosomes over an enhancer. *EMBO J* 10: 2569-2576.

Ruh MF and Ruh TS (1988): Specificity of chromatin acceptor sites for steroid hormone receptors. In: *Steroid Receptors and Disease: Cancer, Autoimmune, Bone and Circulatory Disorders.* Sheridan, P. J., Blum, K., Trachtenberg, M. C., eds., New York: Marcel Dekker.

Schreck R, Zorbas H, Winnacker EL and Baeuerle PA (1990): The NF-κB transcription factor induces DNA bending which is modulated by its 65 kD subunit. *Nucleic Acids Res* 18: 6497-6502.

Schuchard M, Subramaniam M, Ruesnick T and Spelsberg TC (1991): Nuclear matrix localization and specific DNA binding by receptor binding factor 1 of the avian oviduct progesterone receptor. *Biochemistry* 30: 9516-9522.

Seyfred MA and Gorski J (1990): An interaction between the 5' flanking distal and proximal regulatory domains of the rat prolactin gene is required for transcriptional activation by estrogens. *Mol Endocrinol* 4: 1226-1234.

Tazi J and Bird A (1990): Alternative chromatin structure at CpG islands. *Cell* 60: 909-920.

Travers AA (1989): DNA conformation and protein binding. *Annu Rev Biochem* 58: 427-452.

Tsao Y-P, Wu H-Y and Liu LF (1989): Transcription-driven supercoiling of DNA: direct biochemical evidence from *in vitro* studies. *Cell* 56:111-118.

Udvardy A, Maine E and Schedl P (1985): The 87A7 Chromomere: identification of novel chromatin structures flanking the heat shock locus that may define the boundaries of higher order domains. *J Mol Biol* 185: 341-358.

Wang W, Carey M and Gralla JD (1992): Polymerase II promotor activation: closed complex formation and ATP-driven start site opening. *Science* 255: 450-453.

Weintraub H (1984): Histone-H1-dependent chromatin superstructures and the suppression of gene activity. *Cell* 38: 17-27.

Weintraub H, Cheng PF and Conrad K (1986): Expression of transfected DNA depends on DNA topology. *Cell* 46: 115-122.

Weisbrod S (1983): Active chromatin. *Nature* 297: 289-295.

Weiss MA, Ellenberger T, Wobbe CR, Lee JP, Harrison SC and Struhl K (1990): Folding transition in the DNA-binding domain of GCN4 on specific binding to DNA. *Nature* 347: 575-578.

Workman JL, Taylor ICA and Kingston RE (1991): Activation domains of stably bound GAL4 derivatives alleviate repression of promotors by nucleosomes. *Cell* 64: 533-544.

Wu H-Y, Shyy S, Wang JC and Liu LF (1988): Transcription generates positively and negatively supercoiled domains in the template. *Cell* 53: 433-440.

Yamamoto KR (1985): Steroid receptor regulated transcription of specific genes and gene networks. *Annu Rev Genet* 19: 209-215.

Zaret KS and Yamamoto KR (1984): Reversible and persistent changes in chromatin structure accompany activation of a glucocorticoid-dependent enhancer element. *Cell* 38: 29-38.

FUNCTIONAL INTERACTION OF THE ESTROGEN RECEPTOR WITH THE TISSUE-SPECIFIC, HOMEODOMAIN TRANSCRIPTION FACTOR, PIT-1

Richard A. Maurer
Department of Physiology and Biophysics
University of Iowa

Richard N. Day
Department of Medicine
University of Virginia

Yasuhiko Okimura
Barbara E. Nowakowski
Department of Physiology and Biophysics and
Molecular Biology Ph.D. Program
University of Iowa

INTRODUCTION

The prolactin gene has provided a useful system for analysis of the mechanisms which permit estrogen to stimulate the transcription of specific genes. A least two factors have aided studies of the estrogenic regulation of prolactin gene expression. One advantage of the prolactin system concerns the fact that prolactin is a major product of the pituitary. The relatively high level of prolactin gene expression has facilitated biochemical approaches to the analysis of prolactin production. The

STEROID HORMONE RECEPTORS
V. K. Moudgil, Editor
© 1993 Birkhäuser Boston

relatively high level expression has contributed to the preparation of a number of useful reagents including antibodies to prolactin and cDNAs encoding prolactin. A second factor concerns the availability of cell culture systems in which estrogen can stimulate prolactin gene expression in vitro. The GH clonal pituitary tumor cell lines which were developed by Armen Tashjian and Gordon Sato (Tashjian et al, 1968) have been particularly valuable for these studies. These cells produce both growth hormone and prolactin. A number of different clonal strains are available, with the GH3 cells perhaps the most widely used. While a number of studies have utilized primary cultures of rat pituitary cells, the GH3 cells offer a number of advantages. In particular, these cells have proven to be particularly useful for gene transfer experiments examining the DNA sequences and transcription factors required for expression of the prolactin gene.

Utilizing a variety of tools including biochemistry, molecular biology and cell biology and relying heavily on the GH3 cell culture system, studies from a number of laboratories have defined several steps involved in the ability of estrogen to increase production of prolactin. These studies have demonstrated that estradiol increases the transcription of the prolactin gene. The DNA sequences required for estrogen effects on prolactin expression have been mapped and more recently the transcription factors involved in mediating this response have been explored.

Estrogen effects on prolactin biosynthesis, prolactin mRNA and transcription of the prolactin gene

Initial studies of estrogen effects on prolactin gene expression involved analysis of prolactin synthesis. Studies using incorporation of radiolabeled amino acids demonstrated that estrogen treatment results in increased de novo synthesis of prolactin (MacLeod et al, 1969; Maurer and Gorski, 1977; Yamamoto et al, 1975). Estrogen has been observed to stimulate prolactin synthesis in both monolayer cultures of dispersed pituitary cells (Lieberman et al, 1978; Vician et al, 1979) and the GH3 pituitary tumor cell line (Haug and Gautvik, 1976). Thus estrogen appears to have a direct effect on prolactin production by the pituitary.

Estrogen effects on prolactin synthesis are mediated by changes in prolactin mRNA levels. Early studies used cell-free translation systems to estimate prolactin mRNA concentrations (Seo et al, 1979a; Shupnik et al, 1979; Stone et al, 1977; Vician et al, 1979) and demonstrated that estrogen treatment resulted in several-fold increases in prolactin mRNA concentrations. The increases in the concentration of prolactin mRNA as estimated by cell-free translation were similar to previously observed changes in prolactin synthesis. Subsequently, these findings were confirmed by hybridization analysis using radiolabeled prolactin cDNAs (Maurer, 1980; Ryan et al, 1979; Seo et al, 1979b). Thus, these studies provide strong evidence that estrogen induced increases in prolactin synthesis are mediated by increases in the cellular content of prolactin mRNA.

Changes in prolactin mRNA concentrations could be due to changes in transcription or to changes in mRNA degradation. To examine possible changes in transcription of the prolactin gene, a nuclear run-on assay was utilized. This approach involves the quantitation of de novo synthesis of prolactin mRNA sequences by isolated nuclei. The run-on assay was developed by McKnight and Palmiter (McKnight and Palmiter, 1979) and depends on the observation that RNA synthesis by isolated nuclei involves primarily chain elongation with relatively little initiation of new RNA chains (Reeder and Roeder, 1972; Zylber and Penman, 1971). Because most RNA synthesis by isolated nuclei is simply extension of chains initiated in vivo, quantitation of specific mRNA sequences synthesized in vitro provides an estimate of mRNA synthesis occurring at the time the cells were homogenized to prepare nuclei. For studies of estrogen effects on prolactin gene transcription, rats were treated with estradiol and then pituitaries removed at varying times after treatment and nuclei were isolated. Isolated nuclei were allowed to synthesize RNA in the presence of [a-32P]UTP and newly synthesized 32P-RNA was isolated. To quantitate the synthesis of prolactin mRNA sequences, total 32P-RNA synthesized in vitro was hybridized to a cloned rat prolactin cDNA (Gubbins et al, 1979; Gubbins et al, 1980) which was immobilized on nitrocellulose disks. An estimate of non-specific hybridization was obtained by hybridization to filters containing immobilized pBR322 DNA and this value was subtracted from the values obtained for filters containing the prolactin cDNA. Using this system it was demonstrated that the synthesis of prolactin mRNA was tissue specific, mediated by RNA polymerase II and that the prolactin mRNA transcripts represented the sense strand of the prolactin gene (Maurer, 1981). This assay was then used to examine estrogen effects on transcription of the prolactin gene (Maurer, 1982). The results demonstrated that estradiol stimulates transcription of the prolactin gene in as little as 20 minutes (Figure 1). Unlike some other well studied estrogen responsive genes such as the ovalbumin gene, basal transcription of the prolactin gene is maintained quite well in the absence of estrogen

134

(at a rate of several hundred parts per million of total transcription). Following treatment with estradiol, there is a rapid, several-fold increase in the synthesis of prolactin mRNA sequences. This occurs in the same time frame in which tight nuclear binding of estrogen receptors is detected (Figure 1). Rapid effects of estradiol to stimulate the transcription of the prolactin gene have also been detected in cultured

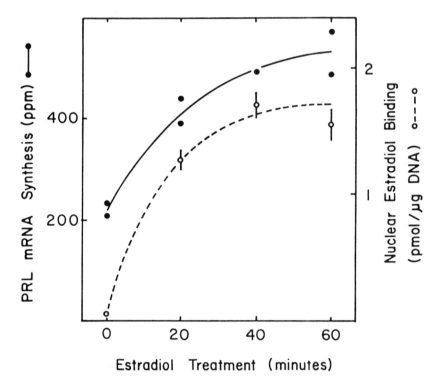

FIGURE 1. Effects of varying times of estrogen treatment on the transcription of the rat prolactin gene. Ovariectomized rats were treated with estradiol and at the indicated times after treatment, the pituitaries were removed and nuclei prepared. Nuclei were incubated in the presence of [a-32P]UTP and unlabeled ATP, CTP and GTP and then incorporation of radioactivity into prolactin mRNA sequences and total RNA was determined. Prolactin mRNA synthesis is expressed as ppm which is the cpm in prolactin sequences/cpm in total RNA x 106 (•). Nuclear estradiol receptors (o) were quantitated by an estradiol exchange assay (Anderson et al, 1972; Mulvihill and Palmiter, 1977). Reprinted with permission from Maurer RA (1982): Estradiol regulates the transcription of the prolactin gene. *J Biol Chem* 257: 2133-2136.

GH3 cells (Waterman et al, 1988). The rapid kinetics of the estrogenic activation of prolactin transcription is consistent with a direct effect of estrogen receptor activation on increased transcription of the prolactin gene. Furthermore, as it has been demonstrated that estrogenic stimulation of prolactin gene transcription is independent of protein synthesis (Shull and Gorski, 1984) it seems very likely that the estrogen stimulation of prolactin transcription is a primary response to the receptor. The magnitude of estrogen effects on transcription is similar to the magnitude of estrogen effects on prolactin mRNA and suggests that changes in transcription are likely the principal mechanism involved in regulation of prolactin mRNA concentrations.

Identification of DNA Sequences Which Interact with the Estrogen Receptor and Which are Required for Estrogen Responses of the Prolactin Gene

Early studies which examined the subcellular localization of steroid receptors led to the prediction that steroid-receptors may interact with specific DNA sequences to alter transcription (Gorski et al, 1968; Jensen and DeSombre, 1968; Jensen et al, 1968). Initial studies of receptor binding failed to demonstrate limited capacity, high affinity binding sites in either isolated nuclei (Chamness et al, 1974; Higgins et al, 1973; Williams and Gorski, 1972) or total DNA (Yamamoto and Alberts, 1974). The failure to detect specific binding of steroid receptors in these studies was attributed to the presence of a large excess of non-specific binding sites which masked binding to a small number of high affinity sites. It required advances in the ability to isolate specific genomic sequences by recombinant DNA technology to test this hypothesis. Following the cloning of chromosomal genes for steroid responsive genes, it became possible to test binding to specific DNA sequences and selective binding was initially demonstrated for the glucocorticoid receptor (Govindan et al, 1982; Payvar et al, 1981; Pfahl, 1982; Scheidereit et al, 1983) and the progesterone receptor (Bailly et al, 1983; Compton et al, 1983; Mulvihill et al, 1982). Thus, at the time studies of estrogen receptor binding to the prolactin gene were initiated there was evidence for specific DNA binding by the glucocorticoid receptor and the progesterone receptor, but not the estrogen receptor.

Initial studies examining the interaction of estrogen receptor with the prolactin gene used a competition DNA binding assay (Maurer, 1985). This assay tests the ability of soluble DNA fragments to compete with immobilized calf thymus DNA for binding the estrogen receptor. The

136

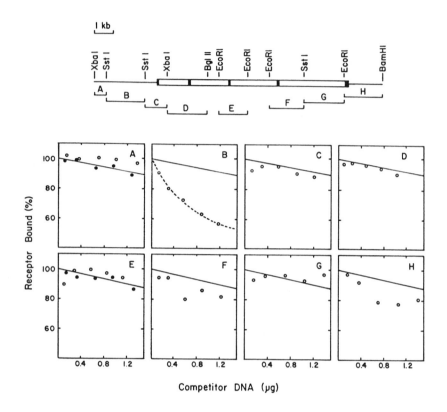

FIGURE 2. Analysis of estrogen receptor binding to specific DNA fragments from the rat prolactin gene. The restriction map (top) identifies the DNA fragments used in a competition assay to identify fragments which specifically interact with estrogen receptor. The filled boxes of the map indicate exons of the prolactin gene and the open boxes indicate introns. Cytosols in which the estrogen receptor was labeled by incubation with [³H]estradiol were incubated with calf thymus DNA-cellulose and soluble DNA consisting of either prolactin gene fragments (o) or calf thymus DNA (•). The 100% binding level was determined in the absence of competitor DNA. Reprinted (with permission from Mary Ann Liebert, Inc.) from Maurer, R.A. (1985): Selective binding of the estradiol receptor to a region at least one kilobase upstream from the prolactin gene. *DNA* 4: 1-9.

assay detects receptor binding to the immobilized DNA by quantitation of [3H]estradiol labeled receptor. The use of radiolabeled estradiol to quantitate receptor binding has the advantage that crude preparations containing receptor can be used for the assay. To calibrate the assay, a preparation of sonicated, calf thymus total DNA is used as the competitor DNA. DNA fragments with high affinity for receptor should compete for the receptor better than total calf thymus DNA. Fragments of the prolactin gene were prepared and used to test for selective binding to the estrogen receptor (Figure 2). Eight different fragments spanning more than 10 kilobase pairs (kbp) of DNA including most of the prolactin gene and a significant amount of 5'- and 3'- flanking sequence were tested. Only a single fragment (Figure 2, panel B) demonstrated substantially better competition for receptor binding than the standard, total calf thymus DNA. Thus, this fragment which is located between 2.6 and 0.6 kbp upstream of the transcription initiation site of the prolactin gene apparently contains a relatively high affinity binding site for the estrogen receptor.

Although the DNA competition studies provided evidence for selective interaction of the receptor with specific DNA sequences, the fact that crude extracts were used raised the possibility that the receptor interacts with another protein which is actually responsible for the selective binding. To explore this issue further, highly purified preparations of receptors were tested for binding to fragments of the prolactin gene using a nitrocellulose filter binding assay (Maurer and Notides, 1987). This assay depends on the fact that double stranded DNA is not retained by a nitrocellulose filter unless it is associated with a protein. To prepare DNA for the assay, deletion mutations of the 5' flanking region of the rat prolactin gene were subcloned into pUC13. Plasmids containing the prolactin fragment were digested with a restriction enzyme which excised the prolactin fragment and the DNA termini were radiolabeled. This procedure labels both the prolactin fragment and the pUC13 DNA to approximately equal specific activity. The pUC13 DNA can then serve as a monitor for non-specific DNA binding. The DNA fragments were incubated with highly purified estrogen receptor, filtered through nitrocellulose and the DNA fragments retained on the nitrocellulose filter analyzed by agarose gel electrophoresis and autoradiography (Figure 3). The results demonstrate rather strong selective binding of the estrogen receptor to the same 2 kbp SstI fragment from the 5' flanking region which was found to contain high affinity receptor binding sites by the DNA competition assay. The fact that selective binding is observed with highly purified receptor argues

138

strongly that this is an intrinsic property of the receptor and is probably not due to the interaction of the receptor with some other protein.

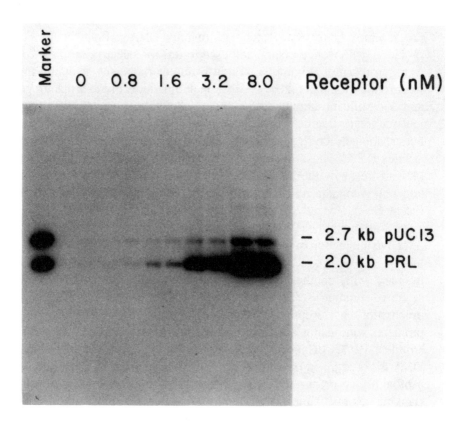

FIGURE 3. The estrogen receptor preferentially binds to a 2.0 kilobase fragment from the 5' flanking region of the rat prolactin gene. Duplicate samples of radiolabled DNA containing a 2.0 kilobase Sst I fragment of the rat prolactin gene (positions -2.6 to -0.6 kilobases with respect to the transcription initiation site) and a 2.7 kilobase pUC13 DNA fragment were incubated with the indicated concentration of estrogen receptor and then filtered through nitrocellulose. DNA retained on the filter was eluted and analyzed by agarose gel electrophoresis. Reprinted with permission (American Society for Microbiology) from Maurer RA and Notides AC (1987): Identification of an estrogen-responsive element from the 5'-flanking region of the rat prolactin gene. *Mol Cell Biol* 7: 4247-4254.

Interestingly, receptor binding to the prolactin fragment showed a strong, nonlinear concentration dependence. For instance, as receptor concentration doubled from 1.6 to 3.2 nM, there was a much more than 2-fold increase in receptor binding to the prolactin fragment. This non-linear concentration dependence likely reflects an important role for formation of receptor dimers and is consistent with the view that receptor dimers may be the physiologically relevant species for binding to DNA (Notides et al, 1981).

Analysis of deletion mutants reveals a substantial decrease in receptor binding with deletion from -1713 to -1532 (Figure 4). Thus, the upstream boundary of the receptor binding likely lies within this region. A similar approach using 3' deletion analysis demonstrated that the downstream boundary of the receptor binding site is located between -1496 and -1575.

To determine if the DNA sequences which bind the receptor are required to mediate a transcriptional response to estrogen, transfection studies were performed. For these studies the 5' flanking region of the rat prolactin gene was coupled to the chloramphenicol acetyl transferase (CAT) marker gene and the fusion gene was transfected into GH3 cells. Initial studies demonstrated that a construct containing 2.4 kbp of the 5' flanking region and promoter of the rat prolactin gene were sufficient to permit a 2 to 3-fold response to estradiol, which is similar to the transcriptional activation of the endogenous gene. This response was promoter specific, as expression of the strong promoter from the long terminal repeat of the Rous sarcoma virus was not estrogen dependent. Primer extension studies demonstrated that the transfected gene utilized the correct transcription initiation site. To determine the sequences necessary for estrogenic regulation, 5' deletions of the prolactin gene were prepared, linked to the CAT gene and transfected into GH3 pituitary tumor cells (Figure 5). In good agreement with the receptor binding experiments, constructs containing at least 1.7 kilobases of 5' flanking sequence showed responses to estrogen, while shorter constructs did not respond to estrogen. Thus the sequences which were found to be necessary for receptor binding by the nitrocellulose filter assay were also found to be required for responses to estrogen. The ability of the upstream region of the prolactin gene to function as an estrogen dependent enhancer sequence was also tested (Figure 6). Either a relatively large fragment (representing sequences -2057 to -870) or a much smaller fragment (positions -1713 to -1495) were found to confer estrogen responsiveness to the thymidine kinase promoter. Thus, both of these fragments function as an estrogen-dependent enhancer sequence.

140

Because a rather small DNA fragment (219 base pairs, positions -1713 to -1495) was found to be sufficient to serve as an estrogen responsive enhancer, it was feasible to more precisely map the DNA sequences which interact with the estrogen receptor. This was performed by an exonuclease III protection assay. The results suggested that DNA

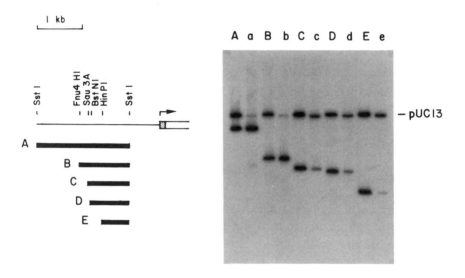

FIGURE 4. Deletion analysis of DNA elements required for estrogen receptor binding to the rat prolactin gene. A physical map is shown indicating the location of the various test fragments. For analysis of DNA binding by the nitrocellulose filter binding assay, a mixture of the radiolabeled test fragment and linearized pUC13 was incubated with the receptor. For lanes A, B, C, D and E the corresponding labeled DNAs (as indicated on the map) were electrophoresed directly on the gel. For lanes a, b, c, d and e, the labeled DNAs were incubated with 8.0 nM estrogen receptor, filtered through nitrocellulose and the bound DNA examined by electrophoresis. Reprinted with permission (American Society for Microbiology) from Maurer RA and Notides AC (1987): Identification of an estrogen-responsive element from the 5'-flanking region of the rat prolactin gene. *Mol. Cell. Biol.* 7: 4247-4254.

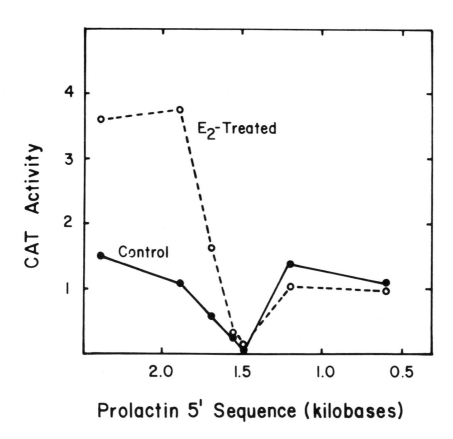

FIGURE 5. Deletion analysis of DNA sequences required for estrogen responsiveness of the rat prolactin gene. Fusion genes containing varying amounts of the 5' flanking region of the prolactin gene linked to the chloramphenicol acetyl transferase gene were transfected into GH3 cells. The cells were incubated in the absence (•) or presence (o) of 10 nM estradiol (E2). After 2 days of exposure to the estradiol, cell extracts were prepared and assayed for chloramphenicol acetyl transferase activity. Reprinted with permission (American Society for Microbiology) from Maurer RA and Notides AC (1987): Identification of an estrogen-responsive element from the 5'-flanking region of the rat prolactin gene. *Mol Cell Biol* 7: 4247-4254.

sequences which interact with the estrogen receptor included positions -1587 to -1563. This region of the prolactin gene contains the sequence TGTCACTATGTCC. This is an imperfect palindrome which is similar to the palindromic estrogen responsive element GGTCANNNTGACC found in the Xenopus vitellogenin gene (Klein-Hitpass et al, 1986) and the chicken vitellogenin gene (Jost et al, 1984). These estrogen responsive genes provided the initial evidence that the estrogen receptor likely interacts with a sequence with a two-fold axis of symmetry. These findings were consistent with a model involving interaction of receptor dimers with DNA, as was suggested by early studies of receptor cooperativity (Notides et al, 1981). Subsequently biochemical studies provided direct evidence for a role for receptor dimers in the interaction with palindromic DNA sequence (Klein-Hitpass et al, 1989; Kumar and Chambon, 1988).

To determine if the imperfect palindrome at -1587 to -1563 which binds the receptor is required for estrogenic stimulation of transcription, additional mutations and transfection studies were performed (Kim et al, 1988). Disruption of the receptor binding site at -1587 to -1563 completely eliminated the ability of the distal enhancer of the prolactin gene to function as an estrogen-dependent enhancer (Figure 7). Thus the imperfect palindrome which binds the estrogen receptor is indeed required for an estrogen response.

The Pituitary-Specific, Homeodomain Transcription Factor, Pit-1 is Required for Estrogen Responsiveness of the Prolactin Gene

It is clear from the studies outlined above that prolactin gene transcription is activated by estrogen through an interaction of the estrogen receptor with a specific DNA element 1.5 kbp upstream from the site of initiation, a region commonly referred to as the distal enhancer. A functional interaction between members of the steroid receptor family, as well as between these receptors and other transcription factors, in the activation of gene expression has been reported (Cato et al, 1988; Cato and Ponta, 1989; Edwards et al, 1989; Meyer et al, 1989). For example, glucocorticoid receptor can act in combination with number of transcription factors to activate an adjacent promoter (Schule et al, 1988a; Schule et al, 1988b; Strahle et al, 1988). Thus, the interaction between the estrogen receptor and other transcription factors that bind cis-elements within the 5' flanking region of the prolactin gene was potentially important for activation of transcription.

The role of transcription factors in prolactin gene expression has been the subject of intensive study. A tissue-specific transcription factor that interacts with the 5'-flanking region of the prolactin gene was proposed based upon gene transfer studies and DNA footprint analysis (Bodner and Karin, 1987; Cao et al, 1988; Nelson et al, 1988; Nelson et al, 1986; Schuster et al, 1988). This protein was cloned by two laboratories and termed Pit-1 (Ingraham et al, 1988) and GHF-1 (Bodner et al, 1988), respectively. The transcription factor Pit-1/GHF-1, hereafter called Pit-1, is a member of the family of homeodomain proteins that are

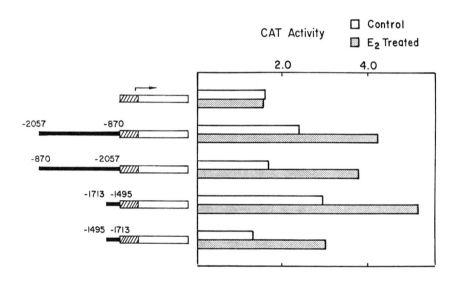

FIGURE 6. An upstream region of the rat prolactin gene can function as an estrogen-dependent enhancer sequence. Fragments from the 5' flanking region of the prolactin gene were placed upstream of the herpes simplex thymidine kinase promoter (hatched) which was linked to the chloramphenicol acetyl transferase gene. The fusion genes were transfected into GH3 cells which were incubated in the absence or presence of 10 nM estradiol as indicated. Cell extracts were prepared 2 days after transfection and chloramphenicol acetyl transferase activity was determined. Reprinted with permission (American Society for Microbiology) from Maurer RA and Notides AC (1987): Identification of an estrogen-responsive element from the 5'-flanking region of the rat prolactin gene. *Mol Cell Biol* 7: 4247-4254.

important activators of gene expression during development (Herr et al, 1988). There are four DNA elements to which the Pit-1 protein binds in the prolactin proximal promoter region and four similar elements were identified in the distal enhancer (Ingraham et al, 1988; Mangalam et al, 1989; Nelson et al, 1988). We investigated the functional relationship between cis-acting DNA elements that bind the pituitary specific factor, Pit-1, and the elements which bind the estrogen receptor within the distal enhancer region of the rat prolactin gene.

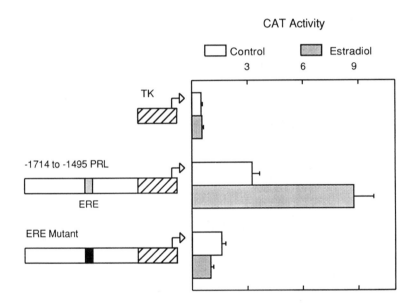

FIGURE 7. Mutation of the estrogen receptor binding site at -1587 to -1563 of the rat prolactin gene eliminates estrogen responsiveness. The -1712 to -1494 region of the rat prolactin gene was placed upstream of the herpes simplex thymidine kinase promoter which was linked to the chloramphenicol acetyl transferase gene. Fusion genes were transfected into GH3 cells and the cells received either no treatment or 10 nM estradiol as indicated. Cell extracts were prepared two days later and chloramphenicol acetyl transferase activity determined. Values are means ± standard errors. Replotted (with permission from The Endocrine Society) from Kim KE, Day RN and Maurer RA (1988): Functional analysis of the interaction of a tissue-specific factor with an upstream enhancer element of the rat prolactin gene. *Mol Endocrinol* 3: 1374-1381.

Both the distal enhancer and proximal promoter regions of the prolactin gene confer responses to cAMP, thyrotropin-releasing hormone and calcium and the purpose of this redundancy in function is not known (Day and Maurer, 1989; Day and Maurer, 1990). It is clear that the Pit-1 binding sites are important for basal, tissue-specific expression of the prolactin gene, and a role for the Pit-1 binding sites in mediating multihormonal responsiveness has recently been demonstrated (Hoggard et al, 1991; Yan and Bancroft, 1991; Yan et al, 1991). Previous results from our laboratory showed synergy between the estrogen response and response to cAMP conferred by the distal enhancer region (Day and Maurer, 1989). This raised the possibility that transcription factors binding to the distal region may interact with the estrogen receptor to modulate enhancer activity and estrogen responsiveness.

To assess the importance of these distal enhancer sequences in conferring response to estrogen, a series of 5' deletions starting from position -1769 bp was prepared and tested (Day et al, 1990). The deleted prolactin fragments were placed upstream of the herpes simplex thymidine kinase (TK) promoter which was linked to the bacterial choramphicol acetyltransferase reporter gene. The fusion genes were transfected into GH3 cells and responses to treatment with 10 nM estradiol were determined (Figure 8). A deletion that removed a portion of the D4 Pit-1 binding site, the most distal Pit-1 binding site (Nelson et al, 1988) had little effect on the ability of estrogen to stimulate expression of the fusion gene. Further deletion to position -1665, slightly reduced the fold induction for estradiol. Deletion to position -1625, which destroyed three of the four Pit-1 binding sites, further reduced estrogen responsiveness to a level near that of the TK promoter alone. It should be noted that this deletion is well upstream of the estrogen receptor binding site. These results demonstrated that responsiveness to estradiol requires a portion of the distal enhancer which includes Pit-1 binding sites. Thus, it is clear that although the estrogen receptor binding site (the imperfect palindrome at -1587 to -1563) is required for estrogen responsiveness, it is not sufficient to mediate an estrogen response. In these studies, the response of the distal enhancer region to stimulation with cAMP was also examined (data not shown). Interestingly, the estrogen and cAMP responses were found to have similar sequence requirements. Thus, the distal enhancer region of the prolactin gene appears to be a complex response unit. This region of DNA appears to bind multiple factors and the interaction of these factors is likely required

146

to mediate hormone regulation. These findings concerning a complex response unit within the distal enhancer of the prolactin gene are quite similar to the identification of a complex glucocorticoid response unit in the phosphoenolpyruvate carboxy kinase gene (Imai et al, 1990).

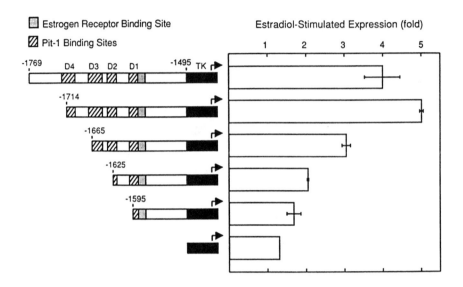

FIGURE 8. Deletional analysis of the prolactin distal enhancer region. A series of 5' deletions of the prolactin distal enhancer were placed upstream of the herpes simplex thymidine kinase (TK) promoter (-105 to +54) which was linked to the bacterial chloramphenicol acetyl transferase (CAT) reporter gene. The fusion genes were transfected into GH3 cells which were maintained in estrogen-depleted medium. The cells were treated with no addition (control) or 10 nM estradiol (E2) and then collected after 24 h for analysis of CAT activity. The results are plotted as ratio of activity of estradiol-treated cultures to control cultures. The upstream limit for each 5' deletion is indicated as well as the relative location of Pit-1 binding sites (hatched boxes) and the estrogen receptor binding site (shaded box). Values are from triplicate independent transfections corrected for transfection efficiency ± standard error. Replotted (with permission from The Endocrine Society) from Day RN, Koike S, Sakai M, Muramatsu M and Maurer RA (1990): Both Pit-1 and the estrogen receptor are required for estrogen responsiveness of the rat prolactin gene. *Mol Endocrinology* 4: 1964-1971.

Gene transfer studies using expression vectors encoding Pit-1 and the rat estrogen receptor were then used to directly examine the potential interaction between these trans-acting factors in conferring enhancer activity and estrogen responsiveness. Previous studies demonstrated that coexpression of the Pit-1 protein with reporter genes driven by the prolactin or GH promoters in cells of nonpituitary origin resulted in activation of both promoters (Fox et al, 1990; Ingraham et al, 1988; Mangalam et al, 1989). However, other studies (Simmons et al, 1990) suggested that Pit-1 alone was not capable of activating the distal enhancer. When the prolactin distal enhancer was cotransfected with an expression vector encoding Pit-1 in the COS-1 monkey kidney cell line, no activation of reporter gene activity was observed (Figure 9). Similarly, cotransfection of the distal enhancer reporter gene and the rat estrogen receptor expression vector did not result in significant activation and only a modest response to estradiol was detected. However, when both the expression vectors encoding Pit-1 and estrogen receptor were cotransfected with the distal enhancer reporter gene, a significant response to estrogen was observed. This response was not observed when estrogen receptor was cotransfected with an expression vector encoding Pit-1 containing an in-frame deletion of the DNA binding domain (Pit-1 MUT). These results demonstrated that both Pit-1 and estrogen receptor must be present for the prolactin distal enhancer to confer estrogen responsiveness, and that Pit-1 must be able to bind to its cognate DNA sites to allow for estrogen receptor activation of the distal enhancer region.

The observation that the prolactin distal enhancer region, which contains four Pit-1 binding sites (Mangalam et al, 1989; Nelson et al, 1988), was not activated by expression of Pit-1 could have been the result of an inappropriate context due to the viral TK promoter. To examine this possibility, the effect of varying concentrations of the Pit-1 expression vector on the activity of the distal enhancer linked to the homologous prolactin promoter region (-306 to +34 bp) was studied in the COS-1 cells. The effect of coexpression of Pit-1 and estrogen receptor on the activity of these reporter genes was also assessed. The activity of the -306 prolactin promoter increased moderately with increasing concentrations of Pit-1 expression vector to approximately 5-fold induction at 10 mg of DNA (Figure 10). The distal enhancer coupled to the prolactin promoter had nearly the identical response to increasing amounts of Pit-1, again demonstrating that expression of Pit-1 was not sufficient to activate the distal enhancer region, even when linked to the homologous promoter. As with the previous study, cotransfection with expression vectors encoding the estrogen receptor and Pit-1 activated

148

the distal enhancer and the response to the estrogen receptor was dependent upon amount of cotransfected Pit-1 expression vector. At the lowest concentration of Pit-1 expression vector tested (0.3 μg), modest activation in the presence of estrogen receptor was observed and the activity of this reporter gene increased markedly with increasing Pit-1 expression vector concentrations. These results confirmed that Pit-1 alone

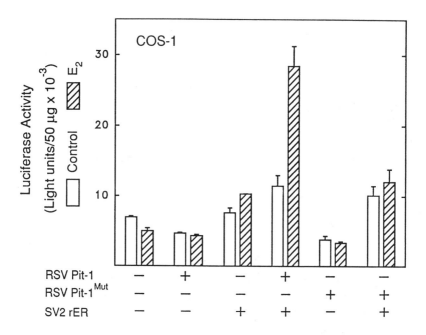

FIGURE 9. Interaction of Pit-1 and estrogen receptor in activation of the prolactin distal enhancer region. COS-1 cells were cotransfected with a reporter gene containing the prolactin distal enhancer linked to TK luciferase and expression vectors encoding Pit-1 (RSV Pit-1), the mutant Pit-1 (RSV Pit-1Mut), or rat estrogen receptor (SV2 rER), either alone or in combination. The cells were treated with no addition (control) or 10 nM estradiol (E2) and collected after 24 h for analysis of luciferase activity. Results are from triplicate independent transfections ± standard error. Reprinted (with permission from The Endocrine Society) from Day RN, Koike S, Sakai M, Muramatsu M and Maurer RA (1990): Both Pit-1 and the estrogen receptor are required for estrogen responsiveness of the rat prolactin gene. *Mol Endocrinol* 4: 1964-1971.

cannot activate the distal enhancer region, and demonstrated that activation of the region by estrogen receptor was dependent upon the concentration of Pit-1. In contrast, the Pit-1 expression vector alone was able to activate the prolactin proximal promoter reporter gene, although this activation was considerably less than that reported for similar studies in HeLa cells (Mangalam et al, 1989). In this regard, it is possible that additional factors that are not expressed or that are limiting in the Simian kidney cell line are required for Pit-1 activation of the prolactin proximal promoter.

To examine this possibility, activation of the prolactin promoter region and the distal enhancer coupled to the prolactin promoter by Pit-1 and estrogen receptor was tested in the rat embryonic fibroblast cell line, RAT-1. In this cell line, the prolactin promoter was activated approximately 20-fold by cotransfection of Pit-1 (Figure 11). Coupling the distal enhancer sequence to the prolactin promoter did not alter the response to expression of Pit-1, but as observed in the previous experiments, this construct was further activated by coexpression of Pit-1 and the estrogen receptor. A truncated prolactin promoter (-39 to +34), in which all of the Pit-1 binding sites were deleted, was not activated by cotransfection of Pit-1. However, the distal enhancer region linked to the truncated prolactin promoter was activated by cotransfection with the estrogen receptor and Pit-1.

It is apparent from these experiments that expression of Pit-1 in both COS-1 and RAT-1 cell lines is sufficient to activate the prolactin proximal promoter region, although this response varies with cell type. However, both Pit-1 and estrogen receptor must be present in these heterologous cell lines for substantial activation of the prolactin distal enhancer region. It is possible that differences in cellular context of transcription factors may contribute to the relative activity of the distal and proximal regions in homologous and heterologous cell lines. Also, the proximal and distal enhancer reporter gene constructs tested result in different relationships between Pit-1 binding sites and the transcription initiation site. Although, by definition, enhancers should function in a positionally independent manner, in many cases enhancers are not immune to positional effects. Thus, both positional effects and differences in binding of transcription factors other than Pit-1 may account for the differential ability of Pit-1 to stimulate expression from constructs containing the distal or proximal enhancer regions.

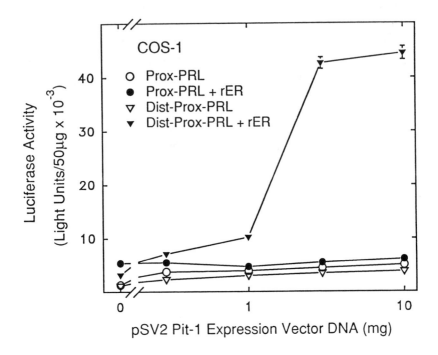

FIGURE 10. Effects of varying concentrations of Pit-1 expression vector on prolactin proximal promoter and distal enhancer activity in COS-1 cells. COS-1 cells were cotransfected with the reporter gene containing the prolactin proximal promoter region (-306 prolactin, circles) or the distal enhancer linked to the proximal promoter (-1765 to -1495 plus -306 to +34 prolactin, triangles) in the presence (closed symbols) or absence (open symbols) of an estrogen receptor expression vector (SV2 rER) and with increasing concentrations of Pit-1 expression vector. Total DNA was kept constant by adding varying amounts of pUC19 DNA. The COS-1 cells were maintained in estrogen containing medium and collected after 24 h for analysis of luciferase activity. Results are from triplicate independent transfections ± standard error. Reprinted with permission (The Endocrine Society) from Day RN, Koike S, Sakai M, Muramatsu M and Maurer RA (1990): Both Pit-1 and the estrogen receptor are required for estrogen responsiveness of the rat prolactin gene. *Mol Endocrinol* 4: 1964-1971.

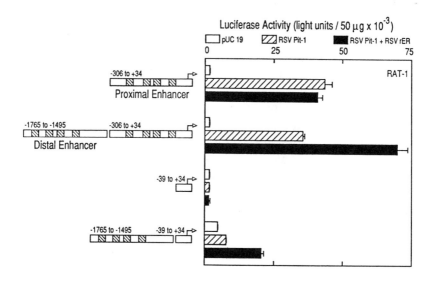

FIGURE 11. Response of the prolactin proximal promoter region and distal enhancer: proximal promoter region reporter gene to the expression of Pit-1 plus or minus estrogen receptor in RAT-1 cells. RAT-1 cells were transfected with reporter genes containing either the prolactin proximal enhancer region (-306 to +34) or minimal promoter (-39 to +34) with or without the distal enhancer region (-1765 to -1495) in combination with the expression vector for Pit-1 plus or minus the expression vector for estrogen receptor. Total DNA was kept constant with pUC 19 DNA. The RAT-1 cells were maintained in estrogen containing medium and collected after 24 h for analysis of luciferase activity. Results are from triplicate independent transfections ± standard error. Reprinted with permission (The Endocrine Society) from Day RN, Koike S, Sakai M, Muramatsu M and Maurer RA (1990): Both Pit-1 and the estrogen receptor are required for estrogen responsiveness of the rat prolactin gene. *Mol Endocrinol* 4: 1964-1971.

CONCLUSIONS

The studies described in this chapter provide evidence that Pit-1 and the estrogen receptor act in concert to activate the prolactin distal enhancer and to confer responsiveness to estradiol. While it appears that Pit-1 is sufficient to activate the prolactin proximal promoter region in at least some situations, Pit-1 alone is not sufficient to permit activation by the prolactin distal enhancer. Thus for both basal activity of the distal enhancer and estrogen responsiveness there is a requirement for the presence of both Pit-1 and the estrogen receptor. These findings are consistent with reports demonstrating that cooperativity between steroid receptors and other transcription factors is required for activation of weak, non-palindromic steroid response elements (Schule et al, 1988b). This cooperativity between distinct cis-acting elements may allow for a range of responses to trans-acting proteins. It should also be noted that the interaction of steroid receptors with other transcription factors can have dramatic effects to alter the nature of the transcriptional response to steroids (Diamond et al, 1990). It is possible that combinatorial interaction of steroid receptors with other factors may create very different response units based on a relatively small number of individual factors.

The interaction of multiple factors to permit hormonal responsiveness may be particularly important for genes such as the prolactin gene which are expressed in a tissue-specific manner. The requirement for interaction between the estrogen receptor and the tissue-specific factor, Pit-1, likely ensures that estrogen receptor does not stimulate expression of the prolactin gene in tissues other than the pituitary. Similarly, the finding that accessory factors which interact with the glucocorticoid receptor are required for glucocorticoid regulation of the phosphoenolpyruvate carboxykinase gene (Imai et al, 1990) may reflect a similar requirement for liver-specific regulation of this important enzyme. Thus, the interaction of steroid receptors with other transcription factors may both permit a broad range of phenotypic responses as well as restrict responsiveness to an appropriate range of receptor-containing cells.

At the present time the mechanisms responsible for the functional interaction of Pit-1 and the estrogen receptor in activation of the prolactin distal enhancer are unknown. A model illustrating possible mechanisms which might account for the functional cooperation of Pit-1 and the estrogen receptor is shown in figure 12. The fact that one of the Pit-1 binding sites (the D1 site) is immediately adjacent to the estrogen receptor binding site, raises the possibility of physical interactions between these factors. Physical interactions between Pit-1 and the estrogen receptor might facilitate or stabilize binding of Pit-1 and the estrogen receptor to the prolactin distal enhancer. This may be particularly important for interaction of the estrogen receptor with the imperfect palindrome which is found in the receptor binding site of the prolactin gene. Presumably the estrogen receptor interacts more weakly with this site than with a perfect palindrome. It should be noted that there is no structural basis to expect that the estrogen receptor can interact with Pit-1. However, recent studies suggest the interaction of steroid receptors with other transcription factors (Diamond et al, 1990; Yang-Yen et al, 1990). It should be possible to directly test for a possible physical interaction between the estrogen receptor and Pit-1. Alternatively the estrogen receptor and Pit-1 might bind cooperatively to a common target protein. As shown in figure 12, this target protein might be an adapter

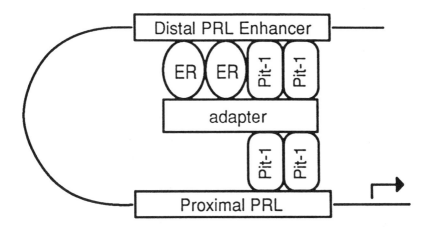

FIGURE 12. Model for functional cooperativity of Pit-1 and the estrogen receptor. A schematic diagram indicates Pit-1 binding to both the proximal region of the prolactin gene as well as the distal enhancer. The model depicts physical interaction between the estrogen receptor and Pit-1. This interaction would either facilitate or stabilize estrogen receptor binding to the imperfect palindrome. The model also indicates that both the estrogen receptor and Pit-1 might contact an adapter protein which is responsible for looping of the intervening sequences and formation of a complex involving both the distal and proximal regions of the prolactin gene.

which was important to facilitate looping resulting in interactions between the distal enhancer and the proximal region of the prolactin gene. One of the most popular models for explaining the properties of enhancers is a looping model in which the enhancer is brought into proximity with the transcription initiation complex by looping out of the intervening DNA (Griffith et al, 1986; Wang and Giaever, 1988). Experimental evidence demonstrating that an enhancer only needs to be brought into the vicinity of a promoter in order to function has been obtained (Müller et al, 1989). An adapter bound to both estrogen receptor and Pit-1 might facilitate such interactions. Rather than contacting an adapter, both Pit-1 and the estrogen receptor might simultaneously contact a component of the transcription complex as has been suggested for other enhancers (Carey et al, 1990). It should be noted that mechanisms other than those shown in figure 12 may also account for the functional cooperativity of the estrogen receptor and Pit-1. For instance, Pit-1 and the estrogen receptor might interact to produce changes in chromatin structure. DNase I hypersensitive sites have been detected in the region of the prolactin distal enhancer (Durrin et al, 1984). An alternative model for enhancer function involves transcription factor tracking along DNA after entry at an enhancer element (Geyer and Corces, 1992; Kadesch and Berg, 1986). In this model, the presence of both the estrogen receptor and Pit-1 would be required to facilitate the binding and tracking of RNA polymerase or specific transcription factors. At the present time there is no compelling evidence in favor of any one of these models. Further information concerning the mechanisms which permit the functional cooperativity of Pit-1 and the estrogen receptor should be important for both understanding the specific regulation of the prolactin gene as well as the general mechanisms supporting the function of enhancers.

REFERENCES

Anderson J, Clark JH and Peck EJ Jr (1972): Oestrogen and nuclear binding sites. Determination of specific sites by (3H)oestradiol exchange. *Biochem J* 126: 561-567.

Bailly A, Atger M, Atger P, Cerbon M-A, Alizon M, Hai MTV, Logeat F and Milgrom E (1983): The rabbit uteroglobin gene. Structure and interaction with the progesterone receptor. *J Biol Chem* 258: 10384-10389.

Bodner M, Castrillo J-L, Theill LE, Deerinck T, Ellisman M and Karin M (1988): The pituitary-specific transcription factor GHF-1 is a homeobox-containing protein. *Cell* 55: 505-518.

Bodner M and Karin M (1987): A pituitary-specific trans-acting factor can stimulate transcription from the growth hormone promoter in extracts of nonexpressing cells. *Cell* 50: 267-275.

Cao Z, Barron EA and Sharp ZD (1988): Prolactin upstream factor I mediates cell-specific transcription. *Mol Cell Biol* 8: 5432-5438.

Carey M, Lin Y-S, Green MR and Ptashne M (1990): A mechanism for synergistic activation of a mammalian gene by GAL4 derivatives. *Nature* 345: 361-364.

Cato ACB, Heitlinger E, Ponta H, Klein-Hitpass L, Ryffel GU, Bailly A, Rauch C and Milgrom E (1988): Estrogen and progesterone receptor-binding sites on the chicken vitellogenin II gene: synergism of steroid hormone action. *Mol Cell Biol* 8: 5323-5330.

Cato ACB and Ponta H (1989): Different regions of the estrogen receptor are required for synergistic action with the glucocorticoid and progesterone receptors. *Mol Cell Biol* 9: 5324-5330.

Chamness GC, Jennings AW and McGuire WL (1974): Estrogen receptor binding to isolated nuclei. A nonsaturable process. *Biochemistry* 13: 327-331.

Compton JG, Schrader WT and O'Malley BW (1983): DNA sequence preference of the progesterone receptor. *Proc Natl Acad Sci USA* 80: 16-20.

Day RN, Koike S, Sakai M, Muramatsu M and Maurer RA (1990): Both Pit-1 and the estrogen receptor are required for estrogen responsiveness of the rat prolactin gene. *Mol Endocrinol* 4: 1964-1971.

Day RN and Maurer RA (1989): The distal enhancer region of the rat prolactin gene contains elements conferring response to multiple hormones. *Mol Endocrinol* 3: 3-9.

Day RN and Maurer RA (1990): Pituitary calcium channel modulation and regulation of prolactin gene expression. *Mol Endocrinol* 4: 736-742.

Diamond MI, Miner JN, Yoshinaga SK and Yamamoto KR (1990): Transcription factor interactions: selectors of positive or negative regulation from a single DNA element. *Science* 249: 1266-1272.

Durrin LK, Weber JL and Gorski J (1984): Chromatin structure, transcription and methylation of the prolactin gene domain in pituitary tumors of Fischer 344 rats. *J Biol Chem* 259: 7086-7093.

Edwards DP, Kuhnel B, Estes PA and Nordeen SK (1989): Human progesterone receptor binding to mouse mammary tumor virus deoxyribonucleic acid: dependence on hormone and nonreceptor nuclear factor(s). *Mol Endocrinol* 3: 381-391.

Fox SR, Jong MTC, Casanova J, Ye Z-S, Stanley F and Samuels HH (1990): The homeodomain protein, Pit-1/GHF-1 , is capable of binding to and activating cell-specific elements of both the growth hormone and prolactin gene promoters. *Mol Endocrinol* 4: 1069-1080.

Geyer PK and Corces VG (1992): DNA position-specific repression of transcription by a Drosophila zinc finger protein. *Genes & Develop.* 6: 1865-1873.

Gorski J, Toft D, Shyamala G, Smith D and Notides A (1968): Hormone receptors: studies on the interaction of estrogen with the uterus. *Rec Progr Hormone Res* 24: 45-80.

Govindan MV, Spiess E and Majors J (1982): Purified glucocorticoid receptor-hormone complex from rat liver cytosol binds specifically to cloned mouse mammary tumor virus long terminal repeats in vitro. *Proc Natl Acad Sci USA* 79: 5157-5161.

Griffith J, Hochschild A and Ptashne M (1986): DNA loops induced by cooperative binding of lambda repressor. *Nature* 322: 750-752.

Gubbins EJ, Maurer RA, Hartley JL and Donelson JE (1979): Construction and analysis of recombinant DNAs containing a structural gene for rat prolactin. *Nuc Acids Res* 6: 915-930.

Gubbins EJ, Maurer RA, Lagrimini M, Erwin CR and Donelson JE (1980): Structure of the rat prolactin gene. *J Biol Chem* 255: 8655-8662.

Haug E and Gautvik KM (1976): Effects of sex steroids on prolactin secreting rat pituitary cells in culture. *Endocrinology* 99: 1482-1489.

Herr W, Sturm RA, Clerc RG, Corcoran LM, Baltimore D, Sharp PA, Ingraham HA, Rosenfeld MG, Finney M, Ruvkun G and Horvitz HR (1988): The POU domain: a large conserved region in the mammalian pit-1, oct-1, oct-2, and Caenorhabditis elegans unc-86 gene products. *Genes Dev* 2: 1513-1516.

Higgins SJ, Rousseau GG, Baxter JD and Tomkins GM (1973): Nuclear binding of steroid receptors: comparison in intact cells and cell-free systems. *Proc Natl Acad Sci USA* 70: 3415-3418.

Hoggard N, Davis JRE, Berwaer M, Monget P, Belayew BPA and Martial JA (1991): Pit-1 binding sequences permit calcium regulation of human prolactin gene expression. *Mol Endocrinol* 5: 1748-1754.

Imai E, Stromstedt PE, Quinn PJ, Carlstedt-Duke J, Gustafsson JA and Granner DK (1990): Characterization of a complex glucocorticoid response unit in the phosphoenolpyruvate carboxykinase gene. *Mol Cell Biol* 10: 4712-4719.

Ingraham HA, Chen R, Mangalam HJ, Elsholtz HP, Flynn SE, Lin CR, Simmons DM, Swanson L and Rosenfeld MG (1988): A tissue-specific transcription factor containing a homeodomain specifies a pituitary phenotype. *Cell* 55: 519-529.

Jensen EV and DeSombre ER (1968): Estrogen-receptor interaction. Estrogenic hormones affect transformation of specific receptor proteins to a biochemically functional form. *Science* 182: 126-134.

Jensen EV, Suzuki T, Kawashima T, Stumpf WE, Jungblut PW and DeSombre ER (1968): A two-step mechanism for the interaction of estradiol with rat uterus. *Proc Natl Acad Sci USA* 59: 632-638.

Jost J-P, Seldran M and Geiser M (1984): Preferential binding of estrogen-receptor complex to a region containing the estrogen-dependent hypomethylation site preceding the chicken vitellogenin II gene. *Proc Natl Acad Sci USA* 81: 429-433.

Kadesch T and Berg P (1986): Effects of the position of the simian virus 40 enhancer on expression of multiple transcription units in a single plasmid. *Mol Cell Biol* 6: 2593-2601.

Kim KE, Day RN and Maurer RA (1988): Functional analysis of the interaction of a tissue-specific factor with an upstream enhancer element of the rat prolactin gene. *Mol Endocrinol* 3: 1374-1381.

Klein-Hitpass L, Schorpp M, Wagner U and Ryffel GU (1986): An estrogen-responsive element derived from the 5' flanking region of the Xenopus vitellogenin A2 gene functions in transfected human cells. *Cell* 46: 1053-1061.

Klein-Hitpass L, Tsai SY, Greene GL, Clark JH, Tsai M-J and O'Malley BW (1989): Specific binding of estrogen receptor to the estrogen response element. *Mol Cell Biol* 9: 43-49.

Kumar V and Chambon P (1988): The estrogen receptor binds tightly to its responsive element as a ligand-induced homodimer. *Cell* 55: 145-156.

Lieberman ML, Maurer RA and Gorski J (1978): Estrogen control of prolactin synthesis in vitro. *Proc Natl Acad Sci USA* 75: 5946-5949.

MacLeod RM, Abad A and Eidson LL (1969): In vivo effect of sex hormones on the in vitro synthesis of prolactin and growth hormone in normal and pituitary tumor-bearing rats. *Endocrinology* 84: 1475-1483.

Mangalam HJ, Albert VR, Ingraham HA, Kapiloff M, Wilson L, Nelson C, Elsholtz H and Rosenfeld MG (1989): A pituitary POU domain protein, Pit-1, activates both growth hormone and prolactin promoters transcriptionally. *Genes Dev* 3: 946-958.

Maurer RA (1980): Immunochemical isolation of prolactin messenger RNA. *J Biol Chem* 255: 854-859.

Maurer RA (1981): Transcriptional regulation of the prolactin gene by ergocryptine and cyclic AMP. *Nature* 294: 94-97.

Maurer RA (1982): Estradiol regulates the transcription of the prolactin gene. *J Biol Chem* 257: 2133-2136.

Maurer RA (1985): Selective binding of the estradiol receptor to a region at least one kilobase upstream from the rat prolactin gene. DNA 4: 1-9.

Maurer RA and Gorski J (1977): Effects of estradiol-17β and pimozide on prolactin synthesis in male and female rats. *Endocrinology* 101: 76-84.

Maurer RA and Notides AC (1987): Identification of an estrogen-responsive element from the 5'-flanking region of the rat prolactin gene. *Mol Cell Biol* 7: 4247-4254.

McKnight GS and Palmiter RD (1979): Transcriptional regulation of the ovalbumin and conalbumin genes by steroid hormones in chick oviduct. *J Biol Chem* 254: 9050-9058.

Meyer M-E, Gronemeyer H, Turcotte B, Bocquel M-T, Tasset D and Chambon P (1989): Steroid hormone receptors compete for factors that mediate their enhancer function. *Cell* 57: 433-442.

Müller HP, Sogo JM and Schaffner W (1989): An enhancer stimulates transcription in trans when attached to promoter via a protein bridge. *Cell* 58: 767-777.

Mulvihill ER, LePennec J-P and Chambon P (1982): Chicken oviduct progesterone receptor: location of specific regions of high affinity binding in cloned DNA fragments of hormone-responsive genes. *Cell* 24: 621-632.

Mulvihill ER and Palmiter RD (1977): Relationship of nuclear estrogen receptor levels to induction of ovalbumin and conalbumin mRNA in chick oviduct. *J Biol Chem* 252: 2060-2068.

Nelson C, Albert VR, Elsholtz HP, Lu LI-W and Rosenfeld MG (1988): Activation of cell-specific expression of rat growth hormone and prolactin genes by a common transcription factor. *Science* 239: 1400-1405.

Nelson C, Crenshaw EB III, Franco R, Lira SA, Albert VR, Evans RM and Rosenfeld MG (1986): Discrete cis-active genomic sequences dictate the pituitary cell type-specific expression of rat prolactin and growth hormone genes. *Nature* 322: 557-562.

Notides AC, Lerner N and Hamilton DE (1981): Positive cooperativity of the estrogen receptor. *Proc Natl Acad Sci USA* 78: 4926-4930.

Payvar F, Wrange O, Carlstedt-Duke J, Okret S, Gustafsson JA and Yamamoto KR (1981): Purified glucocorticoid receptors bind selectively in vitro to a cloned DNA fragment whose transcription is regulated by glucocorticoids in vivo. *Proc Natl Acad Sci USA* 78: 6628-6632.

Pfahl M (1982): Specific binding of the glucocorticoid-receptor complex to the mouse mammary tumor proviral promoter region. *Cell* 31: 475-482.

Reeder RH and Roeder RG (1972): Ribosomal RNA synthesis in isolated nuclei. *J Mol Biol* 67: 433-441.

Ryan R, Shupnik MA and Gorski J (1979): Effect of estrogen on preprolactin messenger ribonucleic acid sequences. *Biochemistry* 18: 2044-2048.

Scheidereit C, Geisse S, Westphal HM and Beato M (1983): The glucocorticoid receptor binds to defined nucleotide sequences near the promoter of mouse mammary tumour virus. *Nature* 304: 749-752.

Schule R, Muller M, Kaltschmidt C and Renkawitz R (1988a): Many transcription factors interact synergistically with steroid receptors. *Science* 242: 1418-1420.

Schule R, Muller M, Otsuka-Murakami H and Renkawitz R (1988b): Cooperativity of the glucocorticoid receptor and the CACCC-box binding factor. *Nature* 332: 87-90.

Schuster WA, Treacy MN and Martin F (1988): Tissue specific trans-acting factor interaction with proximal rat prolactin gene promoter sequences. *EMBO J* 7: 1721-1733.

Seo H, Refetoff S, Martino E, Vassart G and Brocas H (1979a): The differential stimulatory effect of thyroid hormone on growth hormone synthesis and estrogen on prolactin synthesis due to accumulation of specific messenger ribonucleic acids. *Endocrinology* 104: 1083-1090.

Seo H, Refetoff S, Scherberg N, Brocas H and Vassart G (1979b): Isolation of rat prolactin messenger ribonucleic acid and synthesis of complementary deoxyribonucleic acid. *Endocrinology* 105: 1481-1487.

160

Shull JD and Gorski J (1984): Estrogen stimulates prolactin gene transcription by a mechanism independent of pituitary protein synthesis. *Endocrinology* 114: 1550-1557.

Shupnik MA, Baxter LA, French LR and Gorski J (1979): In vivo effects of estrogen on ovine pituitaries: prolactin and growth hormone biosynthesis and messenger ribonucleic acid translation. *Endocrinology* 104: 729-735.

Simmons DM, Voss JW, Ingraham HA, Holloway JM, Broide RS, Rosenfeld MG and Swanson LW (1990): Pituitary cell phenotypes involve cell-specific Pit-1 mRNA translation and synergistic interactions with other classes of transcription factors. *Genes Dev* 4: 695-711.

Stone RT, Maurer RA and Gorski J (1977): Effects of estradiol-17β on preprolactin messenger ribonucleic acid activity in the rat pituitary gland. *Biochemistry* 16: 4915-4921.

Strahle U, Schmid W and Schutz G (1988): Synergistic action of the glucocorticoid receptor with transcription factors. *EMBO J* 7: 3389-3395.

Tashjian AH Jr, Yasumura Y, Levine L, Sato GH and Parker ML (1968): Establishment of clonal strains of rat pituitary tumor cells that secrete growth hormone. *Endocrinology* 82: 342-352.

Vician L, Shupnik MA and Gorski J (1979): Effects of estrogen on primary ovine pituitary cell cultures: stimulation of prolactin secretion, synthesis and preprolactin messenger ribonucleic acid activity. *Endocrinology* 104: 736-743.

Wang JC and Giaever GN (1988): Action at a distance along a DNA. *Science* 240: 300-304.

Waterman ML, Adler S, Nelson C, Greene L, Evans M and Rosenfeld MG (1988): A single domain of the estrogen receptor confers deoxyribonucleic acid binding and transcriptional activation of the rat prolactin gene. *Mol Endocrinol* 2: 14-21.

Willliams D and Gorski J (1972): Kinetic and equilibrium analysis of estradiol in uterus: a model of binding-site distribution in uterine cells. *Proc Natl Acad Sci USA* 69: 3464-3468.

Yamamoto K, Kasai K and Ieiri T (1975): Control of pituitary functions of synthesis and release of prolactin and growth hormone by gonadal steroids in female and male rats. *Jap J Physiol* 25: 645-658.

Yamamoto KR and Alberts BM (1974): On the specificity of the binding of the estradiol receptor protein to deoxyribonucleic acid. *J Biol Chem* 249: 7076-7086.

Yan G-z and Bancroft C (1991): Mediation by calcium of thyrotropin-releasing hormone action on the prolactin promoter via transcription factor Pit-1. Mol. *Endocrinology* 5: 1488-1497.

Yan G-z, Pan WT and Bancroft C (1991): Thyrotropin-releasing hormone action is mediated by the POU protein Pit-1. *Mol. Endocrinol* 5: 535-541.

Yang-Yen HF, Chambard JC, Sun YL, Smeal T, Schmidt TJ, Drouin J and Karin M (1990): Transcriptional interference between c-Jun and the glucocorticoid receptor: mutual inhibition of DNA binding due to direct protein protein interaction. *Cell* 62: 1205-1215.

Zylber EA and Penman S (1971): Products of RNA polymerases in HeLa cell nuclei. *Proc Natl Acad Sci USA* 68: 2861-2865.

INSIGHTS INTO THE GENOMIC MECHANISM OF ACTION OF 1,25-DIHYDROXYVITAMIN D$_3$

J. Wesley Pike
Department of Biochemistry
Ligand Pharmaceuticals, Inc.

INTRODUCTION

Steroid, thyroid, and vitamin hormones are known to exert profound regulatory control over complex gene networks. Many, if not all, of these actions occur at the level of the cellular genome (O'Malley, 1992; Yamamoto, 1985). The products of these modulated genes control processes essential to cellular growth and differentiation as well as influence mechanisms integral to the maintenance of intracellular and extracellular homeostasis. The actions of these signals are mediated by unique intracellular receptors (Jensen et al, 1968; Jensen and DeSombre, 1972). Indeed, the presence of these receptor proteins in cells and tissues represents a principal determinant of response to a particular hormone. These soluble signal-transducing proteins are members of an enlarging gene family of latent transcription factors that acquire unique gene regulating capacities upon activation by their respective hormonal ligands (Evans, 1988). While hormone interaction has been well characterized, the events that follow association of the ligand with its receptor remain less well understood.

STEROID HORMONE RECEPTORS
V. K. Moudgil, Editor
© 1993 Birkhäuser Boston

The vitamin D metabolite 1,25-dihydroxyvitamin D_3 (1,25(OH)$_2$D$_3$) also regulates a variety of biological processes. These include cell growth, differentiation, immune response, as well as the numerous events that are associated with mineral metabolism (Haussler, 1986; DeLuca, 1988; Minghetti and Norman, 1988; Pike, 1991; Suda et al, 1990; Manolagas, et al, 1985). The mechanism, like that of the steroid and thyroid hormones, involves a specific receptor for 1,25(OH)$_2$D$_3$ (VDR) (Haussler and Norman, 1969). While the exact details of how the receptor elicits transcriptional modulation is unclear, many of the protein products that participate in the ensuing biological responses are well known. They include proteins such as the calbindin gene family, ostecalcin (OC), matrix gla protein, osteopontin (OP), and collagen. The purpose of this chapter is to document the recent experiments that have established that the vitamin D receptor is a member of the steroid receptor superfamily of transcription factors. In addition, the evidence contained in this chapter will also show that 1,25(OH)$_2$D$_3$ acts directly through this receptor to modify the expression of at least several genes responsive *in vivo* to the vitamin D hormone.

THE STEROID RECEPTOR SUPERFAMILY

A. *Structural aspects of the steroid receptor family.* The recent molecular cloning of the receptors for most, if not all, of the currently known hormonal ligands has revealed each to be part of an enlarging family of gene products (Evans, 1988; O'Malley, 1990). A salient feature of each of these receptors is a highly homologous central or amino-terminal DNA binding core (Evans, 1988; O'Malley, 1990). The region is highly basic in nature, contains nine positionally conserved cysteine residues, and is structurally comprised of two finger loops. Eight of the nine cysteine sulfur residues are believed to tetrahedrally coordinate two zinc atoms, and are essential for the proper folding of each of the proteins into an active conformation. This hypothesis has been confirmed through extended x-ray adsorption fine structural analysis of the recombinantly expressed glucocorticoid receptor fragment containing the DNA binding domain (Freedman et al, 1988). More recently, the three dimensional solution structures for this region of the glucocorticoid (Hard et al, 1990) and estrogen (Schwabe et al, 1990) receptors have been determined by nuclear magnetic resonance spectroscopy. These experiments suggest that the molecular organization of this region is similar between the two receptors, and is comprised of two important α-helical stretches located on the carboxyl-terminal sides of each finger loop. The high degree of homology within this region between all the steroid receptors predicts that the overall three-dimensional organization may be very similar.

The carboxyl terminal portion of the protein is less highly conserved among the receptors, containing several localized regions of moderate conservation and predominant regions of little homology (Evans, 1988; O'Malley, 1990). While this region is known to confer capacity for high affinity ligand binding to the receptor gene family, internal deletion analysis has not pinpointed significantly the location of the hydrophobic hormone binding pocket. Thus, a broad region of over several hundred amino acids appears to be required presently for this binding function. This region also contributes an important dimerization function to the steroid receptor family (Kumar and Chambon, 1988; Fawell et al, 1990). The interaction of receptors on their cognate cis-acting DNA elements is conferred through the formation of either homo- or heterodimers (see below). Receptor dimers exhibit a considerable increase in affinity for appropriate DNA binding sites over that of their monomeric forms, largely due to the stability achieved by the duplex on corresponding specific DNAs. The dimerization domain within the receptors appears to be bipartite, again spanning a majority of the carboxyl terminus and colocalizing to two of the conserved regions within this portion of the receptors. Additional functional activities within this complex region have also been mapped. They include such activities as nuclear transfer signals (Picard and Yamamoto, 1988) and transactivation domains (Evans, 1988; O'Malley, 1990). As the functions of dimerization, nuclear transfer, and transactivation are all modulated by ligand binding *in vivo*, it is not surprising from a structural point of view that these functional regions might be interconnected.

B. *Cloning and structure of the vitamin D receptor.* The development of monoclonal antibodies directed toward the vitamin D_3 receptor (VDR) in 1982 provided the essential key by which the receptor's rare structural gene was recovered from a cDNA library. Partial cDNAs for the chicken VDR were initially identified and confirmed (McDonnell et al, 1987), whereupon cDNAs for human (Baker et al, 1988) and rat (Burmester et al, 1988) also were obtained. Definitive identity of the cloned VDR was obtained through recombinant expression of the cDNA in a receptor-free mammalian cell background (Baker et al, 1988). The expressed protein was of the appropriate size and immunoreactivity, and bound $1,25(OH)_2D_3$ with normal receptor affinity and specificity.

Inspection of the VDR's primary sequence initially revealed that indeed the protein belonged to the steroid, thyroid, and retinoic acid receptor gene family (Evans, 1988). Its basic structural organization, similar to that of other family members, is comprised of a distinct N-terminal DNA recognition domain and a large C-terminal $1,25(OH)_2D_3$ binding domain. The former region contains two finger structures each folded about a zinc atom tetrahedrally coordinated through the sulfhydryl moieties of four conserved cysteine residues (Hard et al, 1990; Schwabe et al, 1990). This motif appears to be reminiscent of the helix-loop-helix DNA binding motif found in a number of other proteins that carry out related transcriptional regulating functions (Johnston and McKnight, 1989). A model taken from the molecular organization of the estrogen receptor structure (Schwabe et al, 1990) is observed for the DNA binding domain of the VDR in Figure 1. Perhaps more importantly, the cloning of this protein has provided a direct opportunity to determine its function. Indeed, studies on separate structural regions of the VDR have confirmed their participation in both hormone and DNA binding functions as well as transactivation (McDonnell et al, 1989, and see below). Ongoing studies are aimed at further dissection of this interesting macromolecule. Thus, the concept that $1,25(OH)_2D_3$ likely functions through a steroid hormone-like mechanism has been strongly substantiated by our preliminary observations about the receptor.

TRANSCRIPTIONAL RESPONSE TO 1,25-DIHYDROXY-VITAMIN D$_3$

The genes for OC and OP have emerged thus far as the most informative with respect to the modulatory effects of $1,25(OH)_2D_3$. OC is one of the most abundant noncollagenous macromolecules found in adult bone. It is a highly conserved γ-carboxyglutamic acid-containing protein that is believed to be produced exclusively by osteoblasts (Hauschka et al, 1975; Price et al, 1976). The function of OC remains unknown, although its appearance is associated with mineral deposition (Malone et al, 1982), and its capacity to bind hydroxyapatite (Gerstenfeld et al, 1986) and to act as a chemoattractant for bone resorbing cells (Glowacki and Lian, 1989) suggest a role for the protein as an index of bone turnover (Hauschka et al, 1975). Indeed, serum levels of OC are widely considered as clinically diagnostic for osteoblast function and active bone remodelling (Price and Nishimoto, 1980; Lian and Gundberg, 1988; Price, Parthemore and Deftos, 1980; Gundberg et al, 1983; Delmas et al, 1973). Despite the lack of clarity regarding OC function in bone, this protein is highly regulated by a variety of hormones (Price and Baukol, 1980; Lian, Coutts, and Canalis, 1985; Lian, Carnes and Glimcher, 1987). Early work identified $1,25(OH)_2D_3$ as a major upregulator of this protein, and critical investigation suggested that control was exerted at the level of the gene (Pan and Price, 1984). Based

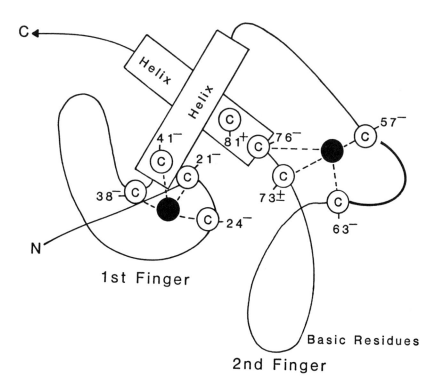

FIGURE 1. Model of the VDR DNA binding domain derived from the NMR structure of the estrogen receptor defined in Schwabe et al, 1990. C and N represent the carboxy and amino terminal orientation of the sequence. The nine positionally conserved cysteines (C) are indicated. The position of the two zinc atoms is indicated by a solid circle. Numbers indicate the amino acid residue within the VDR protein. The superscript + or - indicates the transcriptional activity of a mutant form of the VDR bearing a cysteine to serine mutation at that site (see Sone et al, 1991a). The two helical structures are indicated with a rectangle.

168

upon this background, we and others embarked upon an examination of the promoter for this gene in anticipation that hormonal vitamin D might induce this gene and that ultimately a direct role for the receptor might be demonstrated.

A. *Organization of the osteocalcin gene.* Probes derived from the amino acid sequence of OC enabled recovery of human genomic sequence comprising the gene for human OC (Celeste et al, 1986). Determination of the structural organization of this gene revealed four exons interspersed by three introns that together spanned approximately a kilobase of DNA (Figure 2). Exon 1 contained an in-frame ATG initiator codon as well as codons that direct synthesis of a highly hydrophobic signal sequence responsible for transmembrane transfer. As anticipated, exon 4 contained the TGA termination codon, 3' noncoding sequences, and a canonical AATAAA polyadenylation sequence. The subsequent cloning and structural determination of the rat gene for OC indicated an overall organization highly conserved with respect to the human gene (Lian, Steward, Puchacz et al, 1989).

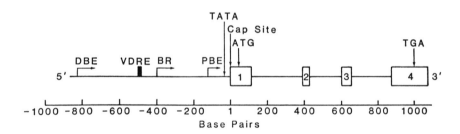

FIGURE 2. Structural and functional organization of the human osteocalcin gene. Features of the control region of the gene promoter are shown, including the cap site, TATA box, putative GC boxes, basal repressor region (BR), VDRE locus, and distal basal enhancer (DBE). The start site is +1 from the TATA box. The exons are designated 1 through 4. The nucleotide base scale is indicated below the diagram.

Visual inspection of the 5' portions of both the human and rat genes for OC suggested the presence of putative TATA box sequences as well as consensus sequences for a variety of important transcription factors (Lian, Stewart, Puchacz et al, 1989; Celeste et al, 1986). These observations, together with the previous studies demonstrating that the OC gene was highly regulated by retinoic acid, $1,25(OH)_2D_3$ and glucocorticoids (Price and Baukol, 1980; Nishimoto et al, 1987; Jowell et al, 1987), led us to examine the molecular basis for promoter activity. Determination of activity was made at the level of transcriptional control by using chimeric genes containing the OC promoter and upstream 5' flanking regions fused to the chloramphenicol acetyltransferase (CAT) structural gene and an SV40 polyadenylation signal. Plasmids were introduced by DNA-mediated gene transfer into cultured cells of osteoblastic lineage such as ROS 17/2.8, UMR-106, and MC3T3-E1, and basal as well as hormone-inducible activities of the reporter examined.

B. *Basal functional activity of the OC promoter in vitro.* An OC chimeric gene containing approximately 1300 base pairs of upstream 5' flanking human promoter sequence exhibited strong basal activity when evaluated in ROS 17/2.8 cells (McDonnell et al, 1989). This DNA exhibited strong promoter activity on the reporter response. These osteosarcoma-derived cells express high levels of endogenous OC (Price and Baukol, 1980), and thus the correlation between endogenous expression of OC and its promoter activity correlated well. In contrast, subsequent examination of this fusion gene in NIH3T3 and CV-1 fibroblasts as well as in the pre-osteoblastic cell line MC3T3-E1, none of which produce endogenous OC, revealed an extremely low basal level of activity (Pike, 1990; Uchida et al, 1991). Studies with the rat OC promoter led to similar results; high levels of basal activity in ROS 17/2.8 cells and low levels of activity in OC-non producing cells such as UMR 106 (Yoon et al, 1988). These observations suggest that a factor(s) important for the endogenous expression of OC may be acting similarly upon the exogenously added foreign chimeric genes, and that this factor(s) together with the DNA sequence elements with which it interacts might be identified. Moreover, it is possible that this activator(s) represents a major determinant of osteoblast-specific expression of OC.

Critical regions that contributed to basal expression of the human OC promoter in ROS 17/2.8 cells were identified by utilizing a series of promoter sequences that contained the TATA box and increasing upstream sequence (Kerner et al, 1989). The results of these experiments, as illustrated in Figure 2, suggest that important regions lie immediately proximal to the region of initiation as well as greater than 400 bp upstream. Studies with the rat OC promoter similarly have identified a potentially functional region near the TATA sequence that appears as a CAAT box (Markose et al, 1990). Certain sequence homologies exist within this region between the rat and human genes. Nevertheless, the

CAAT box in the rat gene appears as a CAAT sequence whereas in the human promoter, this sequence is replaced by CAAAT and bounded on both sides by GC boxes. Indeed, selective mutagenesis of each of these three sites followed by transfection into ROS 17/2.8 cells suggests that only the two GC boxes in the human promoter contribute to basal activity (K. Ozono and J. W. Pike, unpublished data). Distal basal elements include an AP-1 site within the human promoter at -500 (Ozono et al, 1990) and an apparent unmapped element within the rat promoter approximately a kilobase upstream (Markose et al, 1990; Terpening et al, 1991). These elements clearly contribute to the capacity of certain steroid

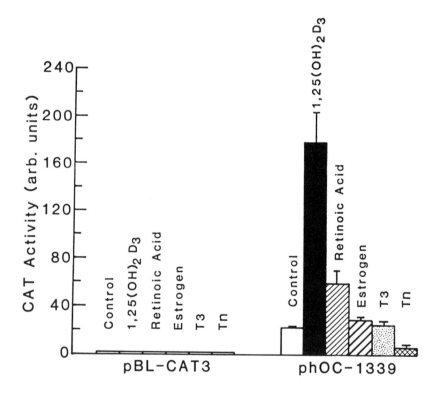

FIGURE 3. Transcriptional response of the human osteocalcin promoter to various hormones. A plasmid comprised of the human osteocalcin promoter (-1339 to +10, phOC-1339) fused to CAT was introduced into the ROS 17/2.8 cell line by transfection methods and examined for response to $1,25(OH)_2D_3$, retinoic acid, estrogen, thyroid hormone (T3), and triamcinolone (TN). CAT activity (arbitrary units) was determined 48 hours following treatment. Error bars are indicated (SEM).

hormones to induce both promoters (see below). Despite these studies, the contribution of these elements to tissue-specific expression of OC remain unknown.

C. *Hormonal responsiveness of the OC promoter.* The same 1300 bp of the 5' flanking region of the human gene was introduced transiently into ROS 17/2.8 cells, and its activity in the presence of a variety of hormones evaluated (McDonnell et al, 1989; Morrison et al, 1989). As observed in Figure 3, enhanced activity was identified when the cells were treated with $1,25(OH)_2D_3$ and retinoic acid, and suppression of the gene was observed when glucocorticoids were utilized (McDonnell et al, 1989; Pike, 1990; Kerner et al, 1989; Morrison et al, 1989). Each result was consistent with the effects of these hormones on endogenous gene expression (Price and Baukol, 1980; Nishimoto et al, 1987; Jowell et al, 1987). Indeed, similar results with $1,25(OH)_2D_3$ were observed with the rat OC promoter (Terpening et al, 1991; Demay et al, 1990). Thus, it is clear that $1,25(OH)_2D_3$ as well as retinoic acid alter the level of expression of OC directly through an action at the level of the gene's promoter.

That OC was regulated at the level of its promoter was also evaluated *in vivo* utilizing transgenic mouse strains (Kesterson et al, 1993). Several mouse strains homozygous for the OC promoter CAT transgene were treated with strontium chloride to suppress circulating levels of $1,25(OH)_2D_3$ and a subset of these "inhibited" mice were then further treated by injection with 25 ng of exogenous $1,25(OH)_2D_3$ per day for three days to restore circulating levels of the hormone. Calvaria, femur, and brain tissues were removed and evaluated for CAT activity, and these activities were contrasted with those derived from dietary normal animals. The results suggest that indeed, removal of the hormone from circulation led to a significant suppression of tissue transgene activity, and restoration of the hormone resulted in recovery and in several cases increased activity of the reporter transgene. These *in vivo* studies confirm the *in vitro* observation that the OC promoter contains elements that are solely responsible for mediating the induction conferred by $1,25(OH)_2D_3$.

D. *Mapping the vitamin D and retinoic acid responsive elements.* The precise location of the vitamin D-responsive element (VDRE) within the OC gene promoter was mapped functionally using unidirectional and/or internal deletion analysis. As previously, the individual constructions were introduced into ROS 17/2.8 cells by transfection, and the activity of the construction assessed in the presence and absence of the vitamin D hormone. Our studies revealed that the region that conferred vitamin D response was located in the human gene between

-512 and -485 relative to the start site of transcription (Kerner et al, 1989; Ozono et al, 1990). Deletion of this sequence in the human promoter led to loss of vitamin D response. Perhaps more definitively, transfer of these sequences in either orientation to heterologous viral promoters not normally inducible by vitamin D, such as those of the mouse mammary tumor virus LTR or the Herpes simplex virus thymidine kinase gene conferred a novel new responsiveness to the hormone (Pike, 1990; Kerner et al, 1989). Interestingly, retinoic acid responsiveness also mapped to the same DNA sequence (Pike, 1990; Schule et al, 1990), suggesting that the action of both hormones was mediated through the same cis-acting element. Similar studies using the rat gene promoter were carried out, leading to the identification of a structurally similar element lying between -458 and -433 (Demay et al, 1990; Terpening et al, 1991). These results suggest that indeed DNA sequences identified within the OC gene promoter are capable of directing the response to vitamin D, and that these sequences retain properties typical of hormone-responsive enhancers (Beato, 1989).

    E. *Structural organization of the VDRE.* The human OC VDRE is comprised of two direct but imperfect repeats separated by three nucleotide pairs, as documented in Figure 4 (Ozono et al, 1990). Studies in the rat OC gene (Terpening et al, 1991) as well as more recent examination of the mouse osteopontin gene (Noda et al, 1990) have identified cognate VDREs that are structurally identical to that found in the human gene. The consensus sequences for the two hexanucleotide half-sites of these three VDREs are G G G/T T G/C A and G G G/T T G/T/A A. These direct repeats, as illustrated in Figure 5, are clearly reminiscent of the structural motif and sequence of elements that mediate both thyroid hormone and retinoic acid response in other genes (Glass et al, 1988; Umesono et al, 1988; de The et al, 1990). Thus, it is perhaps not surprising that retinoic acid acts on the OC gene through this element. However, the observation that thyroid hormone does not stimulate the activity of this promoter (Figure 3) also indicates the importance of promoter context in the hormonal response mechanism. Recent systematic evaluation of these elements for determinants that confer hormonal specificity for vitamin D, retinoic acid, thyroid hormone, and other hormones suggests that the spacing between the hexanucleotide halfsites is a critical factor, although it is clearly not the only specifier (Umesono et al, 1991). Thus, the "3-4-5" rule, as hypothesized by the Evans group (Umesono et al, 1991) predicts that a maximal vitamin D hormone response is registered using two directly repeated A G G T C A hexanucleotides when three nucleotides are located in between.

Similarly, maximal thyroid hormone and retinoic acid response is obtained when the two repeats are interspersed with 4 and 5 base pairs, respectively. A final prediction of this rule is that perhaps additional

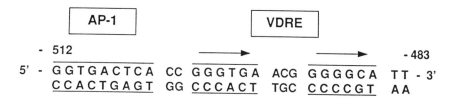

FIGURE 4. Structural organization of the human osteocalcin VDRE. The AP-1 and VDRE sites are indicated over the nucleotide sequence. Arrows indicate the two direct repeats within the VDRE. The location of this locus within the OC promoter is indicated by the nucleotides numbers at the 5' and 3' ends of the sequence.

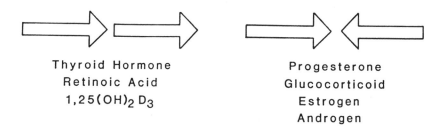

FIGURE 5. Dual nature of response elements that mediate hormonal action. Thyroid, retinoic acid, and vitamin D receptors interact predominately through response elements represented by directly repeated hexamers. Progesterone, glucocorticoid, estrogen, and androgen receptors function predominantly via palindromic hexamers.

174

unknown hormones may prefer repeated halfsites spaced by different base separations. Despite these findings, it is clear that the specific sequence of the repeated hexanucleotides is also influential in determining hormonal specificity.

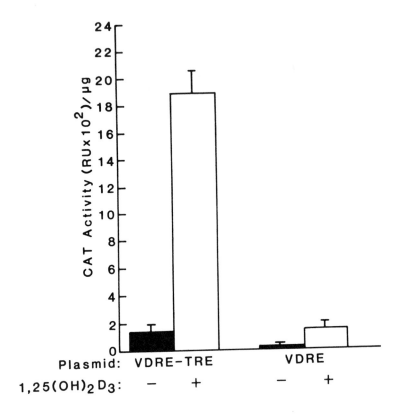

FIGURE 6. Effects of the AP-1 site on vitamin D inducibility of gene expression. The natural osteocalcin promoter (VDRE-TRE) as well as a version of the promoter containing a deletion of the AP-1 site (VDRE) were examined in osteoblastic ROS 17/2.8 cells (VDR positive) for basal activity and activation by $1,25(OH)_2D_3$. Reporter activity (CAT) was determined and quantitated as relative units (RU) per ug of protein. Basal activity is reduced in the absence of the AP-1 site. While both constructs are inducible by the vitamin D hormone, the magnitude of hormone inducibility of the wildtype promoter is clearly substantially elevated.

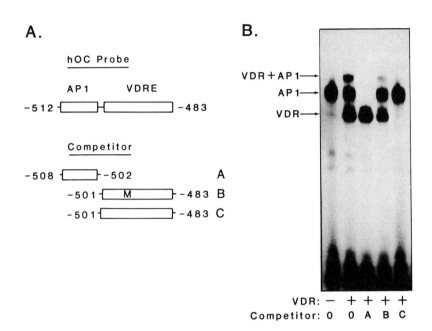

FIGURE 7. AP-1 and the VDR colocalize on the human osteocalcin VDRE locus. Bandshift analysis was utilized to evaluate the interaction of the VDR and AP-1 protein complex with the VDRE/AP-1 probe. **A.** The radiolabelled hOC probe as well as competitor cold probes A, B, and C are indicated. **B.** Bandshift analysis utilizing the DNA probes in A. Extracts containing AP-1 protein alone (VDR-) or supplemented with recombinant VDR (VDR+) were incubated with hOC probes in the absence (O) or presence of the 100 fold excess cold probe (A, B, or C) and then subjected to analysis. AP-1 alone reveals a single protein-DNA interaction. The addition of recombinant VDR to the AP-1 extract results in two additional protein-DNA interactions. Cold mutant VDRE competitor B exhibits little effect on the complexes. However, competition with cold probe A and C both eliminate the most retarded band as well as their respective VDR or AP-1 binding proteins. These data as well as footprint data not shown indicate that the most highly retarded band represents the simultaneous binding of both AP-1 and VDR.

F. *The AP-1 site within the human OC promoter synergizes with the VDRE.* Immediately juxtaposed to the VDRE in the human OC is a unique consensus DNA sequence for the AP-1 proto-oncogene family (see Figure 4). This site is crucial for vitamin D hormone activation, although the mechanism by which this contribution to hormonal response is manifested remains undefined. The structure of this responsive region indicates that it is composite in nature (Diamond et al, 1990). That this site represents a functional component of OC expression is derived from several lines of evidence. First, administration of phorbol esters such as TPA to cultured cells containing either transfected or integrated OC promoters enhances reporter function independent of vitamin D activation (Ozono et al, 1991). This observation suggests that the role of the AP-1 site is to enhance osteocalcin response. Consistent with this interpretation is the observation that deletions (Figure 6) or internal point mutagenesis of the AP-1 site (Ozono et al, 1991), at nucleotide bases known to compromise the ability of c-Fos/c-Jun heterodimers to interact, results in a down regulation of osteocalcin promoter activity. Indeed, comparison of both basal and $1,25(OH)_2D_3$ inducible levels of the mutant and intact promoters indicate the presence of a functional synergism when both cis-acting sites and their transactivating proteins are present.

The mechanism by which this synergism occurs remains to be clarified. However, *in vitro* DNA binding studies have revealed that AP-1 protein complexes such and Fos/Jun or osteoblastic AP-1 complexes of unknown identity, and the VDR (see below for VDR DNA binding studies) can simultaneously bind to this dual element (Ozono et al, 1991). Thus, as identified in Figure 7, incubation of both the VDR and extracts containing AP-1 to the AP-1/VDRE DNA under conditions of probe excess reveal protein-DNA complexes following bandshift analysis comprised of VDR alone, AP-1 alone, and a highly retarded complex of both VDR and AP-1. Thus, synergism is likely to arise through complex protein-protein interactions at the site of DNA binding, although alternative possibilities must also be considered. Importantly, while synergism is apparent in the current studies, inhibition of $1,25(OH)_2D_3$ response might also arise in the context of the osteoblast if an inhibitor of the Fos/Jun heterodimer were to be expressed at an appropriate time during this bone cell's development. Interestingly, in the rat OC gene promoter, an upstream element that remains to be identified appears to represent a functional homolog of the human AP-1 sequence (Yoon et al, 1988; Terpening et al, 1991).

ROLE OF THE VITAMIN D RECEPTOR IN OC
GENE ACTIVATION

A. *The vitamin D receptor is essential to 1, 25(OH)$_2$D$_3$ activation of OC.* Strong activation of the OC promoter by 1,25(OH)$_2$D$_3$ was obtainable in cells such as ROS 17/2.8 and MC3T3-E1 that naturally express the VDR. In contrast, no hormonal response was evident when the OC promoter was transfected into cells that did not contain the VDR. While these data supported the contention that the VDR mediated the observed hormonal response, no direct evidence of this fact could be concluded from the studies. Evidence for the involvement of the VDR in OC transcription was obtained, therefore, by introducing an expression vector containing the receptor cDNA into VDR-negative cells (McDonnell et al, 1989). Under these conditions, 1,25(OH)$_2$D$_3$ response was promptly restored. This restoration was dependent upon receptor concentration, and required synthesis of a normal receptor; receptors that were not full-length, did not bind 1,25(OH)$_2$D$_3$ or contained mutations within the DNA binding domain were inactive (McDonnell et al, 1989; Sone et al, 1989; Sone et al, 1991a). These studies provide the most definitive evidence to date that the VDR is an essential component that acts in trans to mediate 1,25(OH)$_2$D$_3$ action.

In addition to establishing the relevance of the VDR in transactivation, the development of the above assay also permitted an evaluation of receptor structural requirements necessary for functional activation of OC transcription. Deletion and point mutagenesis revealed that the DNA binding domain was essential but not sufficient for transactivation (McDonnell et al, 1989; Sone et al, 1991a). Also revealed was the critical nature of conserved amino acids in maintaining the structural integrity of the DNA binding domain. Thus, mutagenesis of basic amino acid residues or eight of the nine conserved cysteine residues led to loss of proper folding, DNA binding capability, and capacity to activate transcription (Sone et al, 1991a). Likewise, the ligand binding domain was essential to transactivation capability, suggesting that a transactivation domain was located in the VDR within that region. In contrast, deletion of the 21 amino acids lying amino terminal to the DNA binding domain did not compromise any of the apparent functions of the receptor (Sone et al, 1991a).

B. *Characteristics of binding of the VDR to VDRE DNA in vitro.* Direct interaction of mammalian cell derived VDR with VDRE DNA *in vitro* has been demonstrated (Ozono et al, 1990; Terpening et al, 1991; Demay et al, 1990; Noda et al, 1990; Sone et al, 1991a). Indeed, the

VDR exhibits sufficiently high affinity for these elements that protein-DNA complexes can be identified following resolution by bandshift assays as well as through VDRE affinity chromatography. Determination of the affinity of this interaction by incubating receptor with increasing amounts of labelled DNA probe has revealed an equilibrium dissociation constant of approximately 0.2 nM (Sone et al, 1991a, Figure 8). This affinity measurement is consistent with that of other steroid receptor-DNA interactions. Interestingly, the *in vitro* interaction of the VDR with VDRE DNA is modulated by the presence of hormone (Sone et al, 1991a; Liao et al, 1990; Sone et al, 1990). Thus,

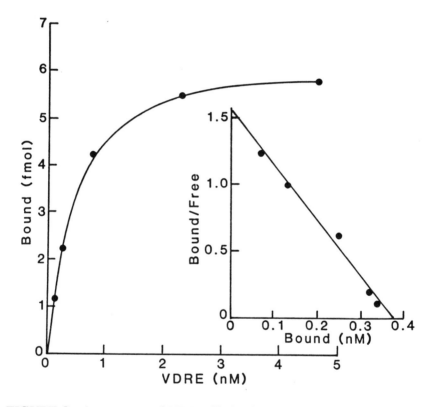

FIGURE 8. Assessment of VDR affinity for the VDRE. Purified VDR was incubated with increasing concentrations of VDRE DNA probe and then subjected to bandshift analysis to separate VDR-DNA complexes from free DNA. Complexes and free DNA was excised from the gel and quantitated. The inset represents a Scatchard plot of the saturation data. The equilibrium dissociation constant calculated from the plot is 0.24 nM with an r value of -0.99.

while incubation of unoccupied receptor with VDRE DNA leads to a poor yield of VDR-DNA complexes following bandshift analysis, addition of $1,25(OH)_2D_3$ to the incubation prior to assay dramatically increases receptor DNA binding. Therefore, the capacity of $1,25(OH)_2D_3$ to activate the VDR and increase transcriptional response *in vivo* is mimicked by this specific action *in vitro*.

In addition to hormone, recent experiments suggest that high affinity binding of the VDR to DNA requires a heterologous protein factor (Sone et al, 1991a; Liao et al, 1990; Sone et al, 1990; Sone et al,1991b). We observed that when the VDR was produced by *in vitro* translation from synthetic transcripts, or prepared by recombinant expression in either bacteria or yeast, the protein would not bind DNA with high affinity. Indeed, chromatographic studies using concatemerized VDRE DNA linked to an inert support revealed that even in the presence of $1,25(OH)_2D_3$, these sources of VDR bound with weak affinity to specific DNA as non cooperative monomers (Sone et al, 1991a). Single VDRE halfsites and the VDR DNA binding domain alone were the only requirements for this weak interaction. In contrast, the addition of mammalian cell extract to recombinantly produced VDR strongly enhanced the latter's capacity to bind VDRE DNA, and a fully intact response element was essential for this high affinity interaction. Additional investigation revealed that the interaction between the VDR and this trypsin- and temperature-sensitive protein factor designated nuclear accessory factor (NAF) did not require the presence of DNA. Thus, immobilized VDR could be utilized to recover NAF from a soluble protein mixture, supporting the idea that VDR-NAF heterodimeric complexes formed in solution.

The experiments discussed above collectively suggest that the binding of the VDR to appropriate DNA elements requires the presence of a protein partner and that formation of the heteromer, while likely influenced by DNA, does not require its presence. Recent experiments with retinoic acid and thyroid hormone receptors have led to similar conclusions (Glass et al, 1990; Burnside et al, 1990). If these heterodimeric complexes for thyroid hormone, retinoic acid, and vitamin D prove to be units that function *in vivo*, they will contrast clearly with those for the sex steroids and the glucocorticoids, where the functional receptor units are homodimeric in nature (Kumar and Chambon, 1988; Fawell et al, 1990). These observations may reflect the differing structural nature of the responsive elements; the former are comprised of direct repeats whereas the latter are palindromic in nature.

Two additional observations about the interaction of the VDR and NAF with DNA have been identified. First, studies using VDRs containing large deletions of the DNA or ligand binding domains have revealed that dimerization occurs through the steroid binding region of the molecule, and does not appear to require the DNA binding core (Sone et al, 1991a). Thus, the carboxy terminal portion of the receptor is sufficient for interaction with NAF. More detailed studies have in fact localize the dimerization region of the VDR to the amino acid segments that remain conserved within the receptor family (Kerner and Pike, unpublished data). Second, and perhaps most important, $1,25(OH)_2D_3$ has been demonstrated to enhance the formation of VDR/NAF heterodimers in solution (Sone et al,, 1991a). Thus, incubation of yeast-produced VDR with $1,25(OH)_2D_3$, followed by incubation with increasing concentrations of protein extract containing NAF results in an enhancement of salt-extractable NAF when compared to an identical experiment carried out in the absence of the vitamin D hormone. Determination of the equilibrium dissociation constant for both conditions revealed an increase in affinity of the VDR for NAF of 10-fold when $1,25(OH)_2D_3$ was present. These results suggest that at least one role of the vitamin D hormone may be to alter the equilibrium between VDR and NAF monomers and the VDR-NAF heterodimer such that an increase in DNA binding becomes evident. While formation of the heteromer, DNA binding, and transactivation may be synonymous, this hypothesis remains to be proven.

PROPERTIES OF NAF AND OTHER STEROID RECEPTOR ACCESSORY FACTORS

Several protein factors have been characterized very recently that contribute to the specific DNA binding capacity of vitamin D, thyroid hormone, and retinoic acid (RAR) receptors (Liao et al, 1990; Glass et al, 1990; Burnside et al, 1990). It is important to note, however, that although these proteins clearly function to promote cooperative association of the respective receptor with its cognate response element, evidence linking these macromolecules to function *in vivo* remains inconclusive. The forthcoming identity of these protein partners will aid in experiments designed to test functional relevance. Moreover, the likely possibility that at least a subset of these accessory factors may belong to the steroid receptor gene superfamily suggests an added complexity - that their function may also be modified by a hormonal ligand.

A. *Distribution of NAF.* We have determined the distribution of NAF in tissues and in cultured cells by employing a complementation bandshift assay wherein extracts that contain NAF or NAF-like components facilitate yeast-derived VDR binding (Sone et al, 1991a; Liao et al, 1990; Sone et al, 1991b). NAF is detectable in both mouse liver and kidney extracts. This experiment indicates that NAF is not coexpressed with the VDR, as the latter is not produced in liver. Thus, it is likely that NAF retains an inherent function of its own. NAF is also expressed widely in cultured cells, including fibroblasts, osteoblasts, hepatocytes, lymphocytes, and cancer cells. Again, while many of these cells synthesize endogenous VDR, several of them do not. To date, the only cell types that do not express NAF activity are bacteria and yeast. These studies indicate that NAF is widely distributed and not coexpressed together with the VDR. Thus, it is likely that in addition to the potential role it plays in $1,25(OH)_2D_3$ activation, NAF protein(s) retains a function independent of this activation.

B. *Properties of NAF.* In addition to distribution, we have investigated other physical and functional features of NAF. We examined the requirement for NAF in VDR binding to a variety of VDREs. Natural VDREs derived from the human and rat OC genes and the osteopontin gene were investigated. VDR binding to each of these DNA sequences required NAF (Sone et al, 1991b). Thus, it is clear that heterodimer formation is a general feature of VDR DNA binding and not a specific function of the element evaluated. The molecular mass of NAF was determined both by crosslinking VDR and NAF as well as by assessing the activity of renatured protein fractions derived from SDS-polyacrylamide gel electrophoresis (Sone et al, 1991a; Sone et al, 1991b). In both experiments, NAF behaves as a 53 to 55 KDa protein. Finally, chromatographic analysis has revealed that NAF binds to a series of ion exchange supports and immobilized ligands that include DEAE- and phosphocelluloses, calf thymus DNA, and VDRE DNA. Like the VDR, NAF exhibits a weak affinity for immobilized VDRE DNA when applied alone, but strong affinity when it is combined with the VDR. These behavioral attributes of NAF support the idea that NAF may be a member of the steroid receptor gene superfamily.

C. *The retinoid X receptors display NAF activity.* Three highly related but separate receptors for all trans retinoic acid have been cloned and identified as members of the steroid receptor superfamily of genes (O'Malley, 1990). More recently, a class of receptors that exhibit low homology to the retinoic acid receptors in both the DNA and ligand binding domains have been identified (Mangelsdorf et al, 1990). These proteins, as documented in Figure 9, are termed the retinoid X receptors (RXRs) and are comprised of α, β and γ forms (Mangelsdorf et al, 1992). They bind and are activated specifically by the retinoic acid metabolite 9-cis retinoic acid (Heyman et al, 1992). Indeed, 9-cis retinoic acid may be the first natural small molecule hormone to be identified since the discovery of $1,25(OH)_2D_3$. Interestingly, in addition to their ability to form functional homodimers on selected response elements, the RXR family of proteins appear to form heterodimeric complexes with the thyroid hormone receptors, RAR, and VDR (Yu et al, 1991; Leid et al, 1992; Zhang et al, 1992; Kliewer et al, 1992). Indeed, the gene for an accessory protein that heterodimerized with both the RAR and the thyroid hormone receptor was cloned by two independent groups by separate methodologies and shown to be RXRβ (Yu et al, 1991; Leid et al, 1992). Based upon this as well as additional research, the RXR family may represent a pool of common protein partners for the action of several

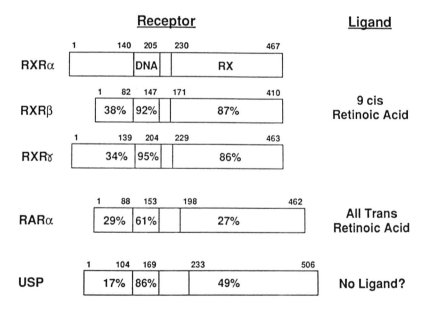

FIGURE 9. The RXR family of genes. The three RXR genes (α, β, and γ) are indicated, and compared to one RAR (α) and the Drosophila RXR homolog ultraspiracle (USP). The relative homology within the specific domain regions of the receptors is indicated as a percentage of RXRα. Numbers above the receptor rectangles indicate the amino acid landmarks of the individual proteins. Ligands for the illustrated receptors are indicated to the right.

liganded receptor systems as summarized in Figure 10. As observed in Figure 11, recombinantly expressed RXRα, β, and γ bind to the VDR to form VDR heterodimers. In contrast, addition of similar extracts of RARα have no effect on VDR binding to the VDRE (data not shown). These experiments, as well as those previously reported (Yu et al, 1991; Leid et al, 1992; Zhang et al, 1992; Kliewer et al, 1992) suggest that members of the RXR family contain unique properties with regard to their capacity to form complexes with other members of the steroid receptor family. At present, however, little functional evidence exists to

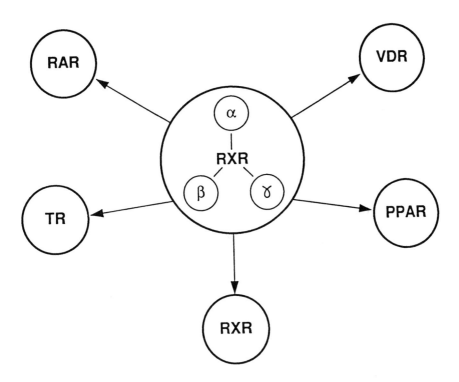

FIGURE 10. The central role of the RXR family in mediating transactivation by other receptors. The RXR family of receptors exhibit ligand-dependent activity on specific genes. In addition, however, they appear in their unliganded state to form heterodimeric partners with heterologous receptor-ligand complexes whose identities are indicated. The affect of the RXR ligand on the action of these heterodimers remains unclear.

184

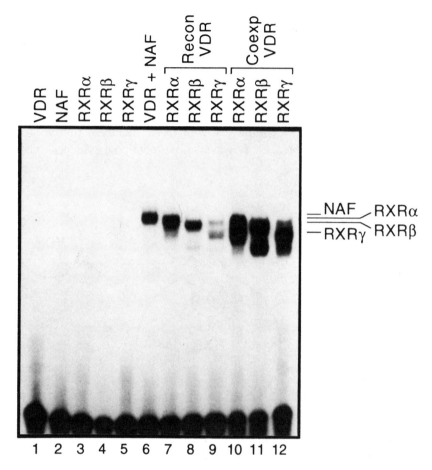

FIGURE 11. VDR and the RXRs form heterodimers *in vitro*. Bandshift analysis was utilized to demonstrate the requirement of the RXRs in high affinity VDR binding to the OC VDRE. Yeast recombinant VDR, RXRα, RXRβ, or RXRγ were incubated independently or in the various combinations indicated with the VDRE probe and subjected to bandshift analysis. "NAF" represents a Hela cell extract containing an endogenous factor (most likely one of the RXRs) shown previously to partner with the VDR. Reconstitution of separately expressed yeast extracts containing VDR and RXRα, RXRβ, or RXRγ is indicated by "Recon VDR". Analysis of yeast extracts coexpressing VDR with either RXRα, RXRβ, or RXRγ is indicated by "Coexp VDR". Neither VDR, NAF, or the RXRs independently bind VDRE DNA. However, the addition of NAF, or each of the RXRs strongly promotes VDR binding to VDRE DNA.

suggest that an alteration in the cellular complement of RXR, achieved through transfection of an RXR expression vector, is effective in modifying cellular response to the vitamin D hormone. Moreover, it is also unclear at present whether NAF is identical to RXR. Thus, additional experiments will be required to determine whether the two proteins are identical, as well as to evaluate the *in vivo* significance of these *in vitro* observations.

SUMMARY

The studies outlined in this chapter describe recent advances in our understanding of the molecular mechanisms by which the osteoblast-specific bone protein OC is regulated at the genomic level. The gene is controlled at the basal level by a series of cis-acting elements that function together with their cognate transactivators at appropriate times during osteoblast maturation. Finally, the gene is modulated by a series of steroid hormones that include $1,25(OH)_2D_3$. The action of this hormone is mediated by the VDR. Interestingly, however, *in vitro* studies have suggested an additional complexity in this regulation. Thus, it appears that the VDR requires a non-VDR protein for cooperative interaction with specific DNA. Current evidence suggests that the dimerization partner may be another member of the steroid receptor superfamily, namely the retinoid X receptor subfamily. This increase in complexity may well extend the diversity of response generated by the vitamin D hormone and its receptor protein, particularly if ligands for the non VDR protein partners are found to play a role in functional response.

REFERENCES

Baker AR, McDonnell DP, Hughes MR, Crisp TM, Mangesldorf DJ, Haussler MR, Pike JW, Shine J and O'Malley BW (1988): Cloning and expression of full length cDNA encoding human vitamin D receptor. *Proc Natl Acad Sci USA* 85: 3294-3298.

Beato M (1989): Gene regulation by steroid hormones *Cell* 5: 335-344.

Burmester JK, Maeda N and DeLuca HF (1988): Isolation and expression of rat 1, 25-dihydroxyvitamin D receptor cDNA. *Proc Natl Acad Sci USA* 85: 1005-1009.

Burnside J, Darling DS and Chin WW (1990): A nuclear factor that enhances binding of thyroid hormone receptors to a thyroid hormone response element. *J Biol Chem* 265: 2500-2504.

Celeste AJ, Rosen V, Buecker JL, Kriz R, Wang EA and Wozney JM (1986). Isolation of the human gene for bone gla protein utilizing mouse and rat cDNA clones. *EMBO J* 5: 1885-1890.

de The H, del Mar Vivanco-Ruiz NM, Tiollais P, Stunnenberg H and Dejean A (1990): Identification of a retinoic acid responsive element in the retinoic acid receptor beta gene. *Nature* 343: 177-180.

Delmas PD, Stenner D, Wahmer HW, Mann KG and Riggs BI (1973): Assessment of bone turnover in post menopausal osteoporosis by measurement of serum bone gla-protein. *J Clin Invest* 71: 1316-1323.

Demay MB, Gerardi JM, DeLuca HF and Kronenberg HM (1990): DNA sequences in the rat osteocalcin gene that bind the 1,25-dihydroxyvitamin D_3 receptor and confer responsiveness to 1,25-dihydroxyvitamin D_3. *Proc Natl Acad Sci USA* 87: 369-373.

Diamond MI, Miner JN, Yoshinaga SK and Yamamoto KR (1990): Transcription factor interactions: selectors or positive or negative regulation from a single DNA element. *Science* 249: 1266-1272.

Evans RM (1988): The steroid and thyroid hormone receptor superfamily. *Science* 240: 889-895.

Fawell WE, Lees JA, White R and Parker MG (1990): Characterization and colocalization of steroid binding and dimerization activities in the mouse estrogen receptor. *Cell* 60: 953-962.

Freedman LP, Luisi BF, Korszun ZR, Basavappa R, Sigler PB and Yamamoto KR (1988): The function and structure of the metal coordination sites within the glucocorticoid receptor DNA binding domain. *Nature* 334: 543-546.

Gerstenfeld LC, Chipman SD, Glowacki J and Lian JB (1986): Expression of differentiated function by mineralizing cultures of chicken osteoblasts. *Dev Biol* 122: 49-60.

Glass CK, Franco R, Weinberger C, Albert VR, Evans RM and Rosenfeld MG (1988): A c-erb A binding site in rat growth hormone gene mediates trans-activation by thyroid hormone. *Nature* 329: 738-741.

Glass CK, Devary OV and Rosenfeld MG (1990): Multiple cell type-specific proteins differentially regulate target sequence recognition by the alpha retinoic acid receptor. *Cell* 63: 729-736.

Glowacki J and Lian JB (1989): Impaired recruitment of osteoclast progenitors by osteocalcin-deficient bone implants. *Cell Diff* 21: 247-254.

Gundberg CM, Lian JB, Gallop PM and Steinberg JJ (1983): Urinary γ-carboxyglutamic acid and serum osteocalcin as bone markers: studies in osteoporosis and Paget's disease. *J Clin Endocrinol Metab* 57: 1221-1225.

Hard T, Kellenbach E, Boelens R, Maler B, Dahlman K, Freedman LP, Carlstedt-Duke J, Yamamoto KR, Gustafsson J-A and Kaptein R (1990): Solution structure of the glucocorticoid receptor DNA-binding domain. *Science* 249: 157-160.

Hauschka PV, Lian JB and Gallop PM (1975): Direct identification of the calcium binding amino acid g-carboxyglutamate in mineralized tissues. *Proc Natl Acad Sci USA* 72: 3925-3929.

Heyman RA, Mangelsdorf DJ, Dyck JA, Stein RB, Eichele G, Evans RM and Thaller C (1992): 9-Cis retinoic acid is high affinity ligand for the retinoid X receptor. *Cell* 68: 397-406.

Jensen EV, Suzuki T, Kawashima T, Stumpf WE, Jungblut PW and DeSombre ER (1968): A two step mechanism for the interaction of estradiol with rat uterus. *Proc Natl Acad Sci USA* 59: 632-638.

Jensen EV and DeSombre ER (1972): Mechanism of action of the female sex hormones. *Annu Rev Biochem* 41: 203-230.

Jowell PS, Epstein S, Fallon MD, Reinhardt TA and Ismail F (1987): 1,25-Dihydroxyvitamin D_3 modulates glucocorticoid-induced alteration in serum bone gla protein and bone histomorphometry. *Endocrinology* 120: 531-536.

Johnston PF and McKnight SL (1989): Eukaryotic transcriptional regulatory proteins. *Annu Rev Biochem* 58: 799-839.

Kerner SA, Scott RA and Pike JW (1989): Sequence elements in the human osteocalcin gene confer basal activation and inducible response to hormonal vitamin D_3. *Proc Natl Acad Sci USA* 86: 4455-4459.

Kesterson RA, Stanley L, Finegold MJ, DeMayo F and Pike JW (1993): The human osteocalcin promoter directs bone-specific vitamin D-regulatable expression in transgenic mice. *Mol Endo* 7: 462-467.

Kliewer SA, Umesono K, Mangelsdorf DJ and Evans RM (1992): Retinoid X receptor interacts with nuclear receptors in retinoic acid, thyroid hormone and vitamin D_3 signalling. *Nature* 355: 446-449.

Kumar V and Chambon P (1988): The estrogen receptor binds tightly to its responsive element as a ligand-induced homodimer. *Cell* 55: 145-156.

Leid M, Kastner P, Lyons R, Nakshatro H, Saunders M, Zacharewski T, Chen J-Y, Staub A, Garnier J-M, Mader S and Chambon P (1992): Purification, cloning, and RXR identity of the Hela cell Factor with which RAR or TR heterodimerizes to bind target sequences efficiently. *Cell* 68: 377-395.

Lian JB and Gundberg CM (1988): Osteocalcin: biochemical, considerations and clinical application. *Clin Orthop Relat Res* 226: 267-291.

Lian JB, Coutts MNC and Canalis E (1985): Studies of hormonal regulation of osteocalcin synthesis in cultured fetal rat calvariae. *J Biol Chem* 260: 8706-8710.

Lian JB, Carnes DL and Glimcher M (1987): Bone and serum concentrations of osteocalcin as a function of 1,25-dihydroxyvitamin D_3 circulating levels in bone disorders in rats. *Endocrinology* 120: 2123-2130.

Lian JB, Stewart C, Puchacz E, Mackowiak S, Shalhoub V, Collart D, Sambetti G and Stein G (1989): Structure of the rat osteocalcin gene and regulation of vitamin D-dependent expression. *Proc Natl Acad Sci USA* 86: 1143-1147.

Liao J, Ozono K, Sone T, McDonnell DP and Pike JW (1990): Vitamin D receptor interaction with specific DNA requires a nuclear protein and 1,25-dihydroxyvitamin D_3. *Proc Natl Acad Sci USA* 87: 9751-9755.

Malone JD, Teitelbaum SL, Griffin GL, Senior RM and Kahn AJ (1982): Recruitment of osteoclast precursors by purified bone matrix constituents. *J Cell Biol* 92: 227-230.

Mangelsdorf DJ, Ong ES, Dyck JA and Evans RM (1990): Nuclear receptor that identifies a novel retinoic acid response pathway. *Nature* 345: 224-229.

Mangelsdorf DJ, Borgmeyer U, Heyman RA, Zhou JY, Ong ES, Oro AE, Kakizuka A and Evans RM (1992): Characterization of three RXR genes that mediate the action of 9-cis retinoic acid. *Genes Devel* 6: 329-344.

Markose ER, Stein JL, Stein GS and Lian JB (1990): Vitamin D-mediated modifications in protein-DNA interactions at two promoter elements of the osteocalcin gene. *Proc Natl Acad Sci USA* 87: 1701-1705.

McDonnell DP, Scott RA, Kerner SA, O'Malley BW and Pike JW (1989). Functional domains of the human vitamin D_3 receptor regulate osteocalcin gene expression. *Mol Endo* 3: 635-644.

McDonnell DP, Mangelsdorf DJ, Pike JW, Haussler MR and O'Malley BW (1987): Molecular cloning of complementary DNA encoding the avian receptor for vitamin D. *Science* 235: 1214-1217.

Morrison NA, Shine J, Fragonas J-C, Verkest V, McMenemy L and Eisman JA (1989): 1,25-Dihydroxyvitamin D-responsive element and glucocorticoid repression in the osteocalcin gene. *Science* 246: 1158-1161.

Nishimoto SK, Salka C and Nimni ME (1987): Retinoic acid and glucocorticoids enhance the effect of 1,25-dihydroxyvitamin D_3 on bone γ-carboxyglutamic acid protein synthesis by rat osteosarcoma cells. *J Bone Mineral Res* 2: 571-577.

Noda M, Vogel RL, Craig AM, Prahl J, DeLuca HF and Denhardt D (1990). Identification of a DNA sequence responsible for binding of the 1,25-dihydroxyvitamin D_3 receptor and 1,25-dihydroxyvitamin D_3 enhancement of mouse secreted phosphoprotein 1 (Supp-1 or osteopontin) gene expression. *Proc Natl Acad Sci USA* 87: 9995-9999.

O'Malley BW (1990): The steroid receptor superfamily: more excitement predicted for the future. *Mol Endo* 4: 363-369.

Ozono K, Liao J, Kerner SA, Scott RA and Pike JW (1990): The vitamin D responsive element in the human osteocalcin gene: association with a nuclear proto-oncogene enhancer. *J Biol Chem* 265: 21881-21888.

Ozono K, Sone T and Pike JW (1991): Functional interactions between vitamin D receptors and the AP-1 proto-oncogene family. *J Bone Min Res* 6, Suppl. 1, 677 (Abstr).

Pan LC and Price PA (1984): The effect of transcriptional inhibitors on the bone γ-carboxylic acid protein response to 1,25-dihydroxyvitamin D in osteosarcoma cells. *J Biol Chem* 259: 5844-5847.

Picard D and Yamamoto KR (1988): Two signals mediate hormone-dependent translocation of the glucocorticoid receptor. *EMBO J* 6: 3333-3340.

Pike JW (1990): Cis and trans regulation of osteocalcin gene expression by vitamin D and other hormones. In: Cohn, D.V., Glorieux , F.H., Martin, T. J, eds. *Calcium Regulation and Bone Metabolism*, vol 10, Excerpta Medica, Amsterdam, p. 127-136.

Price PA, Otsuka AS, Posner JW, Kristaponis J and Raman N (1976): Characterization of a γ-carboxyglutamic acid-containing protein from bone. *Proc Natl Acad Sci USA* 73: 1447-1451.

Price PA and Nishimoto SK (1980): Radioimmunoassay for the vitamin D-dependent protein of bone and its discovery in plasma. *Proc Natl Acad Sci USA* 77: 2234-23238.

Price PA, Parthemore JG and Deftos LF (1980): New biochemical marker for bone metabolism: measurement by radioimmunoassay of bone gla-protein in the plasma of normal subjects and patients with bone disease. *J Clin Invest* 66: 878-883.

Price PA and Baukol SA (1980): 1,25-Dihydroxyvitamin D_3 increases the synthesis of the vitamin D-dependent bone protein by osteosarcoma cells. *J Biol Chem* 255: 11660-11663.

Schule R, Umesono K, Mangelsdorf DJ, Bolado J, Pike JW and Evans RM (1990): Jun-Fos and receptors for vitamins A and D recognize a common response element in the human osteocalcin gene. *Cell* 61: 497-504.

Schwabe JWR, Neuhaus D and Rhodes D (1990): Solution structure of the DNA-binding domain of the oestrogen receptor. *Nature* 348: 458-461.

Sone T, Scott RA, Hughes MR, Malloy PJ, Feldman D, O'Malley BW and Pike JW (1989): Mutant vitamin D receptors which confer hereditary resistance to 1,25-dihydroxyvitamin D_3 in humans are transcriptionally inactive *in vivo*. *J. Biol. Chem.* 264: 20230-20234.

Sone T, McDonnell DP, O'Malley BW and Pike JW (1990): Expression of human vitamin D receptor in Saccharomyces cerevisiae: purification, properties, and generation of polyclonal antibodies. *J Biol Chem* 265: 21997-22003.

Sone T, Kerner SA and Pike JW (1991a): Vitamin D receptor interaction with specific DNA: association as a 1,25-dihydroxyvitamin D_3-modulated heterodimer. *J Biol Chem* 266: 23296-23305.

Sone T, Ozono K and Pike JW (1991b). A 55 kilodalton accessory factor facilitates vitamin D receptor DNA binding. *Mol Endo* 5: 1578-1586.

Terpening CM, Haussler CA, Jurutka PW, Galligan MA, Komm BS and Haussler MR (1991): The vitamin D-responsive element in the rat bone gla protein gene is an imperfect direct repeat that cooperates with other cis-elements in 1,25-dihydroxyvitamin D_3-mediated transcriptional activation. *Mol Endo* 5: 373-385.

Uchida M, Ozono K and Pike JW (1991): 24R,25-Dihydroxyvitamin D_3 activates osteocalcin gene transcription through the vitamin D receptor. In: Norman,A. W., Bouillon, R., and Thomasset, M., eds. *Vitamin D, gene regulation, structure-function analysis and clinical application*, de Gruyter, Berlin, p 304-305.

Umesono K, Giguere V, Glass CK, Rosenfeld MG and Evans RM (1988): Retinoic acid and thyroid hormone induce gene expression through a common responsive element. *Nature* 336: 262-265.

Umesono K, Murikami KK, Thompson CC and Evans RM (1991): Direct repeats as selective response elements for the thyroid hormone, retinoic acid, and vitamin D_3 receptors. *Cell* 65: 1255-1266.

Yamamoto KR (1985): Steroid receptor regulated transcription of specific genes and gene networks. *Annu Rev Genet* 19: 209-252.

Yoon D, Turledge SJC, Buenaga RF and Rodan GA (1988): Characterization of the rat osteocalcin gene: stimulation of promoter activity by 1,25-dihydroxyvitamin D_3. *Biochemistry* 27: 8521-8526.

Yu V, Delsert C, Andersen B, Holloway JM, Devary OV, Naar AM, Kim SY, Boutin J-M, Glass CK and Rosenfeld MG (1991): RXRβ: A coregulator that enhances binding of retinoic acid, thyroid hormone, and vitamin D receptors to their cognate response elements. *Cell* 67: 1251-1266.

Zhang X-K, Hoffmann B, Tran PB-V, Graupner G and Pfahl M (1992): Retinoid X receptor is an auxiliary protein for thyroid hormone and retinoic acid receptors. *Nature* 355: 441-446.

MOLECULAR MECHANISMS OF THYROID HORMONE AND RETINOIC ACID ACTION

Magnus Pfahl
Cancer Center
La Jolla Cancer Research Foundation

INTRODUCTION

Understanding of the molecular mechanisms that underlie the large diversity of biological responses to thyroid hormones and the vitamin A derived hormones (retinoids) has been greatly advanced by the cloning of specific intracellular receptors for thyroid hormone (T_3) and retinoic acid (RA) (Sap et al, 1986; Weinberger et al, 1986; Benbrook and Pfahl, 1987; Petkovich et al, 1987; Giguere et al, 1987; Benbrook et al, 1988; Brand et al, 1988; Thompson et al, 1987; Nakai et al, 1988; Giguere et al, 1990; Krust et al, 1989). In mammals, T_3 receptors (TRs) are encoded by two genes, TRα and TRβ, while the retinoid receptors are encoded by at least six genes that fall into two subfamilies, the classical RA receptors (RARs) α, β, and γ, and the retinoid X receptors (RXRs) α, β, and γ (Hamada et al, 1989; Mangelsdorf et al 1990; 1992; Yu et al, 1991; Leid et al, 1992,). From the individual receptor genes, different isoforms can be generated by differential splicing and/or different promoter usage (Lehmann et al, 1991; Leroy et al, 1991; Zelent et al, 1991; Benbrook and Pfahl, 1987; Nakai et al, 1988). TRs, RARs, and RXRs belong to a subgroup of the nuclear receptor superfamily that have closely related DNA binding domains (reviewed in Umesono and Evans, 1989) and that are able to

STEROID HORMONE RECEPTORS
V. K. Moudgil, Editor
© 1993 Birkhäuser Boston

recognize and function from overlapping and related response elements. This subfamily of receptors includes also the vitamin D receptor (VDR), the peroxisome proliferator activated receptor (PPAR) (Issemann and Green, 1990; Dreyer et al, 1992), as well as the COUP and other orphan receptors for which specific ligands are not known. In analogy with the steroid hormone receptors, it had been assumed that nuclear receptors function by binding as homodimers to specific DNA regions (termed response elements) (reviewed in Green and Chambon, 1988; Evans, 1988). This was supported by the observation that natural T_3 or RA response elements (TREs or RAREs), revealed distinct two-fold symmetries and that inverted and direct repeats functioned as highly effective TREs and/or RAREs (Glass et al 1988; Graupner et al, 1989; Umesono et al, 1988; 1991; Näar et al, 1991; Zhang et al, 1991a). However, more recently, our view on how TRs, RARs, and RXRs function has undergone some quite dramatic change. It is now clear that these receptors require auxiliary receptor(s) for efficient DNA binding and gene activation and that RXRs are such auxiliary receptors (Zhang et al, 1992a; Yu et al, 1991; Leid et al, 1992; Kliewer et al, 1992a; Bugge et al, 1992; Marks et al, 1992). In addition, RXRs can also function on their own in the presence of specific ligands and form homodimers (Zhang et al, 1991b; Lehmann et al 1992). Furthermore, TRs as well as RARs can regulate transcription without binding to specific DNA sequences, but by direct interaction with other transcription factors (Yang Yen et al, 1991; Zhang et al, 1991b). In the following paragraphs I review some of our recent work on the molecular mechanisms of thyroid hormone and retinoic acid action.

Heterodimers are Required for Specific DNA Binding and Transcriptional Activation

In contrast to ER and other steroid hormone receptors, it was found that TRs and RARs bound poorly to response elements, while nuclear components enhanced TR and RAR-DNA interaction significantly (Murray and Towle, 1989; Rosen et al, 1991; Zhang et al, 1991b; and references therein). Moreover, TRα was found to bind to the palindromic TRE (TREpal) primarily as a monomer and essentially non-cooperatively. However, in the presence of nuclear extract from various cell lines, we observed a strong enhancement of TRα binding and the formation of a larger complex (Zhang et al, 1991b). While a number of approaches were used by various laboratories to identify the nuclear factors that enhanced TR and/or RAR response element interaction (Yu et al, 1992;

Leid et al, 1992; Zhang et al, 1992a), we analyzed members of the nuclear receptor superfamily for their ability to enhance TR or RAR DNA binding. We reasoned that since the nuclear factor could interact with the same or similar sequences as TR and RAR, it was likely to be a nuclear receptor closely related in its DNA binding domain to TRs and RARs. Indeed, when we mixed a variety of receptors, we observed that one of them, RXRα, dramatically enhanced TR and RAR interaction with the TREpal, a T_3 and RA responsive element (Zhang et al, 1992a). TR-RXR as well as RAR-RXR interaction with DNA was strongly cooperative, while, for instance, TR-RAR binding to DNA was not. Similar results were obtained by others and it could also be shown that TR-RXR and RAR-RXR heterodimer formation occurred in solution in the absence of DNA (Zhang et al, 1992a; Kliewer et al, 1992; Leid et al, 1992; Yu et al, 1991; Bugge et al, 1992; Marks et al, 1992).

Since RXR formed heterodimers with TRs as well as RARs and other receptors, an important question was whether the different heterodimers had distinct response elements specificities. We analyzed DNA binding and transcriptional activation by TR-RXR and RAR-RXR heterodimers with natural TREs and RAREs. The TREs were derived from the α myosin heavy chain (MHC) (Mahdavi et al, 1984) and the malic enzyme (ME) (Desvergne et al, 1991) promoters. The RAREs were derived from the RARβ2 (Hoffmann et al, 1990), the rat cellular retinol binding protein I (CRBPI) (Husmann et al, 1992) and the apolipoprotein AI (ApoAI) (Rottman et al, 1991) gene promoters. We observed that under our *in vitro* assay conditions TR-RXR heterodimers only bound TREs, but not RAREs. In contrast, RAR-RXR heterodimers only bound RAREs, but not TREs. For example, a strong complex is formed between TR-RXR heterodimers and the ME-TRE, while RAR-RXR heterodimers do not bind to this response element under (Fig. 1a). The natural RARE from the hRARβ2 promoter, however, bound only RAR-RXR heterodimers, but not TR-RXR complexes (Fig 1b) (Hermann et al, 1992). In addition, RAREs were activated optimally by RAR-RXR heterodimers, but not by TR-RXR heterodimers, while TREs were only activated by TR- RXR heterodimers (Hermann et al, 1992). For instance, a reporter gene carrying the rCRBPI-RARE required RAR and RXR for optimal induction by RA, while TR alone or in combination with RXR did not activate this reporter gene (Fig. 2) (Husmann et al, 1992; Hermann et al, 1992). The partial activation seen by RARα alone in Fig. 2 can most likely be attributed to heterodimers formed between RARα and endogenous RXR like proteins present in CV-1 cells. RARα is unlikely to function as a monomer or homodimer on the known RAREs, since its

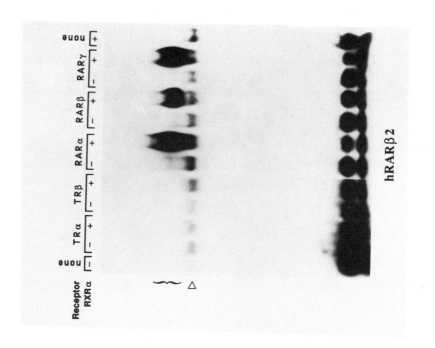

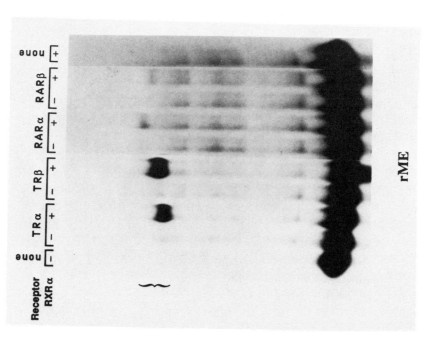

FIGURE 1. RXR Enhances DNA Binding and Specificity.

interaction with these RAREs is very weak in the absence of coreceptors. Very similar results were obtained with a natural TRE containing reporter gene where optimal activation was only observed when TRα was cotransfected with RXR (Hermann et al, 1992).

Our data and similar results obtained by others are summarized in the scheme in Fig. 3. Monomeric TRs, RARs, and RXRs are in equilibrium with TR-RXR and RAR-RXR heterodimers. Only the heterodimers bind effectively to the specific response elements. DNA binding, it should be noted, is independent of ligand binding, such that the heterodimers can have dual functions serving as activators in the presence of specific ligands or repressors in the absence of ligands (Graupner et al, 1989). RXRs were also found to form heterodimers with the v-erbA oncogene, a mutated TRα (Hermann et al, 1993), the VDR (Yu et al, 1991; Leid et al, 1992; Kliewer et al, 1992a; Bugge et al, 1992), and the PPAR (Kliewer et al, 1992b), while the drosophila RXR homologue ultraspiracle is the coreceptor for the ecdysone receptor (Yao et al, 1992).

Ligand Induced RXR Homodimers: A Distinct Response Pathway

From the above observations, it became apparent that RXRs have a very central role in regulating DNA binding and transcriptional activation by RARs, and several other hormone receptors that are controlled by structurally unrelated ligands (Fig. 5). It was, therefore, of interest to investigate what effects RXR specific ligands would have on RXR. As pointed out above, RXR alone does not bind to DNA effectively. Specific DNA binding requires heterodimerization, (Zhang et al, 1992a; Yu et al, 1991; Leid et al, 1992; Kliewer et al, 1992a) unless superficially high concentrations of RXR are used (Mangelsdorf et al, 1991). Using

FIGURE 1. RXR Enhances DNA Binding and Specificity. The effect of RXR on TR and RAR binding to natural response elements was analyzed by gel-shift assay (Zhang et al, 1991b; 1992a). Receptor proteins were synthesized *in vitro* using reticulocyte lysates (Benbrook and Pfahl, 1987; Zhang et al, 1991b). TRα, TRβ, RARα, RARβ, and RARγ were preincubated either with (+) or without (-) equal amounts of *in vitro* synthesized RXRα for 15 minutes at 25° C. Subsequently, the mixtures were incubated with ^{32}P-labeled (A, left panel) ME-TRE or (B, right panel) the βRARE and analyzed by gel retardation. For comparison, binding of RXRα alone, as well as of nonprogrammed reticulocyte lysate, is also shown. Specific binding complexes are indicated by brackets; nonspecific reticulocyte lysate binding is indicated by open triangles.

a gel retardation assay, we made the surprising observation that 9-cis retinoic acid (9-cis RA), strongly enhanced RXR homodimer DNA binding. 9-cis RA, which had been previously shown to bind with high affinity to both RARs and RXRs (Levin et al, 1992; Heyman et al, 1992) was found to induce the formation of RXR homodimers in solution (Zhang et al 1992b). In contrast, RAR DNA binding or homodimer formation was not induced by 9-cis RA. 9-cis RA induced RXR homodimers were found to bind to and activate specific response elements such as the ApoAI-RARE and the CRBPII-RARE, but not the βRARE and the rat CRBPI-RARE (Zhang et al, 1992b; Lehmann et al, 1992). While the ApoAI-RARE is also bound and activated by RAR-RXR heterodimers,

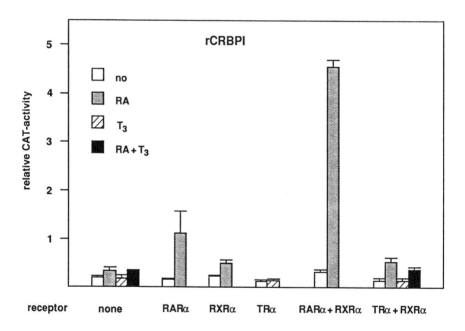

FIGURE 2. RXR is Required for Effective Transactivation By RAR. To measure transcriptional activation from specific response elements by heterodimers, a transient transfection assay was used (Zhang et al, 1992a; 1992b). CV-1 cells were cotransfected with 100 ng reporter construct rCRBPI-RARE-tk-CAT and 5 ng each of the indicated receptor expression vectors. Cells were treated with 10^{-7} M hormone and 16 h later assayed for CAT activity.

the CRABPII-RARE was found to be RXR homodimer specific (Lehmann et al, 1992). Thus, RXR specific ligands, like 9-cis RA, lead to the formation of RXR homodimers that can activate a distinct retinoid response pathway as illustrated in Fig. 4. While 9-cis RA also binds to and activates RARs, we have now been able to define specific retinoids that are RXR selective, in that they only activate RXR homodimers, but not RXR containing heterodimers (Lehmann et al, 1992). Such retinoids can serve to regulate selectively RXR homodimer responsive genes and may, as such, represent valuable tools to examine the roles of RXR in development and disease.

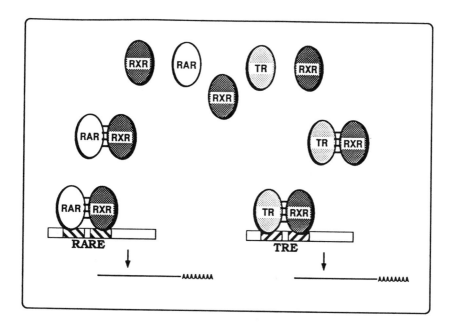

FIGURE 3. Heterodimers as Mediators of T$_3$ and RA Signals. Schematic representative of receptor action. TRs and RARs form heterodimers with RXR in solution. Since only the heterodimers bind with high affinity and specificity to response elements, heterodimerization is likely to be a rate limiting step for DNA binding and transcriptional activation. TR-RXR complexes bind TREs, while RAR-RXR complexes bind to RAREs. DNA binding is independent of hormone.

200

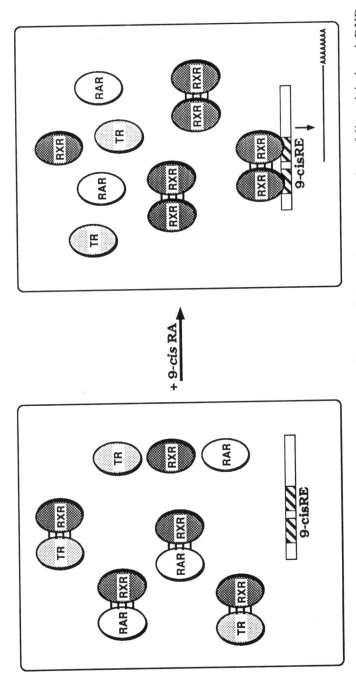

FIGURE 4. RXR Homodimers are Induced by 9-cis RA. Schematic representation of ligand induced RXR homodimer formation. In the absence of ligands, an equilibrium between RXR monomers and RXR heterodimer complexes exist. When 9-cis RA or other RXR specific ligands are present, the equilibrium is shifted towards RXR homodimers. These homodimers bind and activate specific response elements that can be distinct from response elements activated by RAR-RXR heterodimers.

Retinoids that induce RXR homodimer formation can also be expected to limit the availability of RXR molecules for heterodimerization. This is, indeed, the case, although not all heterodimers are affected equally by 9-cis RA. T_3 induction of TRE containing genes by TRα-RXRα heterodimers was found to be strongly repressed in the presence of 9-cis RA or RXR selective retinoids (Lehmann et al, 1993). The repression of the T_3 response appears to be due to sequestering of RXR molecules from TR-RXR heterodimers into RXR homodimers. Thus, a ligand-induced squelching mechanism exists that can regulate the availability of RXR molecules for heterodimerization (See Fig. 5). However not all RXR containing heterodimers are negatively affected by RXR specific retinoids. We found that RAR-RXR heterodimers were not inhibited by 9-cis RA, but in fact, are (in most cases), super-activated. Our data suggest that RARs have a higher affinity for RXRs than TRs. The activity of PPAR-RXR heterodimers also appears to be increased when both PPAR and RXR specific ligands are present (Kliewer et al, 1992b). Thus, 9-cis RA and other retinoids that lead to RXR homodimer formation may inhibit the formation of only certain heterodimers (See also Fig. 5).

COUP Receptors Restrict Retinoid Response Pathways

Although the heterodimers and the ligand-induced RXR homodimers allow for a large degree of diversity and specificity of the retinoid and thyroid hormone responses, the question on how some hormonal responses can be restricted in certain cell types has not yet been understood. When we screened cDNA libraries for RXRα related receptors, two clones -COUPα and COUPβ- identical and closely related to the orphan receptor ear 3 (Miyajima et al, 1988) or COUP-TF (Wang et al, 1989) were obtained (Tran et al, 1992). In transient transfection assays, the COUP receptors dramatically inhibited retinoid receptor activities on several response elements that are activated by RAR-RXR heterodimers or RXR homodimers. COUPα and β bound strongly as a homodimer to these response elements, but did not form heterodimers with RXR (Tran et al, 1992). The elements that were bound effectively by COUP included the palindromic TRE, the CRBPI-RARE as well as the RXR homodimer responsive ApoAI and CRBPII-RAREs. The βRARE was not bound. All response elements that bound effectively COUP homodimers were also inhibited in transient transfection assays by COUP (Tran et al, 1992). These data and similar results reported by others (Cooney et al, 1992; Kliewer et al, 1992c; Windom et al, 1992)

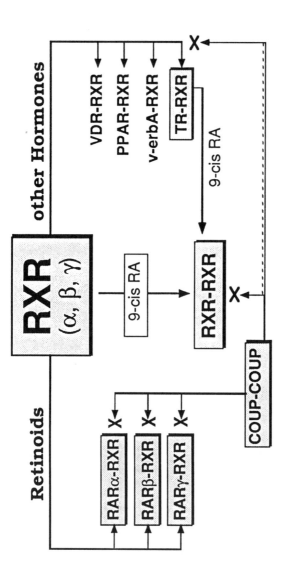

Fig. 5: RXRs as Central Regulators of Multiple Hormonal Response Pathways. RXR forms heterodimers with RARs, that are activated by retinoids. RXR also forms heterodimers with TRs (including v-erbA), VDR, PPAR, and probably other receptors yet to be determined. In the presence of RXR specific retinoids like 9-cis RA, RXR homodimers which recognize a specific subset of RAREs, are formed. The formation of RXR homodimers leads to repression of T_3 responsive genes, while RAR and PPAR containing heterodimers do not appear to be negatively affected by RXR specific ligands. COUP receptors form homodimers that bind with high affinity to several RAREs and can repress RAR/RXR heterodimer as well as RXR homodimer activity. Thus, COUP receptors can restrict retinoid responses to a subset of retinoid responsive genes. COUPs may also antagonize other receptors.

suggest that the COUP receptors are negative regulators that can restrict RA signaling to certain genes. Some other hormonal signals may also be inhibited by COUP (Fig. 5). COUP receptors appear to function as repressors by binding response elements as homodimers but not by heterodimer formation with RXRs as suggested by others (Kliewer et al, 1992c; Windom et al, 1992). Interestingly, the COUP-TF specific binding site in the ovalbumin promoter also functions as an RARE (Tran et al, 1992). Thus, COUP orphan receptors may play an important role in the regulation of cell or tissue specific restriction of RA sensitive programs during development and in the adult, a role consistent with that of the Drosophila COUP homolog *seven-up* (Mlodzik et al, 1990).

Antagonism Between Receptors and the Transcription Factor AP-I

While the basic mechanisms for positive responses to RA and T_3 at the level of gene transcription are well understood, much less is known about negative gene regulation by these hormones. One important mechanism for inhibition of gene transcription involves interaction of RARs and TRs with the cellular oncogenes, c-Jun and c-Fos. c-Jun and c-Fos are constituents of the transcription factor AP-I that mediates signal transduction by peptide hormones, oncogenes and the tumor promoter TPA (Karin, 1990). AP-I binds to specific DNA sites (AP-I sites) and activates gene transcription from adjacent promoters. It has been known for some time that retinoids can antagonize TPA activity and show anticancer activities (Lippman et al, 1987; Huang et al, 1988; Hong et al, 1990). Recent results revealed that this anti-TPA activity of retinoids is mediated by RAR interaction with AP-I. RARs as well as TRs in the presence of their ligands, can interfere with AP-I activity without interacting with the AP-I site (Yang Yen et al, 1991; Zhang et al, 1991b). However, AP-I, when activated by TPA, can also inhibit RAR and TR function. This mutual antagonism appears to result from direct protein-protein interaction between RAR or TR and AP-I. The regulatory mechanism by which the two different families of transcription factors can control each other's activity, is likely to represent a major signaling switch. c-Jun and c-Fos, in general, induce cell proliferation, whereas the nuclear receptors, in general, induce cell differentiation. Since c-Jun and c-Fos mediate the signals of peptide hormones, oncogenes, and tumor promoters, the receptor mediated anti AP-1 activities may represent a major pathway by which retinoids exert their anticancer activities.

204

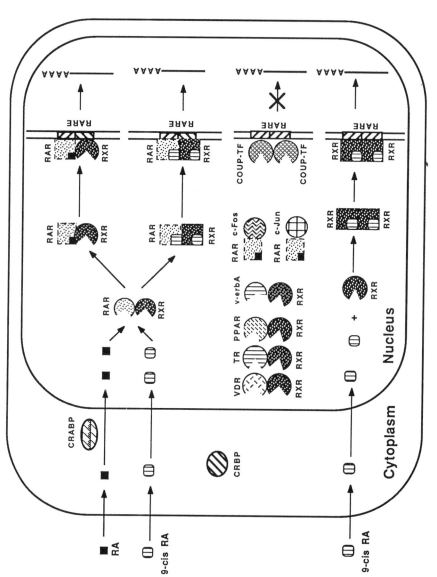

FIGURE 6. Current Model for Retinoid Response Pathways.

FIGURE 6. Current Model for Retinoid Response Pathways. Schematic representation of known retinoid response pathways. RXR heterodimer formation with other receptors and RXR homodimer formation, as well as the action of COUP homodimers, have been discussed above. c-Fos and c-Jun also interact with RARs and this interaction results in mutual inhibition. The details of the receptor AP-1 interaction have not yet been elucidated. While the receptors act in the cell nucleus, cytoplasmic retinol (CRBP) and retinoic acid (CRABP) binding proteins may influence retinoid transport and metabolism.

In conclusion, a number of novel signaling pathways for retinoids and thyroid hormones have recently been elucidated and are summarized in Fig. 6. RXR plays a major role in these pathways and also controls the activities of several receptors that bind structurally unrelated hormones. The advances made in understanding the mechanism of action of these hormones will contribute to the understanding of physiological responses and diseases, while the rational design of more effective hormone-derived therapeutics has become feasible by the design of pathway and receptor selective ligands.

ACKNOWLEDGMENTS

The work reported here was supported by NIH grants CA 50676, CA 51993, and a grant from the Tobacco Related Disease Program of the University of California (2RT0109).

REFERENCES

Benbrook D and Pfahl M (1987): A novel thyroid hormone receptor encoded by a cDNA clone from a human testis library. *Science* 238: 788-791.

Benbrook D, Lernhardt E and Pfahl M (1988): A new retinoic acid receptor identified from a hepatocellular carcinoma. *Nature* 333: 669-672.

Brand N, Petkovich M, Krust A, de The H, Marchio A, Tiollais P and Dejean D (1988): Identification of a second human retinoic acid receptor. *Nature* 332: 850-853.

Bugge TH, Pohl J, Lonnoy O and Stunnenberg HG (1992): RXRα, a promiscuous partner of retinoic acid and thyroid hormone receptors. *EMBO J* 11: 1409-1418.

Cooney AJ, Tsai SY, O'Malley BW and Tsai MJ (1992): Chicken ovalbumin upstream promoter transcription factor (COUP-TF) dimers bind to different GGTCA response elements, allowing COUP-TF to repress hormonal induction of the vitamin D_3, thyroid hormone, and retinoic acid receptors. *Mol Cell Biol* 12: 4153-4163.

Desvergne B, Petty KJ and Nikodem VM (1991): Functional characterization and receptor binding studies of the malic enzyme thyroid hormone response element. *J Biol Chem* 266: 1008-1013.

de Thé H, Vivanco-Ruiz MM, Tiollais P, Stunnenberg H and Dejean A (1990): Identification of a retinoic acid response element in the retinoic acid receptor β gene. *Nature* 343: 177-180.

Dreyer C, Krey G, Keller H, Givel F, Helftenbein G and Wahli W (1992): Control of the peroxisomal β-oxidation pathway by a novel family of nuclear hormone receptors. *Cell* 68: 879-887.

Evans RM (1988): The steroid and thyroid hormone receptor family. *Science* 240: 889-895.

Giguere V, Ong ES, Seigi P and Evans RM (1987): Identification of a receptor for the morphogen retinoic acid. *Nature* 330: 624-629.

Giguere V, Shago M, Zirngibl R, Tate P, Rossant J and Varmuza, S (1990): Identification of a novel isoform of the retinoic acid receptor γ expressed in the mouse embryo. *Mol Cell Biol* 10: 2335-2340.

Glass CK, Holloway JM, Devary OV and Rosenfeld MG (1988): The thyroid hormone receptor binds with opposite transcriptional effects to a common sequence motif in thyroid hormone and estrogen response elements. *Cell* 54: 313-323.

Graupner G, Willis KN, Tzukerman M, Zhang X-k and Pfahl M (1989): Dual regulatory role for thyroid-hormone receptors allows control of retinoic-acid receptor activity. *Nature* 340: 653-656.

Green S and Chambon P (1988): Nuclear receptors enhance our understanding of transcription regulation. *Trends Genet* 4: 309-314.

Hamada K, Gleason SL, Levi B-Z, Hirschfeld S, Apella E and Ozato K (1989): H-2RIIBP, a member of the nuclear hormone receptor superfamily that binds to both the regulatory element of major histocompatibility class I genes and the estrogen response element. *Proc Natl Acad Sci USA* 86: 8289-8293.

Hermann T, Hoffmann B, Zhang X-k, Tran P and Pfahl M (1992): Heterodimeric receptor complexes determine 3,5,3'-triiodothyronine and retinoid signaling specificities. *Mol Endocrinol* 6: 1153-1162.

Hermann T, Hoffmann B, Piedrafita JF, Zhang Z-k and Pfahl M (1993): V-erbA requires auxiliary proteins for dominant negative activity. *Oncogene* 8: 55-65.

Heyman RA, Mangelsdorf DJ, Dyck JA, Stein RB, Eichele G, Evans RM and Thaller C (1992): 9-cis retinoic acid is a high affinity ligand for the retinoid X receptor. *Cell* 68: 397-406.

Hoffmann B, Lehmann JM, Zhang X-k, Hermann T, Graupner G and Pfahl M (1990): A retinoic acid receptor specific element controls the retinoic acid receptor-β promoter. *J Mol Endo* 4: 1734-1743.

Hong WK, Lippmann SM, Itri LM, Karp D, Lee JS, Bvers R, Schantz SP, Kramer AM, Lotan R, Peters LJ, Dimery IW, Brown WB and Goepfert H (1990): Prevention of second primary tumors with isotretinoin in squamous-cell carcinoma of the head and neck. *N Engl J Med* 1323: 795-801.

Huang M, Ye YC, Chen SR, Chai JR, Lu JX, Zhao L, Gu LJ and Wang ZY (1988): Use of all-trans retinoic acid in the treatment of acute promyelocytic leukemia. *Blood* 172: 567-572.

Husmann M, Hoffmann B, Stump DG, Chytil F and Pfahl M (1992): A retinoic acid response element from the rat CRBPI promoter is activated by an RAR/RXR heterodimer. *Biochem Biophys Res Commun* 1187: 1558-1564.

Issemann I and Green S (1990): Activation of a member of the steroid hormone receptor superfamily by peroxisome proliferators. *Nature* 1347: 645-649.

Karin M (1990): The AP-1 complex and its role in transcriptional control by protein kinase C. In: *Molecular Aspects of Cellular Regulation* 6: 143-161.

Kliewer SA, Umesono K, Mangelsdorf DJ and Evans RM (1992): Retinoid X receptor interacts with nuclear receptors in retinoic acid, thyroid hormone and vitamin D$_3$ signaling. *Nature* 1355: 446-449.

Kliewer SA, Umesono K, Noonan DJ, Heyman RA and Evans RM (1992): Convergence of 9-cis retinoic acid and peroxisome proliferator signaling pathways through heterodimer formation of their receptors. *Nature* 358: 771-774.

Kliewer SA, Umesono K, Heyman RA, Mangelsdorf DJ, Dyck JA and Evans RM (1992c): Retinoid X receptor-COUP-TF interactions modulate retinoic acid signaling. *Proc Natl Acad Sci USA* 89: 1448-1452.

Krust A, Kastner PH, Petkovich M, Zelent A and Chambon P (1989): A third human retinoic acid receptor, hRAR-γ. *Proc Natl Acad Sci USA* 86: 5310-5314.

Lehmann JM, Hoffmann B and Pfahl M (1991). Genomic organization of the retinoic acid receptor γ gene. *Nucl Acids Res* 19: 573-578.

Lehmann JM, Jong L, Fanjul A, Cameron JF, Lu XP, Haefner P, Dawson MI and Pfahl M (1992):. Retinoids selective for retinoid X receptor response pathways. *Science* 258: 1944-1946.

Lehmann JM, Zhang XK, Graupner G, Hermann T, Hoffmann B and Pfahl M (1993): Formation of RXR homodimers leads to repression of T₃ response: hormonal cross-talk by ligand induced squelching. *Mol Cell Biol* (in press).

Leid M, Kastner P, Lyons R, Nakshatri H, Saunders M, Zacharewski T, Chen J-Y, Staub A, Garnier J-M, Mader S and Chambon P (1992): Purification, cloning, and RXR identity of the HeLa cell factor with which RAR or TR heterodimerizes to bind target sequences efficiently. *Cell* 68: 377-395.

Leroy P, Krust A, Zelent A, Mendelsohn C, Garnier J-M, Kastner P, Dierich A and Chambon P (1991): Multiple isoforms of the mouse retinoic acid receptor α are generated by alternative splicing and differential induction by retinoic acid. *EMBO J* 10: 59-69.

Levin AA, Sturzenbecker LJ, Kazmer S, Bosakowski T, Huselton C, Allenby G, Speck J, Kratzeisen C, Rosenberger M, Lovey A and Grippo JF (1992): 9-cis retinoic acid steroisomer binds and activates the nuclear receptor RXRα. *Nature* 355: 359-361.

Lippmann SM, Kessler JF and Meyskens FL (1987): Retinoids as preventive and therapeutic anticancer agents. *Cancer Treat Rep* 71: 493-405 (part I)' 493-515 (part 2).

Mahdavi V, Chambers AP and Nadal-Ginard B (1984): Cardiac α- and β-myosin heavy chain genes are organized in tandem. *Proc Natl Acad Sci USA* 81: 2626-2630.

Mangelsdorf DJ, Ong ES, Dyck JA and Evans RM (1990): Nuclear receptor that identifies a novel retinoic acid response pathway. *Nature* 345: 224-229.

Mangelsdorf DJ, Umesono K, Kliewer SA, Borgmeyer U, Ong ES and Evans RM (1991): A direct repeat in the cellular retinol-binding protein type II gene confers differential regulation by RXR and RAR. *Cell* 66: 555-561.

Mangelsdorf DJ, Borgmeyer U, Heyman RA, Zhan JY, Ong ES, Oro AE, Kakizuka A and Evans RM (1992): Characterization of three RXR genes that mediate the action of 9-cis RA retinoic acid. *Genes Dev* 6: 329-344.

Marks MS, Hallenbeck PL, Nagata T, Segars JH, Appella E, Nikodem VM and Ozato K (1992): H-2RIIBP (RXRβ) heterodimerization provides a mechanism for combinatorial diversity in the regulation of retinoic acid and thyroid hormone responsive genes. *EMBO J* 11: 1419-1435.

Miyajima N, Kadowaki Y, Fukushige S-I, Shimizu S-I, Semba K, Yamanashi Y, Matsubara K-I, Toyoshima K and Yamamoto T (1988): Identification of two novel members of erbA superfamily by molecular cloning: the gene products of the two are highly related to each other. *Nucleic Acids Res* 16: 11057-11074.

Mlodzik M, Hiromi Y, Weber U, Goodman CS and Rubin GM (1990): The drosophila *seven-up* gene, a member of the steroid receptor gene superfamily, controls photoreceptor cell fates. *Cell* 60: 211-224.

Murray MB and Towle HC (1989): Identification of nuclear factors that enhance binding of the thyroid hormone receptor to a thyroid hormone response element. *Mol Endocrinol* 3: 1434-1442.

Näar AM, Boutin J-M, Lipkin SM, Yu VC, Holloway JM, Glass CK and Rosenfeld MG (1991): The orientation and spacing of core DNA-binding motifs dictate selective transcriptional responses to three nuclear receptors. *Cell* 65: 1267-1279.

Nakai A, Sakurai A, Bell GI and DeGroot LJ (1988): Characterization of a third human thyroid hormone receptor co-expressed with other thyroid hormone receptors in several tissues. *Mol Endocrinol* 2: 1087-1092.

Petkovich M, Brand NJ, Krust A and Chambon P (1987): Human retinoic acid receptor belongs to the family of nuclear receptors. *Nature* 330: 444-450.

Rosen ED, O'Donnell AL and Koenig RJ (1991): Protein-protein interactions involving erbA superfamily receptors, through the TRAPdoor. *Mol Cell Endocrinol* 78: C83-C88.

Rottman JN, Windom RL, Nadal-Ginard B, Mahdavi V and Karathanasis SK (1991): A retinoic acid-responsive element in the apolipoprotein A1 gene distinguishes between two different retinoic acid response pathways. *Mol. Cell. Biol.* 11: 3814-3820.

Sap J, Munoz A, Damm K, Goldberg Y, Ghysdael J, Leutz A, Beug H and Vennstrom B (1986): The c-erb-A protein is a high-affinity receptor for thyroid hormone. *Nature* 324: 635-640.

Sharp PA (1992): TATA-binding protein is a classless factor. *Cell* 68: 819-821.

Thompson CC, Weinberger C, Lebo R and Evans RM (1987): Identification of a novel thyroid hormone receptor expressed in the mammalian central nervous system. *Science* 237: 1610-1613.

Tran P, Zhang X-k, Gilles S, Hermann T, Lehmann JM and Pfahl M (1992): COUP orphan receptors are negative regulators of retinoic acid response pathways. *Mol Cell Biol* 12: 4666-4676.

Umesono K, Giguere V, Glass CK, Rosenfeld MG and Evans RM (1988): Retinoic acid and thyroid hormone induce gene expression through a common responsive element. *Nature* (London) 336: 262-265.

Umesono K and Evans RM (1989): Determinants of Target Gene Specificity for steroid/thyroid hormone receptors. *Cell* 57: 1139-1146.

Umesono K, Murakami KK, Thompson CC and Evans RM (1991): Direct repeats as selective response elements for the thyroid hormone, retinoic acid and vitamin D_3 receptors. *Cell* 65: 1255-1266.

Wang L-H, Tsai SY, Cook RG, Beattie WG, Tsai M-J and O'Malley BW (1989): COUP transcription factor is a member of the steroid receptor superfamily. *Nature* (London) 340: 163-166.

Weinberger C, Thompson CC, Ong ES, Lebo R, Gruol DJ and Evans RM (1986): The c-erb-A gene encodes a thyroid hormone receptor. *Nature* 324: 641-646.

Windom RL, Rhee M and Karathanasis K (1992): Repression by ARP-1 sensitizes apolipoprotein AI gene responsiveness to RXRα and retinoic acid. *Mol Cell Biol* 12: 3380-3389.

Yao T-P, Segraves WA, Oro AE, McKeown M and Evans RM (1992): Drosophila ultraspiracle modulates ecdysone receptor function via heterodimer formation. *Cell* 71: 63-72.

Yang Yen H-F, Zhang X-k, Graupner G, Tzukerman M, Sakamoto B, Karin M and Pfahl M (1991): Antagonism between retinoic acid receptors and AP-1: implication for tumor promotion and inflammation. *New Biol* 3: 1216-1219.

Yu VC, Delsert C, Andersen B, Holloway JM, Devary OV, Näar AM, Kim SY, Boutin J-M, Glass CK and Rosenfeld MG (1991): RXRβ, a coregulator that enhances binding of retinoic acid, thyroid hormone and vitamin D receptors to their cognate response elements. *Cell* 67: 1251-1266.

Zelent A, Mendelsohn C, Kastner P, Krust A, Garnier J-M, Ruffenach F, Leroy P and Chambon P (1991): Differentially expressed isoforms of the mouse retinoic acid receptor β are generated by usage of two promoters and alternative splicing. *EMBO J* 10: 71-81.

Zhang X-K, Wills KN, Hermann T, Graupner G, Tzukerman M and Pfahl M (1991a): Ligand-binding domain of thyroid hormone receptors modulates DNA binding and determines their bifunctional roles. *New Biol* 3: 1-14.

Zhang X-k, Tran P and Pfahl M (1991b): DNA binding and dimerization determinants for TRα and its interaction with a nuclear protein. *Mol Endocrinol* 5: 1909-1920.

Zhang X-K, Wills KN, Husmann M, Hermann T and Pfahl M (1991c). Novel pathway for thyroid hormone receptor action through interaction with jun and fos oncogene activities. *Mol Cell Biol* 11: 6016-6025.

Zhang X-K, Lehmann J, Hoffmann B, Dawson MI, Cameron J, Graupner G, Hermann T and Pfahl M (1992a): Homodimer formation of retinoid X receptor induced by *9-cis* retinoic acid. *Nature* (London) 358: 587-591.

Zhang X-K, Hoffmann B, Tran P, Graupner G and Pfahl M (1992b). Retinoid X receptor is an auxiliary protein for thyroid hormone and retinoic acid receptors. *Nature* (London) 355: 441-446.

RECEPTOR STRUCTURE AND ASSOCIATED PROTEINS

HEAT SHOCK PROTEINS AND THE CYTOPLASMIC-NUCLEAR TRAFFICKING OF STEROID RECEPTORS

William B. Pratt
Lawrence C. Scherrer
Department of Pharmacology
The University of Michigan Medical School

INTRODUCTION

Transcription factors, such as steroid receptors, must move through the cytoplasm to the nucleus and subsequently within the nucleus in a precisely targeted manner. Experiments carried out twenty years ago in intact cells exposed to metabolic inhibitors suggested that glucocorticoid receptors (GR) that have entered the nucleus (then defined as the low speed particulate fraction of the ruptured cell) are subsequently returned to the cytoplasm in an energy-dependent cycle (Munck et al, 1972; Ishii et al, 1972). This *receptor cycle* has been studied in detail in the laboratory of Allan Munck (Munck and Holbrook, 1984; for review, see Orti et al, 1992), and the shuttling of steroid receptors into and out of the nucleus in intact cells has recently been directly demonstrated by Guiochon-Mantel et al (1991) and Chandran and DeFranco (1992) using immunofluorescence to visualize the receptor protein.

The mechanisms determining such receptor trafficking and its control have not been defined. Indeed, until the 1980s, steroid receptors were often considered as soluble molecules that passively entered nuclei by

STEROID HORMONE RECEPTORS
V. K. Moudgil, Editor
© 1993 Birkhäuser Boston

diffusion and were trapped there by binding to fixed nuclear components, in particular to DNA (for review, see Walters, 1985). However, a variety of observations made with other nuclear proteins indicate that protein movement to and into the nucleus involves an organized system of protein transfer requiring specific signal sequences. It seems rather clear that nuclear proteins of M_r larger than ~40,000 are actively transported across the nuclear envelope through the nuclear pores (Feldherr et al, 1984). Whether the majority of nuclear proteins, including steroid receptors, arrive at nuclear pores via diffusion or via unidirectional transport along cytoskeletal-based movement systems (e.g., Vale, 1987) is as yet unknown. The mechanism of protein movement within the interior of the nucleus is also undefined.

One thing that is clear is that, in those cells where the hormone-free GR is located in the cytoplasm, binding of steroid initiates rapid transfer of the receptor to and within the nucleus. It is highly unlikely this movement reflects a random process, because the receptors must be bound to the appropriate response elements in the genome within a few minutes after steroid addition. Although the mechanism of receptor trafficking is unknown, the receptors must be utilizing a trafficking system that is shared with a wide variety of proteins. Thus, the steroid receptors may be very useful tools for studying the general process of protein transport in the cytoplasm and within the nucleus.

In 1985, it was reported that hormone-free steroid receptors are associated with the 90 kDa heat shock protein hsp90 (Sanchez et al, 1985; Schuh et al, 1985; Catelli et al, 1985), and it was subsequently shown that binding of steroid promotes receptor dissociation from hsp90 with its simultaneous transformation to the DNA binding state (Sanchez et al, 1987; for review, see Pratt, 1987). Recently, coimmunoadsorption studies with progesterone (Kost et al, 1989; Smith et al, 1990a; Renoir et al, 1990) and glucocorticoid (Bresnick et al, 1990; Sanchez et al, 1990b) receptors suggest that the receptor-hsp90 complex is a core unit derived from a larger heteromeric complex that also contains hsp70, hsp56 and some other proteins of unidentified function (for review, see Pratt 1990).

The heat shock protein heterocomplex exists in cytosols independent of steroid receptors (Sanchez et al, 1990a; Perdew and Whitelaw, 1991; Tai et al, 1992) and the three heat shock proteins in the complex are thought to be involved in protein folding/unfolding and protein trafficking in the cell (for review, see Pratt et al, 1992a). It has been shown that hormone-free steroid receptors can be attached to this heterocomplex in an ATP-dependent manner under cell-free conditions (Smith et al, 1990b; Scherrer et al, 1990; Smith et al, 1992; Hutchison et al, 1992b) with accompanying reversal of receptors from the transformed to the untransformed state (Pratt et al, 1992a). For a detailed discussion of hsp90, hsp70, hsp56 and other proteins in the receptor heterocomplex, the stoichiometry of the receptor-heat shock protein heterocomplex, and inactivation of receptor function by hsp90, the reader is referred to a comprehensive review (Pratt, 1993). Recently, we have made the suggestion that this heat shock protein heterocomplex may function as a transport particle (a *transportosome*) (see Fig. 1) to which the steroid receptors remain attached with high affinity while they undergo trafficking within the cell (Pratt, 1992).

In this paper, we will discuss several aspects of receptor localization and movement in the cell with a special view to linkages of phenomena to receptor association with the heat shock protein complex. This paper is not an exhaustive review; rather, our intention is to be speculative and perhaps heuristic.

Receptor Localization and Heat Shock Proteins

Sex hormone receptors. The steroid/thyroid receptor family can be divided into three general classes with respect to their localization in the hormone-free cell (Table I). After their synthesis in the cytoplasm, unliganded progesterone (PR) (Welshons et al, 1984; Perrot-Applanat et al, 1985), estrogen (ER) (King and Green, 1984) and androgen (AR) (Husmann et al, 1990) receptors move directly to the nucleus where they remain in a loosely-bound, inactive state until they are bound by hormone, whereupon they move into high affinity association with nuclear components. The hormone-free receptors are considered to be in loose association with the nucleus because they are recovered almost entirely in the cytosolic fraction upon cell rupture. When cells are broken in the presence of molybdate ion, the sex hormone receptors are recovered entirely in the 9S, hsp90-bound form (see Pratt 1987, 1990). In mammalian systems, the sex hormone receptor complexes also contain hsp70 (Onate et al, 1991; Scherrer et al, 1993) and hsp56 (Renoir et al 1990). In the chicken, hsp70 is clearly present in the PR heterocomplex (Kost et al, 1989; Smith et al 1990a), but because of failure of cross-reactivity of antibody with avian proteins, an analog of hsp56 has not yet been identified.

We have used the term *docking* to describe the relationship between the steroid receptors and the heat shock protein complex (Pratt, 1990). That is, the receptors remained "docked" to the hsp90-containing structure until binding of hormone triggers dissociation from the hsp90 component,

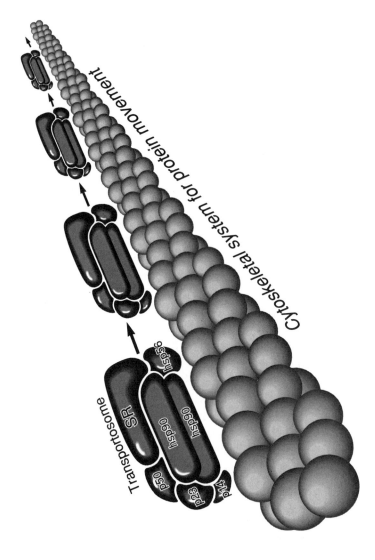

FIGURE 1. The steroid receptor heteroprotein complex as it may function as a transportosome in the intact cell. SR=steroid receptor. Proteins in the complex other than hsp90 and hsp56 are discussed in Pratt (1993a; 1993b). As discussed in the text, several receptors that are nuclear in their hormone-free state (e.g., PR, ER) are associated with hsp70 as well. From Pratt (1992), Fig. 1, with permission.

TABLE 1. Classification of steroid/thyroid receptors by localization of the hormone-free receptor and by hsp90 binding

Class I
  Retinoic acid
  Thyroid hormone
  Vitamin D
	Receptors do *not* form a stable complex with hsp90
	Unliganded receptors move directly to tight nuclear binding sites

Class II
  Glucocorticoid
  Mineralocorticoid
	Receptors form a stable complex with hsp90
	In hormone-free cells, receptor is retained in a *cytoplasmic-nuclear* "docking" complex with hsp90
	hsp90 is required for high affinity steroid binding conformation

Class III Progestin	Receptors form a stable complex with hsp90 In hormone-free cells, receptor is retained in a *nuclear* "docking" complex that is recovered in the cytosolic fraction hsp90 is *not* required for steroid binding conformation

somehow permitting progression of the receptor to high affinity nuclear binding sites where the primary events in transcriptional activation occur (Pratt, 1992). Although receptors recovered from steroid-treated cells are 4S and not associated with hsp90, it is not known whether or not the receptor completely dissociates from the heat shock protein complex in the intact cell, or whether its association with hsp90 is merely *loosened up*, permitting engagement with a nuclear localization signal to mediate its terminal transfer to sites of transcriptional enhancement. In this *docking model*, the receptor is inactive until it is transferred from the docking complex to its ultimate site of action.

The GR. The GR differs from the sex hormone receptors in that it is usually localized to the cytoplasm in hormone-free cells and the receptors shift to the nucleus upon exposure to hormone (Papamichail et al, 1980; Govindan, 1980; Antakly and Eisen, 1984; Fuxe et al, 1985; Wikstrom et al, 1987; Qi et al, 1989). Despite the fact they are cytoplasmic, glucocorticoid receptors, like the nuclear sex hormone receptors, are recovered from hormone-free cells in a heterocomplex with hsp90 and hsp56 (Tai et al, 1986; Renoir et al, 1990; Bresnick et al, 1990; Rexin et al, 1991; Alexis et al, 1992; Rexin et al, 1992). Given that both unliganded receptors that are cytoplasmic and unliganded receptors that are nuclear are recovered in the cytosolic fraction in association with hsp90 and hsp56, it is reasonable to suggest that unliganded receptors remain in association with this heat shock protein heterocomplex while they are transported from their cytoplasmic sites of synthesis to the nucleus.

In most cases, hsp70 has not been found to be a component of the GR heterocomplex (Bresnick et al, 1990; Rexin et al, 1991;; Alexis et al, 1992; Rexin et al, 1992). However, in CHO cells that stably overexpress the mouse GR (Hirst et al, 1990), we found that hsp70 was present in the immunoadsorbed GR heterocomplex (Sanchez et al, 1990b). For reasons that are unknown, the mouse GR is entirely nuclear in hormone-free CHO cells (Sanchez et al, 1990b; Martins et al, 1991). It is interesting that, like the nuclear sex hormone receptors, the nuclear mouse GR in CHO cells is associated with hsp70, whereas the mouse GR in L cells is cytoplasmic and not associated with hsp70 (Sanchez et al, 1990b). This has caused us to speculate that the presence of receptor-associated hsp70 is in some way related to arrival of steroid receptors in the nucleus (Sanchez et al, 1990b).

Members of the hsp70 protein family have been implicated in the translocation of proteins across membranes of the endoplasmic reticulum and mitochondria (Chirico et al, 1988; Deshaies, 1988). It is thought that hsp70 binds to hydrophobic regions of proteins to facilitate their unfolding and that certain proteins have to be maintained in an unfolded state to cross organellar membranes (for review, see Rothman, 1989). The association of steroid receptors that are in a nuclear docking complex with a complex containing hsp70, (as well as hsp90 and hsp56) leads us to speculate that hsp70 could play a role in maintaining steroid receptors in an unfolded state for passage across the nuclear membrane (Sanchez et al, 1990b). In this respect, it is interesting to note that Shi and Thomas (1992) have recently shown an hsp70 requirement for transport of nucleoplasmin into HeLa cell nuclei. Another possibility is that hsp70 associates with the receptor complex at the termini of nuclear transport pathways, being somehow required for the terminal step in the receptor journey to sites of transcriptional activation.

The retinoic acid/thyroid hormone/vitamin D group of receptors. In contrast to the steroid receptors, the evolutionarily more primitive members of the steroid/thyroid hormone family (Class I in Table I) are transported directly from their cytoplasmic sites of synthesis into tight association with the nucleus in hormone-free cells (Samuels et al, 1974; Spindler et al, 1975; Casanova et al, 1984; Nervi et al, 1989). It is thought that in their unliganded state, these receptors are transferred directly to appropriate response elements in the genome, and in some cases, they may produce transcriptional inhibition in their unliganded form. We have found that thyroid hormone (Dalman et al, 1990) and retinoic acid (Dalman et al, 1991b) receptors do not form stable complexes with hsp90. The failure of retinoic acid and thyroid hormone receptors to bind with high affinity to the heterocomplex is consistent with the behavior of these receptors in intact cells where they fail to be retained in docking complexes that are recovered in the cytosolic fraction upon cell rupture.

It is obvious that, if the heat shock protein complex is involved in protein transport, its role is not restricted to steroid receptors, pp60^{src}, and perhaps a few other proteins that attach to it in a particularly stable manner. This is inferred from the fact that hsp90 is a highly abundant and conserved protein and its function in the cell must be a general one that is necessary to the function of many proteins. For example, if hsp90 is important for the transport of steroid receptors to the nucleus, one would expect that it would also be important for the transport of thyroid hormone and retinoic acid receptors. In this respect, it may be important to note that under rapid and gentle conditions of immunoadsorption, we can see some hsp90 coimmunoadsorb with the retinoic acid receptor, implying that a weak complex may exist. Members of the hsp70 family have been shown to interact co-translationally with many proteins (Beckman et al, 1990). We have shown that the GR binds to hsp90 at the termination of receptor translation *in vitro* (Dalman et al, 1989), and we suspect that many proteins may associate with the hsp90 component of the heat shock protein complex at the termination of their translation in the intact cell.

The steroid receptors may have exploited a general role of hsp90 as a protein chaperon component of the protein transport machinery of the cell. In such a model, most proteins would be bound to hsp90 much more weakly than steroid receptors. In binding with high affinity to hsp90 and remaining docked to the heat shock protein complex, the steroid receptors have evolved a method of signal transduction in which their dissociation from this primitive and necessary component of the protein transport system is under hormonal control. Thus, in contrast to thyroid hormone, retinoic acid and vitamin D receptors, steroid receptors are only able to interact with their response elements when the hormone is present. In essence, docking to the heat shock protein complex eliminates any effect of the unbound receptor on transcription, allowing the hormone to turn on the receptor like a switch; that is, from no activity to transcriptional enhancement.

Nuclear Localization Signals and Heat Shock Proteins

The signals. Nuclear proteins larger than relative molecular mass of ~40,000 appear to require a nuclear localization signal (NLS) for passage through the nuclear pores. One of the best characterized nuclear localization signals is that of the SV40 large T antigen (Pro-Lys-Lys-Lys-Arg-Lys-Val) (Kalderon et al, 1984a). When attached to a normally cytoplasmic protein, such as pyruvate kinase, this sequence is sufficient to cause nuclear localization (Kalderon et al, 1984b). Sequences similar to the nuclear localization signal of SV40 large T antigen determine nuclear localization of other karyophilic viral proteins (Roberts et al, 1986).

Using the same type of fusion protein approach employed with SV40 large T antigen, Picard and Yamamoto (1987) identified two nuclear localization signals in the COOH-terminal half of the GR. One signal (NL1) maps to a short segment located just to the COOH-terminal side of the DNA binding domain. NL1 contains a basic sequence (Thr-Lys-Lys-Lys-Ile-Lys-Gly) that is very much like the nuclear localization signal of the SV40 large T antigen. When fused to β-galactosidase, NL1 acts constitutively. A second localization signal (NL2) is located within the hormone binding domain (HBD) of the receptor and is fully hormone-dependent. In the intact wild-type GR, both nuclear localization signals are hormone-dependent, accounting for the cytoplasmic location of the GR in most hormone free cells. However, in the CHO cells overexpressing the mouse GR, NL1 must be constitutively active, permitting total nuclear localization of the unliganded receptor (Sanchez et al, 1990b).

The PR has a constitutive NLS that is homologous to that of the SV40 large T antigen and is located in roughly the same region of the receptor as NL1 of the GR (Guiochon-Mantel, 1989). This signal accounts for the nuclear localization of the unliganded wild-type PR. When it is deleted, the PR is cytoplasmic in the absence of steroid, but the receptor moves to the nucleus and is transcriptionally active in the presence of progesterone because of a second hormone-dependent NLS. In contrast to the GR where the hormone-dependent NL2 is located in the HBD, the hormone-dependent NLS of the PR is located in the DNA binding domain adjacent to the second zinc finger. Similar nuclear localization signals consisting of positively charged amino acids have been identified to the COOH-terminal side of the DNA binding domain in androgen (Simental et al, 1991) and estrogen (Picard et al, 1990) receptors.

Regulation of NLS. The ability of the GR HBD to regulate the activity of NL1 was nicely demonstrated in the fusion protein experiments of Picard and Yamamoto (1987). They showed that β-galactosidase fused to a segment of the rat GR containing the DNA binding domain and NL1 was constitutively nuclear, but a fusion protein that contained the HBD as well as NL1 was cytoplasmic in the hormone-free state and moved to the nucleus upon addition of hormone. It is well established that hsp90 binds to the GR and the PR at a binding site located entirely within the HBD (Pratt et al, 1988; Denis et al, 1988; Howard et al, 1991; Dalman et al, 1991a; Cadepond et al, 1991; Schowalter et al, 1991; Chakraborti and Simons, 1991). To test whether there is a relationship between hsp90 binding and the hormonal regulation of the GR NL1, we transfected COS cells with cDNAs for fusion proteins containing β-galactosidase and portions of the GR (Scherrer et al, 1993). We showed that only those fusion proteins whose cellular localization was shown by Picard and Yamamoto (1987) to be under hormonal control were bound to hsp90. We also found that the presence of the GR HBD prevented reaction of fusion proteins with an antibody to NL1 under conditions where the fusion protein was bound to hsp90 but not under conditions where it was dissociated from hsp90. These observations support a model in which the HBD of the GR confers hormonal control onto the cellular localization of β-galactosidase-GR fusion proteins by conferring hormone-regulated binding to hsp90.

In contrast to the GR, the ER lacks a second NLS within the HBD. Picard et al (1990) demonstrated that the HBD of the ER fails to regulate nuclear localization signals, although it is clear that it can regulate other functions. The constitutive activity of the NLS accounts for the nuclear localization of the ER in the hormone-free cell. When the HBD of the GR is replaced with the HBD of the ER, the resulting GR.ER fusion protein is located in the nucleus in the absence of hormone (Picard et al, 1990). Thus, the GR HBD can inactivate the NL1 of the GR, but why the ER HBD fails to inactivate either its own nuclear localization signal in the wild-type ER or the NL1 of the GR in the GR.ER fusion protein is unknown.

Picard et al (1988) first demonstrated that the HBD of the GR acts as a transferable regulatory cassette that can confer hormonal control onto chimeric proteins. In that work, the transcriptional activating activity of the adenovirus EIA gene product was inactivated by fusion to the HBD of the GR and the inactivation was reversed by hormone. This general inactivating function is common to both GR and ER hormone binding domains and can be conferred to other heterologous fusion proteins (Eilers et al, 1989; Umek et al, 1991; Superti-Furga et al, 1991; Burk and Klempnauer, 1991). In the GR.ER fusion protein studied by Picard et al (1990), where the activity of NL1 was not inactivated by the ER HBD, the transcriptional activating activity of the GR was inactivated and the inactivation was relieved by binding of β-estradiol. As the ER HBD binds hsp90 (Scherrer et al, 1993) and confers hormonal regulation onto fusion proteins without regulating NL1, any general inactivation of fusion protein function is not simply a function of intracellular localization.

Receptor Trafficking Within the Nucleus

Questions. It seems clear that an NLS is required for uptake of receptors into the nucleus, but does the signal function after the receptor has traversed the nuclear pore? At the moment there is no answer to this important question; however, there is some information upon which to base speculation regarding intranuclear pathways of movement. Recently, we have used the technique of confocal scanning microscopy (Martins et al, 1991) to ask several questions regarding unliganded receptors. Is their distribution random, as indicated by a diffuse immunofluorescence throughout the nucleus? Is the distribution nonrandom, with the existence of regions that do not contain receptors? Are unliganded receptors that are retained in the inactive *docking* state located at the nuclear periphery and do they move after hormonal activation? Are the loosely bound docking complexes located near the high affinity acceptor sites, which are presumably the sites where the primary events of transcriptional activation occur?

Confocal observations. For the confocal study we used the stably transfected CHO cell lines, which yield a strong nuclear immuno-fluorescence signal for the unliganded GR (Sanchez et al, 1990b). We found that the unliganded, wild-type receptor was distributed in a mottled, nonrandom manner throughout all planes of the nucleus but was totally excluded from nucleoli (Martins et al, 1991). Because there is no preferential localization of the receptor at the periphery versus the center of the nucleus, the unliganded receptor must move within the nucleus after passage through the nuclear pores. As the intranuclear distribution is nonrandom (i.e., mottled rather than diffuse), either limits are imposed on intranuclear diffusion or, more likely, the receptor is transported to specific sites within the nucleus via an intranuclear transport system. Like cytoplasmic GR, the nuclear mouse GR in the CHO cells is bound to hsp90 (Sanchez et al, 1990b). Thus we would speculate that the GR is bound to the heat shock protein heterocomplex, both while it moves through cytoplasm and while it moves through the nucleus. It is reasonable to propose that the nuclear portion of the receptor movement system contains components that are common to the extranuclear receptor movement system.

In the confocal study, we did not see any change in receptor distribution on exposure to hormone. However, because of the general mottled distribution pattern, one cannot conclude that there is no intranuclear movement as the receptor proceeds from a loosely bound docking state to high affinity association with the nucleus. Perrot-Applanat et al (1986) used a monoclonal antibody against the PR and the protein A-gold technique to show that the unliganded PR in uterine stromal cells was randomly scattered over clumps of condensed chromatin, but after hormone administration, it was mainly detected in border regions between condensed chromatin and nucleoplasm. Such a local redistribution would not be detected in our confocal approach. However, the absence of any gross movement of the overexpressed GR when hormone is added to the CHO cells suggests that the hsp90-containing docking complexes containing unliganded receptor may be located at the termini of the nuclear transport pathways for the receptor. It is possible that the termini of the nuclear transport pathways are located very close to or even at the loci where the primary events in transcriptional activation occur.

Speculations. It seems entirely reasonable to speculate that the steroid hormone receptors bind immediately at the termination of their translation to the hsp90 component of a transport particle (Fig. 1), and if their nuclear localization signal is "engaged" when they are in the unliganded state (e.g., sex hormone receptors, overexpressed GR in CHO cells), the receptors are moved along a pathway that proceeds into and throughout the nucleus. In the absence of hormone, the receptors remain docked at the termini of the nuclear transport system, which are located close to the ultimate sites of receptor action on transcription. In those cells where the GR is cytoplasmic, the nuclear localization signal does not become (for whatever reason) "engaged" with the protein that determines directionality of movement by locking the heat shock protein heterocomplex into the movement system. Thus, these receptors remain docked in the cytoplasm until their interaction with the hsp90 component of the system is "loosened" by binding of hormone to the HBD. Engagement of the NLS with the NLS-binding protein then permits movement to the ends of the pathway in the nucleus where it (like the steroid-activated receptors that had been docked at these sites in their unliganded state) can undergo terminal transfer from the protein transport pathway to sites of transcriptional enhancement. If the transportosome model of Fig. 1 is correct, then we would predict that the NLS-binding protein is itself an integral component of the transport particle.

The Heat Shock Protein Heterocomplex and Protein Movement to the Plasma Membrane

$pp60^{v\text{-}src}$. The first protein shown to be associated with hsp90 was $pp60^{v\text{-}src}$, a membrane-associated tyrosine kinase that induces oncogenic transformation of cells infected with Rous sarcoma virus (for review, see Brugge 1986; Jove and Hanafusa, 1987). As soon as it is translated, $pp60^{v\text{-}src}$ becomes associated with both hsp90 and a 50 kDa protein of unknown function, p50 (Brugge et al, 1981; Courtneidge and Bishop, 1982). The heterocomplex of $pp60^{v\text{-}src}$-hsp90-p50 is very much like the heterocomplex of GR-hsp90-hsp56 in that both heterocomplexes can be formed by reticulocyte lysate in an ATP-dependent manner under cell-free conditions (Scherrer et al, 1990; Hutchison et al, 1992b; Hutchison et al, 1992a) and both heterocomplexes are stabilized by molybdate, vanadate and tungstate (Hutchison et al, 1992c).

Although it is not yet known whether or not p50 is a heat shock protein, it is a component of the cytosolic heat shock protein heterocomplex. This was established when Perdew and Whitelaw (1991) identified a 50 kDa protein (in addition to hsp56, hsp70 and hsp90) that coimmunoadsorbed from rat liver cell cytosol with monoclonal antibodies directed against hsp90. Subsequently, Whitelaw et al (1991) raised an antibody to this p50 and demonstrated by both immunoreactivity and peptide cleavage that this was the same as the 50 kDa phosphoprotein that was the long established component of the $pp60^{src}$-hsp90 heterocomplex. When $pp60^{v\text{-}src}$ or GR heterocomplexes are reconstituted in the cell-free reticulocyte lysate system, the $pp60^{v\text{-}src}$ or GR become attached to a complex that contains both hsp56 and p50, as well as hsp70 and hsp90. When the native GR heterocomplex is isolated from L cells or WEHI cells, however, hsp56 is present but p50 is not (Bresnick et al, 1990; Whitelaw et al 1991). Similarly, results from methionine labeling studies suggest that native pp60^{v-src} heterocomplexes isolated from rat 3Y1 cells contain p50 but not hsp56 (R. Jove, personal communication). As the other proteins are common components of the native heterocomplexes for both pp60^{v-src} and the GR, we would speculate that p50 and hsp56 may in some way be responsible for determining the direction of movement of the two proteins in the cell.

Bidirectional movement of proteins bound to hsp90.   Figure 2 presents a diagram of pp60^{v-src} and GR movement in the cell. As cited above, it is thought that both proteins bind to hsp90 either co-translationally or immediately at the termination of their translation and that they then undergo vectorial movement in the cell while attached to hsp90. In the case of pp60^{v-src} the heterocomplex proceeds to the cell periphery, where it dissociates from hsp90 and is attached to the plasma membrane (Brugge, 1986).   In contrast, the GR and other steroid receptors proceed to the nucleus (and probably within the nucleus) while attached to hsp90.

How is the direction of this protein movement determined? Systems for inward and outward movement of organelles along cytoskeletal scaffolds have been identified (for review, see Vale 1987) and protein transport particles carrying an inward moving protein such as the GR or an outward moving protein like pp60^{v-src} could attach to such a system via a *direction-determining protein (DDP)*, which is the role we speculate above for hsp56 (inward movement) and p50 (outward movement). The observation that heterocomplexes created *in vitro* contain both hsp56 and p50, whereas native heterocomplexes contain either one protein or the other, leads us to suggest that a protein undergoing intracellular trafficking may be attached originally to a complex containing both DDPs, and when one becomes engaged by virtue of interacting with a directional signal (e.g., a nuclear localization signal), the binding affinity of the heterocomplex for the opposite DDP may be decreased such that it dissociates during heterocomplex isolation.

A major question that arises in considering bidirectional movement is whether two signals or one signal are sufficient. In the first model there would be an *outward movement signal* (OMS) in addition to the inward movement signaling role performed by the NLS. This possibility has been raised recently by Chandran and DeFranco (1992) who demonstrated bidirectional movement of the PR into and out of the nuclei of transient heterokaryons, whereas no outward movement of the SV40 large T antigen was seen under the same conditions. The authors felt it likely that the PR must possess a *nuclear export signal that is distinct from* the nuclear import signal (i.e., an NLS). To our knowledge no signal for

228

outward movement has been identified to date, but it is reasonable that some general property of molecules must signal their choice of movement toward the cell periphery. Assuming for a moment the possibility that common elements exist between protein movement systems in the nucleus and those in the cytoplasm, then there could be a similarity between nuclear export signals for proteins, such as steroid receptors, that shuttle back and fourth between the nucleus and the cytoplasm and outward

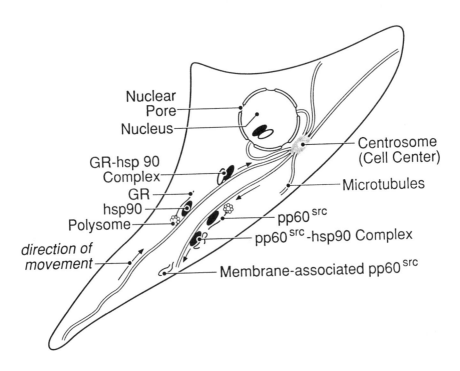

FIGURE 2. Bidirectional movement of proteins bound to hsp90. Binding of GR or pp60^{src} appears to occur at the time of their translation in the cell. It is proposed that the hsp90-containing heat shock protein complexes function as *transportosomes* to which the proteins remain bound during cytoplasmic transport. Systems exist for transport in both directions (i.e., toward the nucleus and away from it) along cytoskeletal protein scaffolds. The direction of movement is determined by targeting signals inherent to the transported protein, such as the nuclear localization signals of steroid receptors. Although not indicated in the diagram, it is proposed that a transport system also exists for carrying the receptor within the nucleus. GR=glucocorticoid receptor. From Pratt (1992), Fig. 3, with permission.

movement signals that direct proteins, such as pp60$^{v\text{-}src}$, directly from their cytoplasmic sites of synthesis through the cytoplasm to the cell periphery.

An alternative to the two signal mechanism is that there is only inward movement signalling which acts as by an override mechanism. That is, all proteins that bind to hsp90 at translation may be bound to a heat shock protein heterocomplex that moves automatically in an outward direction unless they possess an NLS, which by engaging an NLS-binding protein mediates binding of the complex to the inward movement pathway. If this were the case, then one might expect that inactivation of the NLS in a steroid receptor would result in a concentration of the receptor towards the periphery of the cell. This does not seem to be the case with the GR. At a gross level of observation, the unliganded GR in which both NL1 and NL2 are inactive appears to have a rather general distribution throughout the cytoplasm (e.g., Picard and Yamamoto, 1987; Qi et al, 1989). On the other hand, unliganded vitamin D receptors in fibroblasts cultured in the absence of serum and phenol red shift from their normal constitutive nuclear localization to the cell periphery over a period of several days (Barsony et al, 1990). When calcitriol is added, the receptors rapidly migrate to the nucleus.

Regardless of whether two signals are needed for a single molecule to engage in bidirectional movement or whether a NLS alone is sufficient, with the transportosome model, one can easily conceive of a class of transcription factors that shuttle information from the plasma membrane to the nucleus. For example, a protein that regulates transcription could be bound co-translationally to hsp90 in the transportosome complex, and as happens with the glucocorticoid receptor, its NLS could be inactivated by hsp90. A second targeting signal (or simply the lack of dominant NLS function or some general characteristic of the molecule) might then direct its passage to the end of the outward movement pathway. Thus, the protein attached to the heat shock protein complex would move to the inner surface of the plasma membrane where it remains *docked* to the transportosome. A subsequent event, such as phosphorylation by a hormone-bound plasma membrane receptor, could then favor a conformational change that "loosens" its association with the hsp90 component of the transportosome. Now, much as happens with the GR after steroid binding, the nuclear localization signal is derepressed, permitting the heterocomplex-bound protein to shuttle rapidly to the nucleus where it activates transcription.

This kind of hsp90-regulated shuttling mechanism could fill a large gap in our understanding of how the interaction of peptide hormones and growth factors with plasma membrane receptors can directly regulate gene transcription. We have begun to explore this possibility using the c-Raf-1 protein, which is a 74 kDa serine/threonine kinase implicated in normal mitogenic signal transduction and which can be oncogenically activated by mutation (for review, see Li et al, 1991). The c-Raf-1 protein is under tight control by several receptors, including those for PDGF and EGF. In response to hormone interaction with a plasma membrane receptor it undergoes transfer from the plasma membrane to the perinuclear region (Rapp et al, 1988). As we reported at a recent symposium, we have demonstrated that c-Raf-1 can be bound to the heat shock protein heterocomplex by the ATP-dependent, K^+-dependent heterocomplex assembly system in reticulocyte lysate, and as shown with the steroid receptors and $pp60^{v-src}$, its presence in the complex is stabilized by molybdate (Hutchison et al, 1993; Stancato et al, 1993). Thus, the possibility of an hsp90-regulated shuttling mechanism permitting direct communication between membrane signalling and nuclear events is perhaps an entirely reasonable speculation.

Do Steroid Receptors Move to and Within the Nucleus Along Cytoskeletal Pathways?

Although it has not been in any way proven, it is reasonable to assume that receptors shuttle along tracks in the cytoplasm and within the nucleus. Both microtubule- and microfilament-based systems for bidirectional cytoplasmic transport of organelles have been described (Vale 1987) and shuttling of the Nopp140 protein between the cytoplasm and nucleolus occurs on discrete nuclear tracks of unknown composition (Meier and Blobel, 1992). There is certainly considerable precedent for the organized transport of proteins into and out of the nucleus (for review, see Silver, 1991; Nigg et al, 1991) but the mechanisms underlying protein movement are in general poorly understood. Thus, it is not surprising that very little is known about the movement of steroid receptors to and within the nucleus. However, as proteins that undergo targeted movement within the cell, the steroid receptors themselves are powerful tools for probing basic mechanisms of protein transport.

Evidence for localization of receptors to cytoskeleton. In indirect immunofluorescence studies, cytoplasmic steroid receptors have been generally visualized as a diffuse cytoplasmic immunofluorescence, suggesting random distribution of receptors throughout the cytoplasm. In cells that are growing on a culture dish (i.e., cells that are not round), a cytoplasmic signal can appear to be perinuclear simply because the depth of the cytoplasmic space decreases tremendously toward the periphery where the cells are attached to substrate. In the region of the cytoplasm outside the intense perinuclear signal, several investigators have noticed a trabecular appearance to the receptor fluorescence (e.g., Guiochon-Mantel et al, 1989; Wikstrom et al, 1987) consistent with a nonrandom, possibly cytoskeletal, distribution.

Akner et al (1990) published a careful study of the immunocyto-chemical localization of GR in human gingival fibroblasts, showing colocalization of the receptor with cytoplasmic microtubules. Treatment of the cells with colchicine or vinblastine caused redistribution of both the GR and tubulin to the cell periphery, with vinblastine also causing the formation of paracrystals containing both GR and tubulin. A subsequent study by the same group found the GR in human gingival fibroblasts to be colocalized with tubulin using the double immunofluorescence technique and both conventional and confocal microscopy (Akner et al, 1991). Unfortunately, in this study hormone treatment did not induce redistribution of receptor to the nucleus. The reason for this lack of receptor shift is unclear, although in our own work with different cultured cells we have seen a large variation in the percentage of the ligand-tagged GR shifting from the cytoplasmic fraction to the nuclear fraction. From as little as ten percent to essentially all of the receptor undergoes nuclear shift, with the variation apparently determined by the cell type. It is possible that the human gingival fibroblasts studied by Akner et al (1991) fall into the class of cells where the percentage of shift is too low to discern as a shift in immunofluorescence signal.

When bound by hormone, cytosolic steroid receptors in many cells move rapidly to the nucleus. In COS cells, for example, Picard and Yamamoto (1987) reported a half-life of roughly 5 minutes for NL2-mediated nuclear accumulation of a β-galactosidase fusion protein with the GR HBD. Barsony et al (1990) used a rapid microwave fixation technique that allowed temporal resolution within 15-second intervals to follow the calcitriol-dependent shift of vitamin D receptors from the cell periphery into the nucleus of serum-deprived cultured human fibroblasts. They described a fascinating response in which there is first clumping of receptors at the cell periphery within 15-45 seconds, alignment of clumps along fibrils within 30-45 seconds, perinuclear accumulation of clumps within 45-90 seconds, and intranuclear accumulation of clumps within 1-3 minutes. Treatment of cells with wheat germ agglutinin, which blocks protein transport through nuclear pores, also blocked calcitriol-dependent translocation of the receptor. The rapid movement of receptors into the nucleus certainly suggests a process of vectorial transport through the cytoplasm. Barsony et al (1990) noted that the receptor clumps aligned along cytoplasmic fibrils that were directed radially toward the nucleus, again suggesting that cytoskeletal elements may be involved in receptor transport to the nucleus.

Evidence for localization of heat shock proteins to cytoskeleton. As suggested in the transportosome model in Figure 1, it is really the whole heat shock protein complex that is thought to move along a cytoskeletal scaffold. It is totally unclear how such a complex would attach to cytoskeleton in the proposed system. A direct attachment is not implied and there could certainly be a number of intervening proteins. The implication from the model is that either directly, or more likely indirectly, at least a portion of the hsp90 and hsp56 in the cell should be bound to cytoskeleton.

As with steroid receptors, most immunofluorescence studies of hsp90 localization reveal a diffuse fluorescence throughout the cytoplasm (e.g., Carbajal et al, 1986; Lai et al, 1984). However, Collier and Schlesinger (1986) noted that hsp90 had a fibrous distribution toward the periphery of chicken embryo fibroblasts. We have used both the polyclonal antibody used by Collier and Schlesinger (1986) and the AC88 monoclonal anti-hsp90 antibody of Reihl et al (1985) to demonstrate colocalization of hsp90 immunofluorescence with tubulin immunofluorescence in interphase and mitotic cells of various types (Sanchez et al, 1988; Pratt et al, 1989; Redmond et al, 1989). Wikstrom et al (1989) have also shown colocalization of hsp90 and tubulin by immunocytochemical staining of normal human fibroblasts with monoclonal antibodies. However, using a new monoclonal antibody directed against rat hsp90, this laboratory (Akner et al, 1993) found a diffuse, predominantly cytoplasmic staining with a weak, uniform nuclear signal in human gingival fibroblasts. They noted that there was sometimes a more fibrillar pattern in the peripheral region of the cytoplasm, consistent perhaps with cytoskeletal localization. It is important that these observations be repeated in other laboratories using other monoclonal anti-hsp90 antibodies and a variety of cell fixation techniques before it can be concluded that a substantial portion of the hsp90 in the cell is associated with cytoskeleton. Part of the problem in displaying a discrete localization for hsp90 in the cytoplasm is that different monoclonal antibodies may be recognizing only free hsp90 while others are able to react with hsp90 complexed with other molecules.

It is unlikely that there is a direct protein-protein interaction of hsp90 with tubulin itself but cross-linking experiments will be most useful to see if there is binding to microtubule-associated proteins or to other cytoskeletal proteins, such as actin. The latter possibility is raised because the Yahara laboratory has shown that purified hsp90 coprecipitates with rabbit skeletal muscle actin under actin-polymerizing conditions (Koyasu et al, 1986). This binding appears to be regulated by calmodulin in a Ca^{2+}-dependent manner (Nishida et al, 1986). Miyata and Yahara (1991) have now reported that the 8S, heterocomplex form of the GR binds to filamentous actin *in vitro* whereas the transformed, hsp90-free, 4S form of the receptor does not. The authors therefore suggest that the binding of the receptor complex to the actin filaments occurs through the hsp90 moiety. The *in vitro* binding of the 8S GR to actin suggests that GR and hsp90 would localize by immunofluorescence to microfilaments and not to microtubules as was found in the studies cited above.

The cellular localization of hsp56 has been examined by indirect immunofluorescence with two antibodies. Using the EC1 monoclonal antibody in human choriocarcinoma cells, Sanchez et al (1990a) reported a diffuse cytoplasmic fluorescence of hsp56 without any obvious cytoskeletal localization. There was also some nuclear fluorescence. Using a new polyclonal antibody, Ruff et al (1992) have reported that hsp56 is localized to both cytoplasm and nucleus of LLC-PK$_1$ porcine kidney cells, with the cytoplasmic staining being fibrous in nature, suggesting a cytoskeletal localization. Hsp70 has been found to be both cytoplasmic and nuclear in distribution (e.g., Collier and Schlesinger, 1986) and to colocalize with stress fibers in heat-shocked chicken fibroblasts (Schlesinger et al, 1982). In chicken cells, a member of the hsp70 family has been reported to copurify with microtubules and intermediate filaments (Wang et al, 1980; 1981).

There is evidence that both hsp90 and hsp70 themselves undergo cytoplasmic-nuclear trafficking. Welch and Feramisco (1984), for example, showed that the amount of hsp70 in HeLa cell nuclei increased with heat stress. Similarly, Collier and Schlesinger (1986) found that significant amounts of hsp70 moved to the nucleus of chicken embryo fibroblasts during stress and restress. Although hsp90 was primarily localized to the cytoplasm in these cells, when the cells were restressed, most of the hsp90 accumulated in the nucleus within 3 hours. These observations imply that these two heat shock proteins can undergo vectorial movement, consistent with a proposed role in protein trafficking in the cell. However, recent observations by Akner et al (1992) suggest that the nuclear movement of hsp90 in heat stressed human gingival fibroblasts is very slow and does not occur on microtubules or microfilaments.

Speculations. Although it is a reasonable prediction that receptor trafficking occurs along cytoskeletal pathways, it should be clear from the above discussion that direct evidence for cytoskeletal association of both steroid receptors and the relevant proteins in the heat shock protein complex is very limited. If the receptors and/or heat shock proteins are associated with cytoskeleton, they are associated only loosely. Such a loose association might be expected for molecules undergoing trafficking but it makes direct biochemical analysis of interactions difficult. The predicted associations would be very difficult to analyze by most genetic techniques because all of the elements (i.e., heat shock proteins and cytoskeletal proteins) are highly conserved, are utilized for multiple cell functions, and are required for cell survival. Hsp70 and hsp90 are highly abundant proteins and only a fraction of the total protein may be involved in trafficking on cytoskeletal systems at any time. Although these considerations make analysis of receptor trafficking difficult, one must ask what are the alternatives to trafficking on a cytoskeletal pathway? At this time, we see no obvious alternatives.

In our proposed overall model of receptor transport, receptors bound to the heat shock protein complex via hsp90 would move to the termini of the inward movement system. A single NLS would be sufficient for determining direction of movement and progression of the receptor through the nuclear pores to multiple termini located throughout the nucleus (excluding nucleoli). It is reasonable to predict that the termini of the intranuclear transport pathways are located in spatial proximity to sites where transcription occurs, such as sites in the nuclear matrix attached to loops of DNA with inducible genes containing the response elements and appropriate transcription factors. This is inferred from the fact that at the level of confocal microscopy no obvious spatial rearrangement occurs when nuclear receptors are bound by hormone (Martins et al, 1991) and only a very local redistribution of nuclear receptors is observed by electron microscopy (Perrot-Applanat et al, 1986).

In this model, the NLS has derandomized the process of receptor movement from its site of synthesis in the cytoplasm to the termini of the transport system, but what happens now? Given the similarity of nuclear localization signals for a wide variety of proteins unrelated to the receptors, it would seem likely that many nucleophilic proteins may proceed along the same transport pathway. Possibly one could envision as many NLS-binding proteins and even inward transport pathways as there are distinct nuclear localization signals. However, it seems reasonable to us to assume that the first stage in vectorial movement, i.e., the journey to the termini of the protein transport system, is common to most or all nucleophilic proteins (but perhaps not nucleolar proteins) and utilizes a very restricted number of NLS-binding proteins.

After arrival at the nuclear termini of the transport system, the remainder of the receptor journey is totally unclear. However, at some time, nuclear proteins must diverge from one another to eventually arrive at distinct sites of action in the genome. Thus, it seems reasonable to predict that steroid receptors would enter into a very specific protein-protein interaction determining their transfer from docking complexes at the termini of the general protein transport system to more specific "acceptor sites" in the nuclear matrix. It seems also reasonable to predict that the termini of the transport system must be contiguous with the nuclear matrix, to which hormone-activated receptors become bound with high affinity (for review of steroid receptor binding to nuclear matrix, see Barrack and Coffey, 1982; Barrack, 1987). Several approaches have been taken to look for a receptor-protein interaction that occurs after receptor transport and before specific interaction with transcription factors at a subsequent stage of DNA binding. These approaches include cell-free binding of transformed receptors to isolated nuclear matrix (Barrack, 1987), binding of receptors to chromatin and purified receptor binding factors (for review, see Landers and Spelsberg, 1992), and high affinity binding of transformed receptors to cytosolic protein complexes with the hope that such interactions have elements in common with high affinity nuclear binding of receptors (e.g., Nelson et al, 1989; Scherrer and Pratt, 1993).

To date, none of these approaches have been clearly successful in developing a model cell-free system for studying a receptor-protein interaction that is involved in the conversion of steroid receptors from a loosely bound nuclear docking complex to the high affinity nuclear bound state assumed after receptor transformation by hormone in the intact cell. However, some reasonable predictions may be made about the first receptor-protein interaction after hsp90 dissociation that determines their transfer to matrix "acceptor sites". The new protein-protein interaction of the transformed receptors would have to be different from receptor dimerization, as it is difficult to see how dimerization *per se* could contribute to transfer of the receptors to their ultimate nuclear sites of action. The proposed receptor-protein interaction should not require the HBD. This follows from the fact that steroid receptors lacking the HBD often have considerable constitutive transcriptional activating activity. In that the nt[i] mutant GR is bound to hsp90 and undergoes hormone-dependent conversion to a high affinity nuclear bound state, we would predict that the entire amino-terminal half of the receptor is not required. This leaves a predicted meaningful interaction site for conversion to high affinity nuclear binding lying within the region containing the DNA binding domain and the "hinge" segment, which lies between the DNA binding and hormone binding domains. A final speculation is that the role of heat shock proteins in the process may not end with receptor dissociation from hsp90. The unfoldase activity of hsp70 could be important in the transfer of receptors to acceptor sites much as it is thought to be required for the original attachment of receptors to the hsp90 component of the heat shock protein complex (Smith et al, 1992; Pratt et al, 1992b). This might be why hsp70 has been found with nuclear receptor heterocomplexes but not with cytoplasmic GR heterocomplexes.

ACKNOWLEDGMENTS

Work in the authors' laboratory is supported by NIH grants CA28010 and DK31573.

REFERENCES

Akner G, Sundqvist KG, Denis M, Wikstrom AC and Gustafsson JA (1990): Immunocytochemical localization of glucocorticoid receptor in human gingival fibroblasts and evidence for a colocalization of glucocorticoid receptor with cytoplasmic microtubules. *Eur J Cell Biol* 53: 390-401.

Akner G, Mossberg K, Wikstrom AC, Sundqvist KG and Gustafsson JA (1991): Evidence for colocalization of glucocorticoid receptor with cytoplasmic microtubules in human gingival fibroblasts, using two different monoclonal anti-GR antibodies, confocal laser scanning microscopy and image analysis. *J Steroid Biochem Molec Biol* 39: 419-432.

Akner G, Mossberg K, Sundqvist KG, Gustafsson JA and Wikstrom AC (1992): Evidence for reversible, non-microtubule and non-microfilament-dependent nuclear translocation of hsp90 after heat shock in human fibroblasts. *Eur J Cell Biol* 58: 356-364.

Alexis MN, Mavridou I and Mitsiou DJ (1992): Subunit composition of the untransformed glucocorticoid receptor in the cytosol and in the cell. *Eur J Biochem* 204: 75-84.

Antakly T and Eisen HJ (1984): Immunocytochemical localization of glucocorticoid receptor in target cells. *Endocrinology* 115: 1984-1989.

Barrack ER (1987): Steroid hormone receptor localization in the nuclear matrix: interaction with acceptor sites. *J Steroid Biochem Molec Biol* 27: 115-121.

Barrack ER and Coffey DS (1982): Biological properties of the nuclear matrix: steroid hormone binding. *Rec Prog Horm Res* 38: 133-189.

Barsony J, Pike JW, DeLuca HF and Marx SJ (1990): Immunocytology with microwave-fixed fibroblasts shows a 1α,25-dihydroxyvitamin D_3-dependent rapid and estrogen-dependent slow reorganization of vitamin D receptors. *J Cell Biol* 111:2385-2395.

Beckman RP, Mizzen LA and Welch WJ (1990); Interaction of hsp70 with newly synthesized proteins: Implications for protein folding and assembly. *Science* 248: 850-854.

Bresnick EH, Dalman FC and Pratt WB (1990): Direct stoichiometric evidence that the untransformed M_r 300,000, 9S, glucocorticoid receptor is a core unit derived from a larger heteromeric complex. *Biochemistry* 29: 520-527.

Brugge JS (1986): Interaction of the Rous sarcoma virus protein pp60$^{v\text{-}src}$ with the cellular proteins pp50 and pp90. *Curr Top Microbiol Immunol* 123: 1-22.

Brugge JS, Erickson E and Erickson RL (1981): The specific interaction of the Rous sarcoma virus transforming protein, pp60^{src}, with two cellular proteins. *Cell* 25: 363-372.

Burk O and Klempnauer KH (1991): Estrogen-dependent alterations in differentiation state of myeloid cells caused by a v-*myb*/estrogen receptor fusion protein. *EMBO J* 10: 3713-3719.

Cadepond F, Schweizer-Groyer G, Segard-Maurel I, Jibard N, Hollenberg SM, Giguere V, Evans RM and Baulieu EE (1991): Heat shock protein 90 as a critical factor in maintaining glucocorticosteroid receptor in a nonfunctional state. *J Biol Chem* 266: 5834-5841.

Carbajal ME, Duband JL, Lettre F, Valet JP and Tanguay RM (1986): Cellular localization of *Drosophila* 83-kilodalton heat shock protein in normal, heat shocked, and recovering cultured cells with a specific antibody. *Biochem Cell Biol* 64: 816-825.

Casanova J, Horowitz ZD, Copp RP, McIntyre WR, Pascual A and Samuels HH (1984): Photoaffinity labeling of thyroid hormone nuclear receptors: influence of *n*-butyrate and analysis of the half-lives of the 57,000 and 47,000 molecular weight receptor forms. *J Biol Chem* 259: 12084-12091.

Catelli MG, Binart N, Jung-Testas I, Renoir JM, Beaulieu EE, Feramisco JR and Welch WJ (1985): The common 90-kDa protein component of non-transformed '8S' steroid receptors is a heat shock protein. *EMBO J* 4: 3131-3135.

Chakraborti PK and Simons SS (1991): Association of heat shock protein 90 with the 16 kDa steroid binding core fragment of rat glucocorticoid receptors. *Biochem Biophys Res Commun* 176: 1338-1344.

Chandran VR and DeFranco DB (1992): Internuclear migration of chicken progesterone receptor, but not SV40 large tumor antigen, in transient heterokaryons. *Mol Endo* 6: 837-844.

Chirico WJ, Waters MG and Blobel G (1988): 70K heat shock proteins stimulate protein translocation into microsomes. *Nature* 332: 805-810.

Collier NC and Schlesinger MJ (1986): The dynamic state of heat shock proteins in chicken embryo fibroblasts. *J Cell Biol* 103: 1495-1507.

Courtneidge SA and Bishop JM (1982): Transit of pp60$^{v\text{-}src}$ to the plasma membrane. *Proc Natl Acad Sci* USA 79: 7117-7121.

Dalman FC, Bresnick EH, Patel PD, Perdew GH, Watson SJ and Pratt WB (1989): Direct evidence that the glucocorticoid receptor binds hsp90 at or near the termination of receptor translation in vitro. *J Biol Chem* 264: 19815-19821.

Dalman FC, Koenig RJ, Perdew GH, Massa E and Pratt WB (1990): In contrast to the glucocorticoid receptor, the thyroid hormone receptor is translated in the DNA-binding state and is not associated with hsp90. *J Biol Chem* 265: 3615-3618.

Dalman FC, Scherrer LC, Taylor LP, Akil H and Pratt WB (1991a): Localization of the 90-kDa heat shock protein-binding site within the hormone-binding domain of the glucocorticoid receptor by peptide competition. *J Biol Chem* 266: 3482-3490.

Dalman FC, Sturzenbecker LJ, Levin AA, Lucas DA, Perdew GH, Petkovitch M, Chambon P, Grippo JF and Pratt WB (1991b): Retinoic acid receptor belongs to a subclass of nuclear receptors that do not form "docking" complexes with hsp90. *Biochemistry* 30: 5605-5608.

Denis M, Gustafsson JA, Wikstrom AC (1988): Interaction of the M_r=90,000 heat shock protein with the steroid binding domain of the glucocorticoid receptor. *J Biol Chem* 263: 18520-18523.

Deshaies RJ, Koch BD, Werner Washburn M, Craig EA and Shekman R (1988): A subfamily of stress proteins stimulates translocation of secretory and mitochondrial precursor proteins. *Nature* 332: 800-805.

Eilers M, Picard D, Yamamoto KR and Bishop JM (1989): Chimaeras of Myc oncoprotein and steroid receptors cause hormone-dependent transformation of cells. *Nature* 340: 66-68.

Feldherr RD, Kallenbach E and Schultz N (1984): Movement of karyophillic proteins through the nuclear pores of oocytes. *J Cell Biol* 99: 2216-2222.

Fuxe K, Wikstrom AC, Okret S, Agnati LF, Harfstrand A, Yu ZY, Granholm L, Zoli M, Vale W and Gustafsson JA (1985): Mapping of glucocorticoid receptor immunoreactive neurons in the rat tel- and diencephalon using a monoclonal antibody against rat liver glucocorticoid receptor. *Endocrinology* 117: 1803-1812.

Govindan MV (1980): Immunofluorescence microscopy of the intracellular translocation of glucocorticoid-receptor complexes in rat hepatoma (HTC) cells. *Exp Cell Res* 127: 293-297.

Guiochon-Mantel A, Loosfelt H, Lescop P, Sar S, Atger M, Perrot-Applanat M and Milgrom E (1989): Mechanisms of nuclear localization of the progesterone receptor: evidence for interaction between monomers. *Cell* 57:1147-1154.

Guiochon-Mantel A, Lescop P, Christin-Maitre S, Loosfelt H, Perrot-Applanat M and Milgrom E (1991): Nucleocytoplasmic shuttling of the progesterone receptor. *EMBO J* 10: 3851-3859.

Hirst MA, Northrop JP, Danielsen M and Ringold GM (1990): High level expression of wild type and variant mouse glucocorticoid receptors in Chinese hamster ovary cells. *Mol Endo* 4: 162-170.

Howard KJ, Holley SJ, Yamamoto KR and Distelhorst C (1991): Mapping the hsp90 binding region of the glucocorticoid receptor. *J Biol Chem* 265: 11928-11935.

Husmann DA, Wilson CM, McPhaul MJ, Tilley WD and Wilson JD (1990): Antipeptide antibodies to two distinct regions of the androgen receptor localize the receptor protein to the nuclei of target cells in the rat and human prostate. *Endocrinol* 126: 2359-2368.

Hutchison KA, Brott BK, De Leon JH, Perdew GH, Jove R and Pratt WB (1992a): Reconstitution of the multiprotein complex of pp60^{src}, hsp90, and p50 in a cell-free system. *J Biol Chem* 267: 2902-2908.

Hutchison KA, Czar MJ, Scherrer LC and Pratt, WB (1992b): Monovalent cation selectivity for ATP-dependent association of the glucocorticoid receptor with hsp70 and hsp90. *J Biol Chem* 267: 14047-14053.

240

Hutchison KA, Stancato LF, Jove R and Pratt WB (1992c): The protein-protein complex between pp60^{v-src} and hsp90 is stabilized by molybdate, vanadate, tungstate, and an endogenous cytosolic metal. *J Biol Chem* 267: 13952-13957.

Hutchison KA, Scherrer LC, Czar MJ, Stancato LF, Chow Y-H, Jove R and Pratt WB (1993): Regulation of glucocorticoid receptor function through assembly of a receptor-heat shock protein complex. *Ann NY Acad Sci*, 684: 35-48.

Ishii DN, Pratt WB and Aronow L (1972): Steady-state level of the specific glucocorticoid binding component in mouse fibroblasts. *Biochemistry* 11: 3896-3904.

Jove R and Hanafusa H (1987): Cell transformation by the viral *src* oncogene. *Ann Rev Cell Biol* 3: 31-56.

Kalderon D, Richardson WD, Markham AF and Smith AE (1984a): Sequence requirements for nuclear location of similar virus 40 large-T antigen. *Nature* 311: 33-38.

Kalderon D, Roberts BL, Richardson WD and Smith AE (1984b): A short amino acid sequence able to specify nuclear location. *Cell* 39: 499-509.

King WJ and Green GL (1984): Monoclonal antibodies localize oestrogen receptor in the nuclei of target cells. *Nature* 307: 745-747.

Kost SL, Smith DF, Sullivan WB, Welch WJ and Toft DO (1989): Binding of heat shock proteins to the avian progesterone receptor. *Mol Cell Biol* 9: 3829-3838.

Koyasu S, Nishida E, Kadowaki T, Matsuzaki F, Iida K, Harada F, Kasuga M, Sakai H and Yahara I (1986): Two mammalian heat shock proteins, hsp90 and hsp100, are actin-binding proteins. *Proc Natl Acad Sci* USA 83: 8054-8058.

Lai BT, Chin NW, Stanek AE, Keh W and Lanks KW (1984): Quantitation and intracellular localization of the 85 K heat shock protein by using monoclonal and polyclonal antibodies. *Mol Cell Biol* 4: 2802-2810.

Landers JP and Spelsberg TC (1992): New concepts in steroid hormone action: transcription factors, protooncogenes, and the cascade model for steroid regulation of gene expression. *Crit Rev Eukaryot Gene Exp* 2: 19-63

Li P, Wood K, Mamon H, Haser W and Roberts T (1991): Raf-1: a kinase currently without a cause but not lacking in effects. *Cell* 64; 479-482.

Martins VR, Pratt WB, Terracio L, Hirst MA, Ringold GM and Housley PR (1991): Demonstration by confocal microscopy that unliganded overexpressed glucocorticoid receptors are distributed in a nonrandom manner throughout all planes of the nucleus. *Mol Endo* 5: 217-225.

Meier UT and Blobel G (1992): Nopp140 shuttles on tracks between nucleolus and cytoplasm. *Cell* 70: 127-138.

Miyata Y and Yahara I (1991): Cytoplasmic 8S glucocorticoid receptor binds to actin filaments through the 90-kDa heat shock protein moiety. J Biol Chem 266: 8779-8783.

Munck A and Holbrook NJ (1984): Glucocorticoid-receptor complexes in rat thymus cells. Rapid kinetic behavior and a cyclic model. J Biol Chem 259: 820-831.

Munck A, Wira C, Young DA, Mosher KM, Hallahan C and Bell PA (1972): Glucocorticoid-receptor complexes and the earliest steps in the action of glucocorticoids on thymus cells. J Steroid Biochem 3: 567-578.

Nelson K, Baranowska-Kortylewicz J, van Nagell JR, Gallion HH, Donaldson ES, Kenedy DE, McRoberts WP and Pavlik EJ (1989): Intermolecular engagement of estrogen receptors indicated by the formation of a high molecular weight complex during activation. Biochemistry 28: 9741-9749.

Nervi C, Grippo JF, Sherman MI, George MD and Jetten AM (1989): Identification and characterization of nuclear retinoic acid binding activity in human myeloblastic leukemia HL-60 cells. Proc Natl Acad Sci USA 86: 5854-5858.

Nigg EA, Baeuerle PA and Luhrman R (1991): Nuclear import-export: in search of signals and mechanisms. *Cell* 66:15-22.

Nishida E, Koyasu S, Sakai H and Yahara I (1986): Calmodulin-regulated binding of the 90-kDa heat shock protein to actin filaments. *J Biol Chem* 261: 16033-16036.

Onate SA, Estes PA, Welch WJ, Nordeen SK and Edwards DP (1991): Evidence that the heat shock protein-70 associated with progesterone receptors is not involved in receptor-DNA binding. *Mol Endo* 5: 1993-2004.

Orti E, Bodwell JE and Munck A (1992): Phosphorylation of steroid hormone receptors. *Endocrine Rev* 13: 105-128.

Papamichail M, Tsokos G, Tsawdaroglou N and Sekeris CE (1980): Immunochemical demonstration of glucocorticoid receptors in different cell types and their translocation from the cytoplasm to the cell nucleus in the presence of dexamethasone. Exp *Cell* Res 125: 490-493.

Perdew GH and Whitelaw ML (1991): Evidence that the 90-kDa heat shock protein (hsp90) exists in cytosol in heteromeric complexes containing hsp70 and three other proteins with M_r 63,000, 56,000 and 50,000. *J Biol Chem* 266: 6708-6713.

Perrot-Applanat M, Logeat F, Groyer-Picard MT and Milgrom E (1985): Immunocytochemical study of mammalian progesterone receptor using monoclonal antibodies. *Endocrinology* 116: 1473-1484.

Perrot-Applanat M, Groyer-Picard MT, Logeat F and Milgrom E (1986): Ultrastructural localization of the progesterone receptor by an immunogold method: effect of hormone administration. *J Cell Biol* 102: 1191-1199.

Picard D and Yamamoto KR (1987): Two signals mediate hormone-dependent nuclear localization of the glucocorticoid receptor. *EMBO J* 6: 3333-3340.

Picard D, Salser SJ and Yamamoto KR (1988): A movable and regulable inactivation function within the steroid binding domain of the glucocorticoid receptor. *Cell* 54: 1073-1080.

Picard D, Kumar V, Chambon P and Yamamoto KR (1990): Signal transduction by steroid hormones: nuclear localization is differentially regulated in estrogen and glucocorticoid receptors. *Cell Regulation* 1: 291-299.

Pratt WB (1987): Transformation of glucocorticoid and progesterone receptors to the DNA-binding state. *J Cell Biochem* 35: 51-68.

Pratt WB (1990): Interaction of hsp90 with steroid receptors: organizing some diverse observations and presenting the newest concepts. *Mol Cell Endocrinol* 74: C69-C76.

Pratt WB (1992): Control of steroid receptor function and cytoplasmic-nuclear transport by heat shock proteins. BioEssays 14: 841-848.

Pratt WB (1993): Role of heat shock proteins in steroid receptor function. In: *Steroid Hormone Action: Frontiers in Molecular Biology,* Parker M.G., ed. Oxford: Oxford University Press.

Pratt WB, Jolly DJ, Pratt DV, Hollenberg SM, Giguere V, Cadepond FM, Schweizer-Groyer G, Catelli MG, Evans RM and Baulieu EE (1988): A region in the steroid binding domain determines formation of the non-DNA-binding, 9S glucocorticoid receptor complex. *J Biol Chem* 263: 267-273.

Pratt WB, Sanchez ER, Bresnick EH, Meshinchi S, Scherrer LC, Dalman FC and Welsh MJ (1989): Interaction of the glucocorticoid receptor with the M_r 90,000 heat shock protein: an evolving model of ligand-mediated receptor transformation and translocation. *Cancer Res* 49 (Suppl.): 2222s-2229s.

Pratt WB, Hutchison KL and Scherrer LC (1992a): Steroid receptor folding by heat shock proteins and composition of the receptor heterocomplex. *Trends Endocrinol Metab* 3: 326-333.

Pratt WB, Scherrer LC, Hutchison KA and Dalman FC (1992b): A model of glucocorticoid receptor unfolding and stabilization by a heat shock protein complex. *J Steroid Biochem* 41: 223-229.

Qi M, Hamilton BJ and DeFranco D (1989): V-*mos* oncoproteins affect the nuclear retention and reutilization of glucocorticoid receptors. *Mol Endo* 3: 1279-1288.

Rapp UR, Heidecker G, Huleihel M, Cleveland JL, Choi WC, Pawson T, Ihle JN and Anderson WB (1988): *raf* Family serine/threonine protein kinases in mitogen signal transduction. *Cold Spring Harbor Symp Quant Biol* 53: 173-184.

Redmond T, Sanchez ER, Bresnick EH, Schlesinger MJ, Toft DO, Pratt WB and Welsh MJ (1989): Immunofluorescence colocalization of the 90-kDa heat-shock protein and microtubules in interphase and mitotic mammalian cells. *Eur J Cell Biol* 50: 66-75.

Reihl RM, Sullivan WP, Vroman BT, Bauer VJ, Pearson GR and Toft DO (1985): Immunological evidence that the nonhormone binding component of avian steroid receptors exists in a wide range of tissues and species. *Biochemistry* 24: 6586-6591.

Renoir JM, Radanyi C, Faber LE and Baulieu EE (1990): The non-DNA-binding heterooligomeric form of mammalian steroid hormone receptors contains a hsp90-bound 59-kilodalton protein. *J Biol Chem* 265: 10740-10745.

Rexin M, Busch W and Gehring U (1991): Protein components of the nonactivated glucocorticoid receptor. *J Biol Chem* 266: 24601-24605.

Rexin M, Busch W, Segnitz B and Gehring U (1992): Structure of the glucocorticoid receptor in intact cells in the absence of hormone. *J Biol Chem* 267: 9619-9621.

Roberts BL, Richardson WD, Kalderon DD, Cheng SH, Markland W and Smith AE (1986): In *Nucleocytoplasmic Transport*, Peters R., Trendelenburg M., eds. Berlin: Springer-Verlag.

Rothman JE (1989): Polypeptide chain binding proteins: catalysts of protein folding and related processes in cells. *Cell* 59: 591-601.

Ruff VA, Yem AW, Munns PL, Adams LD, Reardon IM, Deibel MR and Leach KL (1992): Hsp56 is localized to the cytoskeleton and throughout the nucleus of LLC-PK$_1$ kidney cells. *J Biol Chem* 267: 21285-21288.

Samuels HH, Tsai JS, Casanova J and Stanley F (1974): In vitro characterization of solubilized nuclear receptors from rat liver and cultured GH₁ cells. *J Clin Invest* 54: 853-865.

Sanchez ER, Toft DO, Schlesinger MJ and Pratt WB (1985): Evidence that the 90-kDa phosphoprotein associated with the untransformed L-cell glucocorticoid receptor is a murine heat shock protein. *J Biol Chem* 260: 12398-12401.

Sanchez ER, Meshinchi S, Tienrungroj W, Schlesinger MJ, Toft DO and Pratt WB (1987): Relationship of the 90-kDa murine heat shock protein to the untransformed and transformed states of the L cell glucocorticoid receptor. *J Biol Chem* 262: 6986-6991.

Sanchez ER, Redmond T, Scherrer LC, Bresnick EH, Welsh MJ and Pratt WB (1988): Evidence that the 90-kilodalton heat shock protein is associated with tubulin-containing complexes in L cell cytosol and intact PtK cells. *Mol Endo* 2: 756-760.

Sanchez ER, Faber LE, Henzel WJ and Pratt WB (1990a): The 56-59 kilodalton protein identified in untransformed steroid receptor complexes is a unique protein that exists in cytosol in a complex with both the 70- and 90-kilodalton heat shock proteins. *Biochemistry* 29: 5145-5152.

Sanchez ER, Hirst M, Scherrer LC, Tang H-Y, Welsh MJ, Harmon JM, Simons SS, Ringold GM and Pratt WB (1990b): Hormone-free glucocorticoid receptors overexpressed in Chinese hamster ovary cells are localized to the nucleus and are associated with both hsp90 and hsp70. *J Biol Chem* 265: 20123-20130.

Scherrer LC and Pratt WB (1992): Energy-dependent conversion of transformed cytosolic glucocorticoid receptors from soluble to particulate-bound form. *Biochemistry* 31: 10879-10886.

Scherrer LC, Dalman FC, Massa E, Meshinchi S and Pratt WB (1990): Structural and functional reconstitution of the glucocorticoid receptor-hsp90 complex. *J Biol Chem* 265: 21397-21400.

Scherrer LC, Picard D, Massa E, Harmon JM, Simons SS, Yamamoto KR and Pratt WB (1993) Evidence that the hormone binding domain of steroid receptors confers hormonal control on chimaeric proteins by determining their hormone-regulated binding to hsp90. *Biochemistry* 32: 5381-5386.

Schlesinger MJ, Aliperti G and Kelley PM (1982): The response of cells to heat shock. *Trends Biochem Sci* 7: 222-225.

Schowalter DB, Sullivan WP, Maihle NJ, Dobson ADW, Conneely OM, O'Malley BW and Toft DO (1991): Characterization of progesterone receptor binding to the 90- and 70-kDa heat shock proteins. *J Biol Chem* 266: 21165-21173.

Schuh S, Yonemoto W, Brugge J, Bauer VJ, Riehl RM, Sullivan WP and Toft DO (1985): A 90,000-dalton binding protein common to both steroid receptors and the Rous sarcoma virus transforming protein, pp60$^{v\text{-}src}$. *J Biol Chem* 260: 14292-14296.

Shi Y and Thomas JO (1992): The transport of proteins into the nucleus requires the 70-kilodalton heat shock protein or its cytosolic cognate. *Mol Cell Biol* 12: 2186-2192.

Silver PA (1991): How proteins enter the nucleus. *Cell* 64: 489-497.

Simental JA, Sar M, Lane MV, French FS and Wilson EM (1991): Transcriptional activation and nuclear targeting signals of the human androgen receptor. *J Biol Chem* 266: 510-518.

Smith DF, Faber LE and Toft DO (1990a): Purification of unactivated progesterone receptor and identification of novel receptor-associated proteins. *J Biol Chem* 265: 3996-4003.

Smith DF, Schowalter DB, Kost SL and Toft DO (1990b): Reconstitution of progesterone receptor with heat shock proteins. *Mol Endo* 4: 1704-1711.

Smith DF, Stensgard BA, Welch WJ and Toft DO (1992): Assembly of progesterone receptor with heat shock proteins and receptor activation are ATP-mediated events. *J Biol Chem* 267: 1350-1356.

Spindler SR, MacLeod KM, Ring J and Baxter JD (1975): Thyroid hormone receptors: binding characteristics and lack of hormonal dependency for nuclear localization. *J Biol Chem* 250: 4113-4119.

Stancato LF, Chow Y-H, Hutchison KA, Perdew, GH, Jove R and Pratt WB (1993): Raf exists in a native heterocomplex with hsp90 and p50 that can be reconstituted in a cell-free system. *J. Biol. Chem. 268:* in press.

Superti-Furga G, Bergers G, Picard D and Busslinger M (1991): Hormone-dependent transcriptional regulation and cellular transformation by Fos-steroid receptor fusion proteins. *Proc Natl Acad Sci* USA 88: 5114-5118.

Tai PK, Maeda Y, Nakao K, Wakim NG, Duhring JL and Faber LE (1986): A 59-kilodalton protein associated with progestin, estrogen, androgen, and glucocorticoid receptors. *Biochemistry* 25: 5269-5275.

Tai PK, Albers MW, Chang H, Faber LE and Schreiber SL (1992): Association of a 59-kilodalton immunophilin with the glucocorticoid receptor complex. *Science* 256: 1315-1318.

Umek RM, Friedman AD and McKnight SL (1991): CCAAT-enhancer binding protein: a component of a differentiation switch. *Science* 251: 288-292.

Vale RD (1987): Intracellular transport using microtubule-based motors. *Ann Rev Cell Biol* 3: 347-378.

Walters MR (1985): Steroid hormone receptors and the nucleus. *Endocrine Rev* 6: 512-543.

Wang C, Asai DJ and Lazarides E (1980): The 68,000-dalton neurofilament-associated polypeptide is a component of nonneuronal cells and of skeletal myofibrils. *Proc Natl Acad Sci* USA 77: 1541-1545.

Wang C, Gomer RH and Lazarides E (1981): Heat shock proteins are methylated in avian and mammalian cells. *Proc Natl Acad Sci* USA 78: 3531-3535.

Welch WJ and Feramisco JR (1984): Nuclear and nucleolar localization of the 72,000-dalton heat shock protein in heat-shocked mammalian cells. *J Biol Chem* 259: 4501-4513.

Welshons WV, Lieberman ME and Gorski J (1984): Nuclear localization of unoccupied oestrogen receptors. *Nature* 307: 747-749.

Whitelaw ML, Hutchison KA and Perdew GH (1991): A 50-kDa cytosolic protein complexed with the 90-kDa heat shock protein (hsp90) is the same protein complexed with $pp60^{v-src}$ hsp90 in cells transformed by the Rous sarcoma virus. *J Biol Chem* 266: 16436-16440.

Wikstrom AC, Bakke O, Okret S, Bronnegard M and Gustafsson JA (1987): Intracellular localization of the glucocorticoid receptor: evidence for cytoplasmic and nuclear localization. *Endocrinology* 120: 1232-1242.

Wikstrom AC, Denis M, Akner G, Bakke O and Gustafsson JA (1989): The association of the glucocorticoid receptor with M_r 90,000 heat shock protein and tubulin. In: *The Steroid/Thyroid Hormone Receptor Family and Gene Regulation*, Carlstedt-Duke J, Eriksson H., Gustafsson J.A., eds. Basel: Birkhauser.

IN VITRO ASSEMBLY OF THE AVIAN PROGESTERONE RECEPTOR

David F. Smith
Department of Pharmacology
University of Nebraska Medical Center

William P. Sullivan
Jill Johnson
David O. Toft
Department of Biochemistry and Molecular Biology
Mayo Foundation

INTRODUCTION

When in the absence of hormone, most steroid receptors exist in an inactive multi-protein complex (see Pratt, 1992 and Smith and Toft, 1993, for review). With the avian progesterone receptor, these complexes contain hsp90, hsp70, and three additional proteins identified by size as p54, p50, and p23 (Smith et al, 1990). That all of these proteins associate with the receptor inside the cell is still speculative as is their functional significance to receptor activity. Until quite recently, research in this area was hampered by the inability to form and study these receptor complexes in vitro. Some success was achieved with the estrogen receptor by Inano et al, (1990, 1992). They found that the estrogen receptor would re-bind to hsp90 under mildly denaturing conditions. An alternative approach has recently been described for

STEROID HORMONE RECEPTORS
V. K. Moudgil, Editor
© 1993 Birkhäuser Boston

reconstitution of hsp90 complexes with the receptors for progesterone (Smith et al, 1990b) and glucocorticoids (Pratt, 1991) using rabbit reticulocyte lysate as the reconstitution medium.

In this report we will review the conditions for reconstitution of the progesterone receptor (PR) in reticulocyte lysate and describe two proteins, p60 and p23, that appear to participate in this process.

Progesterone Receptor Reconstitution

To study the binding of PR to receptor-associated proteins, we use an antibody precipitation technique where the receptor from chick oviduct cytosol is bound to protein A-Sepharose containing anti-receptor antibody. This allows rapid isolation of the receptor, and when conducted under high ionic conditions (0.5 M KCl), the receptor is stripped of its associated proteins. When this stripped, immobilized receptor is suspended in oviduct cytosol, no binding of hsp90 or other associated proteins occurs (Smith et al, 1990b). However, when the receptor is suspended in rabbit reticulocyte lysate, it re-binds to endogenous hsp90 through a process that is time and temperature dependent and requires ATP, Mg^{++}, and K^+ (Smith et al, 1992). Similar requirements were found for the reconstitution of glucocorticoid receptor complexes (Hutchison et al, 1992). Figure 1 illustrates some basic features of this reconstitution. The protein pattern for the isolated stripped receptor is shown in lane 1. The two receptor forms, A and B, are observed at about 80 kDa and 110 kDa, respectively. Although these receptor forms have common antibody epitopes and are isolated together, they have been shown to form separate complexes with hsp90 (Smith et al, 1990a). There is a non-specific contaminant just above receptor A in this experiment and some hsp70 is also apparent that was not stripped from the receptor. Two conditions for re-binding receptor to hsp90 and hsp70 are shown in Fig. 1, lanes 3 and 4. The stripped receptor was incubated in reticulocyte lysate for 30 minutes at 30° in the absence (lane 3) or in the presence (lane 4) of an ATP regeneration system. While both conditions promote hsp90 binding, this binding is clearly superior in the presence of ATP regeneration. In this case, the level of ATP is 2 to 4 mM at the start and changes little during the incubation. When the phosphocreatine/creatine kinase regeneration system is omitted, the original store of ATP is rapidly metabolized to micromolar levels (Smith et al, 1992). The requirement of ATP for reconstitution is shown in lane 5. In this sample, the plant ATPase, apyrase, was added to hydrolyze ATP. This results in a total lack of hsp90 binding and minimal binding of hsp70.

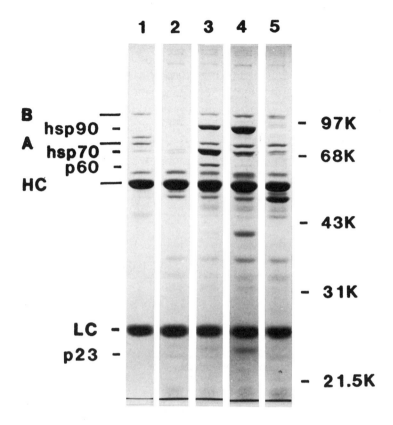

FIGURE 1. ATP dependence of hsp90 binding to progesterone receptor. Data shown are Coomassie-stained SDS-PAGE lanes containing progesterone receptor from 1.5 ml of oviduct cytosol (adjusted to 0.5 M KCl); receptor was immobilized on 20 µl of PR22-protein A resin and incubated in 200 µl of rabbit reticulocyte lysate at 30°C for 30 min., except as noted below. One aliquot of receptor resin was analyzed without lysate treatment (lane 1). In other lanes, resin aliquots were incubated with rabbit reticulocyte lysate before washing and analysis. As a negative control, an aliquot of PR22-protein A resin lacking receptor was incubated with untreated lysate (lane 2). Aliquots of receptor-containing resin were incubated in the following lysates: untreated lysate (lane 3), lysate supplemented with an ATP-regenerating system (lane 4), or lysate pretreated with apyrase, 1 unit/100 µl (lane 5). Receptor forms A and B along with associated proteins and PR22 heavy (HC) and light (LC) chains are indicated on the left; molecular weight markers are indicated on the right. Reprinted with permission of the American Society for Biochemistry & Molecular Biology from Smith et al, 1992.

Some additional proteins should be noted in the comparison of Figure 1, lanes 3 and 4. The optimal conditions for hsp90 binding (lane 4) also result in binding of a 23 kDa protein, p23. This protein has been shown to be related to the p23 that is associated with PR isolated from oviduct cytosol since a monoclonal antibody to chicken p23 cross-reacts with rabbit p23 on Western blots. A protein band at about 40 kDa is also shown in lane 4. This protein has not yet been investigated and whether it is a contaminant or a specific receptor-associated protein remains unknown. Without the ATP regeneration system (lane 3), an additional protein called p60 is bound as well as more extensive binding of hsp70. Control experiments show that p60 is not a contaminant but its binding appears to be specific to the receptor complex.

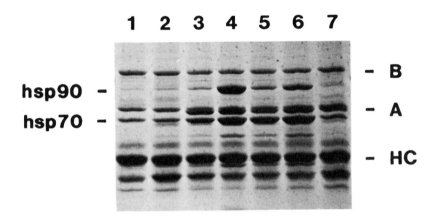

FIGURE 2. Effects of lysate dialysis and nucleotide supplementation on hsp90 binding to progesterone receptor. Lysate was dialyzed overnight at 4°C against buffer containing 10 mM Tris (pH 7.5), 3 mM $MgCl_2$, and 50 mM KCl. Stripped receptor-resin complexes are shown in lane 1, followed by receptor-resin complexes treated as usual for reconstitution but using dialyzed lysate supplemented with the following nucleotides: no nucleotide (lane 2); 1 mM ATP (lane 3); 5 mM ATP (lane 4); 5 mM ADP (lane 5); 5 mM GTP (lane 6); or 5 mM AMP-PNP, a nonhydrolyzable ATP analogue (lane 7). Receptor-associated proteins are indicated on the right. Reprinted with permission of the American Society for Biochemistry & Molecular Biology from Smith et al, 1992.

A further test for the nucleotide dependency of receptor reconstitution is shown in Figure 2. In this experiment the reticulocyte lysate was first dialyzed extensively to remove ATP. This lysate is totally inactive with regard to hsp90 or hsp70 binding (lane 2). However, binding is partially restored by the addition of 1 mM ATP (lane 3) and fully restored by 5 mM ATP (lane 4). Some binding occurs with the addition of 5 mM ADP (lane 5) or GTP (lane 6), but the non-metabolizable analogue, AMP-PNP, does not support binding (lane 7). Since an ATP regeneration system was not used in this experiment, the binding of p60 is also observed. Its binding appears to be proportional to the level of hsp90 binding. This, plus the ATP dependency, has led us to suggest that p60 may be part of an intermediate complex in the assembly of receptor complexes (Smith et al, 1992).

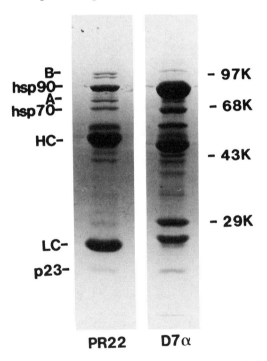

FIGURE 3. Comparison of the protein composition of progesterone receptor complexes and more general complexes of hsp90. Complexes of PR or hsp90 were isolated as described in Fig. 1 using either antibody PR22 (to PR) or D7α (to hsp90). The results shown are from two separate experiments.

Receptor-Associated Protein p23

The above studies show that at least four proteins from reticulocyte lysate contribute to the reconstitution of PR complexes: hsp90, hsp70, p60, and p23. Further characterization of p60 and p23 has been assisted by the development of monoclonal antibodies to these proteins and by the analysis of more general complexes of hsp90. Figure 3 illustrates a comparison of protein complexes isolated from chick oviduct cytosol using a receptor antibody (PR22) or an antibody to hsp90 (D7α). The PR complex includes hsp90, hsp70, and p23. The bands above and below the antibody heavy chain (HC) are non-specific contaminants and the receptor-associated proteins described previously as p50 and p54 are masked by heavy chain.

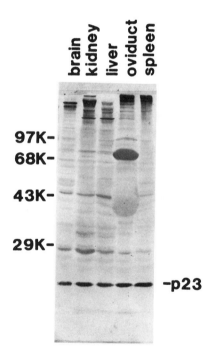

FIGURE 4. The detection of p23 in cytosol extracts of various chicken tissues. Cytosol extracts were prepared from the tissues indicated using four volumes of homogenization buffer. Ten μl of each cytosol were applied to SDS-PAGE and the proteins transferred to PVDF membrane. The detection of p23 was accomplished using antibody R3, a mouse monoclonal IgM, followed by a second antibody-alkaline phosphatase conjugate.

When an antibody to hsp90 is used for immune precipitation, the B and A receptor bands are barely detectable. This is consistent with the fact that the level of hsp90 in cytosol is at least 100-fold that of PR (Catelli et al, 1985). Two major proteins co-purify with hsp90 and migrate between hsp90 and antibody heavy chain. These have been identified by antibody probes to be hsp70 and p60 (Smith et al, 1993). P23 is also observed in the hsp90 complex. In this complex, p23 appears to be more abundant than PR but less abundant than hsp70 and p60. We have isolated hsp90 complexes from a variety of tissue extracts and, in all cases, the co-purification of hsp70 and p60 is observed (Smith et al, 1993). This suggests that the three proteins, hsp90, hsp70, and p60, function together in some cases.

We have recently obtained a monoclonal antibody to p23 which has been useful for analyzing the distribution of this protein and in screening for cDNA clones. Figure 4 illustrates a western blot of several chicken tissue extracts using antibody R3 against p23. Similar quantities of the protein were observed in all tissues that were tested. In addition, p23 has been detected in cytosol extracts from human, rabbit, rat and mouse tissues. Thus, p23 is a ubiquitous protein and its expression is not unique to cells that contain PR. Since one of the common heat shock proteins is in the size range of 24 kDA to 27 kDa (Lindquist and Craig, 1988), it was of interest to see if p23 is related to this protein. This appears not to be the case since western blotting studies using antibodies to p23 and to chicken hsp24 show that p23 is clearly smaller than chicken hsp24 and no antibody cross-reactions were observed (Johnson et al, in preparation). We have recently obtained both peptide sequence and cDNA clones for p23. These indicate that p23 is a unique protein that has not been described previously (Johnson et al, in preparation).

Receptor Associated Protein p60

A monoclonal antibody was recently obtained to chicken p60 that was isolated with hsp90 complexes. As with other receptor-associated proteins, western blotting of cytosol extracts showed the expression of p60 in all chicken tissues that were tested (Figure 5A). Slight variations in the mobility of p60 were observed which may be caused by differing protein composition of the extracts or by the existence of p60 isoforms. The antibody used, F5, was able to recognize a p60 protein from a wide variety of eukaryotes as shown in Figure 5B. The only negative sample was an extract from Saccharomyces cerevisiae (lane 9). We now believe that this is due to a lack of antibody recognition rather than to a lack of

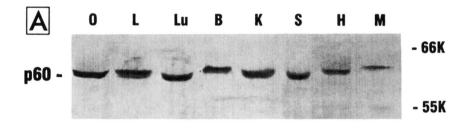

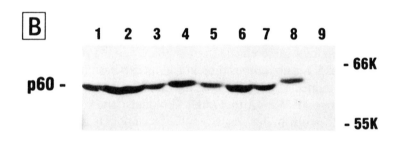

FIGURE 5. **A**. Avian tissue survey for p60 using Mab F5. Crude cytosolic extracts were prepared from eight different chick tissues, separated by SDS-PAGE and transferred to PVDF membrane. The membrane was immuno-stained with monoclonal antibody F5 prepared against protein p60 from oviduct. The tissues surveyed are: O -- oviduct, L -- liver, Lu -- lung, B -- brain, K -- kidney, S -- spleen, H -- heart, M -- breast muscle. Only the region of the immuno-stained membrane containing reactive bands is shown. Molecular weight markers are indicated on the right. **B**. Species survey for antigens crossreactive with Mab F5. Crude cytosolic or total cell extracts were prepared, separated by SDS-PAGE, transferred to PVDF membrane, then immuno-stained with monoclonal antibody F5 against avian p60. The samples surveyed were 1 -- chick oviduct cytosol, 2 -- human HeLa cell extract, 3 -- human MCF7 cell extract, 4 -- rabbit reticulocyte lysate, 5 -- rat liver cytosol, 6 -- rat GH4 cell extract, 7 -- mink CCL cell extract, 8 -- Xenopus liver cytosol, and 9 -- S. cerevisiae cell extract. Migration positions for pre-stained molecular weight markers are indicated on the left. Taken from Smith et al, 1993.

p60 expression in yeast (see below). Since this antibody recognizes p60 from rabbits, it was possible to show by western blotting that the antibody has similar reactivity with the p60 in hsp90 complexes and the p60 in reconstituted receptor complexes.

To more clearly identify p60, efforts were made to obtain sequence information on this protein (Smith et al, 1993). The preparation of highly purified p60 was accomplished by the immuno-isolation of hsp90 complexes, resolution of proteins by SDS-PAGE, and the electroelution of p60 from gel sections. Peptide fragments of p60 were then generated by treatment with trypsin or cyanogen bromide and eight of these were sequenced. A data base search using these sequences revealed one protein having a striking homology to p60. This is a 63 kDa protein from human cells that was reported recently to be upregulated after viral transformation (Honoré et al, 1992). Figure 6 shows the sequence comparison. A range of 50 to 100% homology was obtained for the eight peptides and the human protein which has been referred to as IEF SSP 3521 according to its mobility on 2-d gel analysis. Antibody F5 against chicken p60 cross-reacts specifically with the 63 kDa human protein (Honoré and Smith, unpublished). Figure 6 also shows the sequence of a yeast protein, STI1, which was found by Honoré et al to be related to the human protein. STI1 is a stress-induced protein or heat shock protein that has been described by Nicolet and Craig (1989). Yeast that lack this protein are viable but are unable to grow at extreme temperatures. Thus, it appears that p60 may be another highly conserved heat shock protein. However, it should be noted that p60 is not related to the major heat shock protein in mitochondria that is commonly called hsp60 (Lindquist and Craig, 1988).

DISCUSSION

The original observations of the association of steroid receptors with hsp90 suggested a function of this heat shock protein in maintaining the receptor in an inactive state (Catelli et al, 1985; Sanchez et al, 1985; Schuh et al, 1985). In this state, the receptor does not have DNA-binding activity, but hormone binding causes dissociation of hsp90 as part of a process of receptor activation.

More recent studies indicate that the structures of at least some steroid receptors are much more complex when in their inactive state. As many as seven different receptor-associated proteins have been described, but not all are from the same receptor system or state (Smith and Toft, 1993). Some insight into this complexity may be gained by studying the

```
        1                                                           50
P60c  . .EEAQALKE  RGNQALAAGD  IGAAVRHYTA  AIAL.DAXN  . . . . . . . . . .
P60h  .MEQVNELKE  KGNKALSVGN  IDDALQCYSE  AIKL.DPHNH  VLYSNRSAAY
P60y  MSLTADEYKQ  QGNAAFTAKD  YDKAIELFTK  AIEVSETPNH  VLYSNRSACY

        51                                                          100
P60c  . . . . . . . . . .  . . . . . . . . . .  . . . . . . . . . .  . . . .FLNRF  EE. . . . . . . .
P60h  AKKGDYQKAY  EDGCKTVDLK  PDWGKGYSRK  AAALEFLNRF  EEAKRTYEEG
P60y  TSLKKFSDAL  NDANECVKIN  PSWSKGYNRL  GAAHLGLGDL  DEAESNYKKA

        101                                                         150
P60c  . . . . . . . . . .  . . . . . . . . . .  . . . . . . . . . .  . . . . . . . . . .  . . . . . . . . . .
P60h  LKHEANNPQL  KEGLQNM. . .  .EARLAERK.  .FMNPFNMPN  LYQKLESDPR
P60y  LELDASNKAA  KEGLDQVHRT  QQARQAQPDL  GLTQLFADPN  LIENLKKNPK

        151                                                         200
P60c  . .ALLGDPSL  RELLEQLR. .  . . . . . . . . . .  . . . . . . . . . .  . . . . . . . . . .
P60h  TRTLLSDPTY  RELIEQLRNK  PSDLGTKL.Q  DPRIMTTLSV  LLGVDL. . . .
P60y  TSEMMKDPQL  VAKLIGYKQN  PQAIGQDLFT  DPRLMTIMAT  LMGVDLNMDD

        201                                                         250
P60c  . . . . . . . . . .  . . . . . . . . . .  . . . . . . . . . .  . . . . . . . . . .  . . . . . . . . . .
P60h  . . . .GSMDEE  EEIATP. . . .  . . . . . . . . . .  . . . . .PPPPP  PKKETK.PEP
P60y  INQSNSMPKE  PETSKSTEQK  KDAEPQSDST  TSKENSSKAP  QKEESKESEP

        251                                                         300
P60c  QQDLNENAK  QALKE. . . . .  . . . . . . . . . .  . . . . . . . . . .  . . . . . . .TYIT
P60h  MEEDLPENKK  QALKEKELGN  DAYKKKDFDT  ALKHYDKAKE  LDPTNMTYIT
P60y  MEVDEDDSKI  EADKEKAEGN  KFYKARQFDE  AIEHYNKAWE  LH.KDITYLN

        301                                                         350
P60c  NQAAVYFEKG  DYNXEXELEE  K. . . . . . . . .  . . . . . . . . . .  . . . . . . . . . .
P60h  NQAAVYFEKG  DYNKCRELCE  KAIEVGRENR  EDYRQIAKAY  ARIGNSYFKE
P60y  NRAAAEYEKG  EYETAISTLN  DAVEQGREMR  ADYKVISKSF  ARIGNAYHKL

        351                                                         400
P60c  . .YKDAVHFY  NK. . . . . . .  . . . . . . . . . .  . . . . . . . . . .  . . . . . . . . . .
P60h  EKYKDAIHFY  NKSLAEHRTP  DVLKKCQQAE  KILKEQERLA  YINPDLALEE
P60y  GDLKKTIEYY  QKSLTEHRTA  DILTKLRNAE  KELKKAEAEA  YVNPEKAEEA

        401                                                         450
P60c  . . . . . . . . . .  . . . . . . . . . .  . . . . . . . . . .  . . . . . . . . . .  . . .LLEFPLA
P60h  KNKGNECFQK  GDYPQAMKHY  TEAIKRNPKD  AKLYSNRAAC  YTKLLEFQLA
P60y  RLEGKEYFTK  SDWPNAVKAY  TEMIKRAPED  ARGYSNRAAA  LAKLMSFPEA

        451                                                         500
P60c  LKDQEEQIR.  . . . . . . . . . .  . . . . . . . . . .  . . . . . . . . . .  . . . . . . . . . .
P60h  LKDCEECIQL  EPTFIKGYTR  KAAALEAMKD  YTKAMDVYQK  A. . . . . .LDL
P60y  IADCNKAIEK  DPNFVRAYIR  KATAQIAVKE  YASALETLDA  ARTKDAEVNN

        501                                                         550
P60c  . . . . . . . . . .  . . . . . . . . . .  . . . . . . . . . .  . . . . . . . . . .  . . . . . . . . . .
P60h  DSSCKEAADG  YQRCMMAQYN  R. . .HDSPED  VKRRAMADPE  VQQIMSDPAM
P60y  GSSAREIDQL  YYKASQQRFQ  PGTSNETPEE  TYQRAMKDPE  VAAIMQDPVM

        551                                        590
P60c  . . . . . . . . . .  . . . . . . . . . .  . . . . . . . . . .  . . . . . . . . . .
P60h  RLILEQMQKD  PQALSEHLKN  PVIAQKIQKL  MDVGLIAIR.
P60y  QSILQQAQQN  PAALQEHMKN  PEVFKKIQTL  IAAGIIRTGR
```

FIGURE 6.

assembly of receptor complexes in vitro. It seems that some receptor-associated proteins, such as hsp90 and perhaps p23, bind stoichiometrically to the receptor. On the other hand, our results indicate that p60 only associates with the receptor complex when the in vitro conditions are sub-optimal with regard to ATP. Thus, p60 may be a transient participant in assembly of the receptor complex. There is also a transient element to the association of hsp70. In vitro, the binding of hsp70 is much greater in the absence of an ATP regeneration system than in the presence of continuous ATP.

A model to describe the above observations is shown in Figure 7. This model depicts assembly of the receptor complex as an ATP-dependent process where a key functional element is a three-way complex between hsp90, hsp70, and p60. An abundant complex of these heat shock proteins exists in many tissues and their association as an intermediate in receptor assembly provides an explanation for the appearance of p60 and higher amounts of hsp70 under conditions that may cause a build up of intermediate states. Under optimal conditions, the one or more intermediate states become negligible and a stable complex of inactive receptor is produced that contains less hsp70 and little or no p60. Since little is known about the receptor-associated proteins p54, p50, and p23, these are indicated collectively in the model as a unit labeled X.

This model will certainly change as we gain more information on individual steps and components of the assembly process. An important aspect of the model is the concept of this as a receptor assembly process rather than a static entity. This offers an explanation for the relatively large number of proteins that seem to be involved. Some of these, the heat shock proteins have been clearly implicated as molecular chaperons

FIGURE 6. Sequence comparison of p60 peptides with the 63 kDa human transformation-sensitive protein and the yeast STI1 protein. Gel-purified p60 was digested with trypsin or cyanogen bromide and fragments were separated and sequenced. Comparing the sequences obtained with sequence data bases, a striking similarity was observed with a recently cloned gene coding for a 63 kDa protein from human MRC-5 fibroblasts whose abundance increases in response to SV40 transformation (indicated as p60h). The sequence of a related yeast protein, STI1, is also shown (indicated as p60y). The published sequences are aligned with p60 peptide sequences with sequence identities indicated by shading. X indicates a residue whose determination was ambiguous.

Receptor Assembly

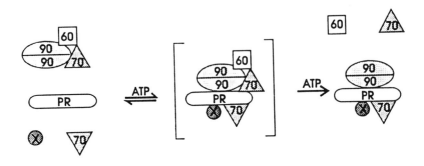

FIGURE 7. Model for the assembly of PR complexes. In this model, binding of hsp90 to receptor is facilitated by a heteromeric complex containing hsp90, hsp70, and p60. Binding of the heteromeric complex to receptor requires ATP. Another hsp70 molecule binds receptor independently. Also, additional proteins, such as p23, bind receptor, but the mechanism for their binding is unknown. Additional ATP is required for the dissociation of hsp70 and p60 from hsp90, thus forming the "native" unactivated PR complex. Assembly steps are reversible and perhaps require excess hsp90 heteromeric complexes to maintain receptor complexes at physiological temperatures.

that participate in the folding and processing of proteins to their native state (Gething and Sambrook, 1992; Georgopoulos, 1992). Once a native state is achieved, the association with chaperoning proteins is lost. The situation may be similar for receptor assembly where p60 and hsp70 may have an early function in receptor folding and assembly which is more or less completed in the intermediate steps. However, if one takes the view that receptors are only truly native when bound to hormone, some chaperoning functions must be sustained until hormone binding occurs. Hsp90 may have the predominant role at this stage with accessory roles by hsp70 or other proteins found in the inactive complex.

REFERENCES

Catelli MG, Binart N, Jung-Testas I, Renoir JM, Baulieu EE, Feramisco JR and Welch WJ (1985): The common 90 kDa protein component of non-transformed "8S" steroid receptors is a heat-shock protein. *EMBO J* 4: 3131-3135.

Gething M-J and Sambrook J (1992): Protein folding in the cell. *Nature* 355: 33-45.

Georgopoulos C (1992): The emergence of the chaperon machines. *TIBS* 17: 295-299.

Honoré B, Leffers H, Madsen P, Rasmussen HH, Vandekerckhove J and Celis JE (1992): Molecular cloning and expression of a transformation-sensitive human protein containing the TPR motif and sharing identity to the stress-inducible yeast protein STI1. *J Biol Chem* 267: 8485-8491.

Hutchison KA, Czar MJ, Scherrer LC and Pratt WB (1992): Monovalent cation selectivity for ATP-dependent association of the glucocorticoid receptor with hsp70 and hsp90. *J Biol Chem* 267: 14047-14053.

Inano K (1990): Reconstitution of the 9 S estrogen receptor with heat shock protein 90. *FEBS Lett.* 267: 157-159.

Inano K, Ishida T, Omata S and Horigome T (1992): In vitro formation of estrogen receptor-heat shock protein 90 complexes. *J Biochem* 112: 535-540.

Lindquist S and Craig EA (1988): The heat-shock proteins. *Annu Rev Genet* 22: 631-677.

Nicolet CM and Craig EA (1989): Isolation and characterization of STI1, a stress-inducible gene from Saccharomyces cerevisiae. *Mol Cell Biol* 9: 3638-3646.

Pratt W (1992): Control of steroid receptor function and cytoplasmic-nuclear transport by heat shock proteins. *BioEssays* 14: 841-848.

Sanchez ER, Toft DO, Schlesinger MJ and Pratt WB (1985): Evidence that the 90-kDa phosphoprotein associated with the untransformed L-cell glucocorticoid receptor is a murine heat shock protein. *J Biol Chem* 260: 12398-12401.

Scherrer LC, Dalman FC, Massa E, Meshinchi S and Pratt WB (1990): Structural and functional reconstitution of the glucocorticoid receptor-hsp90 complex. *J Biol Chem* 265: 21398-21400.

Scherrer LC, Hutchison KA, Sanchez ER, Randall SK and Pratt WB (1992): A heat shock protein complex isolated from rabbit reticulocyte lysate can reconstitute a functional glucocorticoid receptor-hsp90 complex. *Biochemistry* 31: 7325-7329.

Schuh S, Yonemoto W, Brugge J, Bauer VJ, Riehl RM, Sullivan WP and Toft DO (1985): A 90,000-dalton binding protein common to both steroid receptors and the Rous sarcoma virus transforming protein, pp60^{v-src}. *J Biol Chem* 260: 14292-14296.

Smith DF, Faber LE and Toft DO (1990a): Purification of unactivated progesterone receptor and identification of novel receptor-associated proteins. *J Biol Chem* 265: 3996-4003.

Smith DF, Schowalter DB, Kost SL and Toft DO (1990b): Reconstitution of progesterone receptor with heat shock proteins. *Mol Endocrinol* 4: 1704-1711.

Smith DF, Stensgard BA, Welch WJ and Toft DO (1992): Assembly of progesterone receptor with heat shock proteins and receptor activation are ATP mediated events. *J Biol Chem* 267: 1350-1356.

Smith DF, Sullivan WP, Marion TN, Zaitsu K, Madden B, McCormick DJ and Toft DO (1993): Identification of a 60-kilodalton stress-related protein, p60, which interacts with hsp90 and hsp70. *Mol Endocrinol* 13: 869-876.

Smith DF and Toft DO (1993): Steroid receptors and their associated proteins. *Mol Endocrinol* 7: 4-11.

STEROID RECEPTOR ASSOCIATED PROTEINS: HEAT SHOCK PROTEIN 90 AND P59 IMMUNOPHILIN

Marie-Claire Lebeau
Nadine Binart
Françoise Cadepond
Maria-Grazia Catelli
Béatrice Chambraud
Nelly Massol
Christine Radanyi
Gérard Redeuilh
Jack-Michel Renoir
Michèle Sabbah
Ghislaine Schweizer-Groyer
Etienne-Emile Baulieu
INSERM U33, Lab Hormones
Bicêtre, France

INTRODUCTION

Nearly ten years ago, hoping to obtain a monoclonal antibody against "native" 8S chick progesterone receptor prepared in the now classical way in the presence of molybdate, we obtained an antibody which displaced progesterone binding in sucrose gradients (Joab et al, 1984). It also displaced estrogen, androgen and glucocorticosteroid binding, and we now know that it can also displace mineralocorticosteroid receptors

STEROID HORMONE RECEPTORS
V. K. Moudgil, Editor
© 1993 Birkhäuser Boston

(Rafestin-Oblin,1989). In low salt gradients, where the 8S complex was conserved, the peak of bound steroid was shifted by the BF4 antibody, but in high salt, the complex dissociates and the 4S peak was not shifted, indicating that BF4 interacted not with the hormone-binding receptor molecule, but with another protein, which was part of the 8S, but not of the 4S peak. We had discovered a new receptor-associated protein, which turned out to be an abundant, cytosol, non-steroid-binding protein of 90 kDa. We soon found out that it was a well known heat shock protein (hsp 90) (Catelli et al, 1985), which could interact with other intracellular proteins such as oncogenic protein kinases (reviewed in Brugge, 1986, Zimiecki et al, 1986). Chick hsp 90 was cloned using BF4 antibody and then sequenced. Among other features it has a negatively charged region, which was called "zone A" (which is predicted to contain two α helices with a proline containing loop in the middle, and where the negative charges could be aligned by the computer on the phosphate backbone of the B-DNA helix)(Binart et al, 1989). At about the same time, the structure of steroid receptors had been described (Hollenberg et al, 1985, Walter et al, 1985, Gronemeyer et al, 1987), and we hypothesized that zone A could be engaged in an ionic interaction with the DNA binding domain (DBD) of steroid receptors, and that somehow hsp 90 could cap the DBD and maintain the receptor in an inactive conformation (Baulieu, 1987).

Steroid Receptors and hsp 90: Functional Analysis by Deletion Mutants

With Bill Pratt, who came to the laboratory on a sabbatical leave, we started to search for the sequence elements that are necessary for receptor:hsp 90 interaction. We used plasmids from Ron Evans' lab, where wild type or mutant glucocorticosteroid receptor expression is under the control of a constitutive promoter, and transfected them into Cos cells, which contain a relatively low level of endogenous glucocorticosteroid receptor. Cytosols were incubated with hormone and analyzed on sucrose gradients (Pratt et al, 1988). In mutants where various regions of the receptor were deleted, but where the steroid/ligand binding domain (LBD) was present, binding of the radioactive steroid indicated the sedimentation coefficient of the receptor to be approximately 8-9S. When mutants lacking the LBD were examined, the fractions were analyzed on SDS gels followed by Western blots which were revealed with antireceptor antibodies. In this case, the receptor was found in the 4S peak. This indicated that the zone important for interaction of the glucocorticosteroid receptor with hsp 90 is the LBD. Moreover, the mutants containing the LBD were incapable of stimulating CAT activity, while those lacking the LBD were constitutively active.

In order to test which part of the LBD is responsible for this association, mutants were tested where three subregions of the LBD were deleted separately. As soon the LBD was altered, hormone binding activity was lost, but all these mutants were 8S, indicating that no limited deletion in the LBD precluded 8S formation, and that there must be more than one zone of interaction (Cadepond et al, 1991). One should note that all these mutants failed to support transcriptional activity, while a mutant where the LBD was replaced by a portion of β-galactosidase didn't bind to hsp 90 and had some constitutive transcription activity.

The situation appeared to be more complex when we studied the estrogen receptor. The sedimentation coefficients and transcriptional activity of a whole series of mutants from Pierre Chambon's laboratory were tested (Chambraud et al, 1990). The results concerning the LBD were essentially the same as those obtained for the glucocorticosteroid

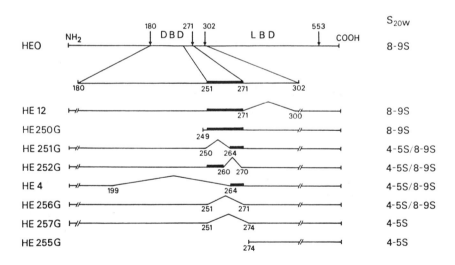

FIGURE 1. Human estrogen receptor: wild type (HEO) or deletion mutants. The positively charged zone is indicated by a black bar, and the sedimentation coefficient of the estrogen receptor in low salt gradients containing molybdate is indicated on the right. LBD: ligand binding domain; DBD: DNA binding domain.

receptor. The N-terminal region of the receptor, including part of the DBD up to amino acid 249 was not necessary either to obtain an 8S complex. However, when mutant HE 255G was compared with HE 250G (Fig.1), a difference of 20 amino acids made a big change in the sedimentation properties of the receptor. While an 8-9 S complex was formed upon deletion of up to 249 amino acids, deletion to 274 led to the formation of only the 4S form. Mutants containing smaller deletions in the positively charged C-terminal part of the DBD were also examined. Deletion of 10-15 amino acids within the zone 251-274 results in the appearance of 4-5 S forms suggesting that the association of hsp 90 with these mutated receptors is destabilized. When the whole 251-274 region is deleted (mutant HE 257G), only the 4-5 S form is obtained. This zone therefore appears to be important for the formation of the 8-9 S complex. It is not however responsible alone for the interaction, since mutants which contain this zone but not the LBD are however 4-5 S. For the estrogen receptor, therefore, both the LBD and a small 251-274 charged region contribute to the formation of the 8-9 S complex. According to Yamamoto's (Picard, D. and Yamamoto, K.R., 1987) and Chambon's (Ylikomi et al, 1992) groups, this zone is included in the nuclear localization signal.

Interaction Between Ligand Binding Domain of Steroid Receptors and hsp 90

Another approach was used in the laboratory to examine if "natural" calf uterus estrogen receptor really associates with hsp 90, and if so, which regions are involved (Sabbah et al, in preparation). Two types of crosslinking agents were used : EDC a zero-length crosslinking agent which creates a peptide bond between adjacent proteins, and DSP, a thiol cleavable homobifunctional reagent, which links two ε-amino groups of lysine residues that are up to 12A° apart (Fig. 2a). The labeled cytosol was crosslinked and then mildly treated with either trypsin or chymotrypsin. After crosslinking with EDC, treating with chymotrypsin which cuts in the middle of the 251-274 zone, and analyzing on a sucrose gradient containing KCl, an 8S peak was obtained, indicating that this region was protected from digestion by the enzyme. On the contrary, trypsin released the LBD and a 4S peak was obtained (Fig. 2b). With DSP, on the other hand, an 8S peak was found in all cases, showing that hsp 90 was probably crosslinked to the LBD, although in this case the enzymatic attack could take place. After reduction of the disulfide

a   **EDC**   1ethyl-3(3dimethylaminopropyl) carbodiimide

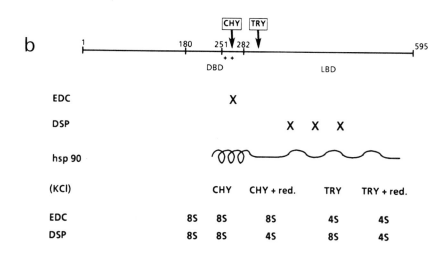

$$CH_3\text{-}CH_2\text{-}N = C = N\text{-}(CH_2)_3\text{-}N \begin{smallmatrix} CH_3 \\ \\ CH_3 \end{smallmatrix} (HCl)$$

(zero length)

DSP   dithio-bis(succinidyl propionate)

$$N\text{-}O\text{-}CO\text{-}(CH_2)_2\text{-}S\text{-}S\text{-}(CH_2)_2\text{-}CO\text{-}O\text{-}N$$

12Å
reversible

b

| CHY | TRY |

1   180   251   282   595

DBD   LBD

EDC   X

DSP   X   X   X

hsp 90

(KCl)	CHY	CHY + red.	TRY	TRY + red.
EDC	8S	8S	8S	4S	4S
DSP	8S	8S	4S	8S	4S

FIGURE 2. a) Crosslinking agents EDC and DSP. b) Calf uterus cytosol was labeled with Tamoxifen azeridine, crosslinked with EDC or DSP, mildly treated by chymotrypsin (CHY) or trypsin (TRY) and sedimented on sucrose gradients containing KCl. The sedimentation coefficient of the labelled fraction is indicated. After DSP crosslinking, treatment with a reducing agent (red) frees the ligand binding domain (LBD). DBD, DNA binding domain.

contained in DSP, the proteolysed, crosslinked estrogen receptor released its LBD (Fig. 2b), which sedimented as a 4S peak.

In general, we noted a weak interaction between hsp 90 and the LBD of estrogen receptor and a strong one with the C-terminal of the DBD (which includes a nuclear localization signal). In the case of glucocorticosteroid receptor, our results and those of Gustaffson and co-workers (Denis et al, 1988), have shown a strong association of hsp 90 at the LBD level with a very weak contribution of the nuclear localisation signal region. In summary (Fig. 3), the LBD could interact with hsp 90, placing it in such a way that the DBD of the receptor is hindered. Upon hormone binding, a change occurs in receptor conformation (most certainly in the LBD) that weakens its affinity for hsp 90, which dissociates from the complex leaving the DBD free to interact with the hormone response elements on the target DNA.

Functions of hsp 90

Finally, after these experiments, and many others from our and other groups, we now believe that hsp 90:

a) protects the receptor from denaturation (which is what a chaperone is supposed to do) (Housley et al, 1990).

b) interacts with the LBD and may stabilize hormone binding [at least for the glucocorticosteroid receptor (Denis, M. and Gustafsson, J-A., 1989)], and could be involved in antihormone action (Groyer et al, 1987).

c) interacts with the C-terminal part of the DBD=NLS whereby it hinders DNA binding (Sabbah et al, in preparation) and may regulate transcriptional activity.

Hsp 90 which is a member of the chaperone family, has been shown to associate with steroid hormone receptors (Joab et al, 1984, Rafestin-Oblin, 1989), oncogenes (Ziemecki et al, 1985), translation factor kinases (Matts and Hurst, 1989, Nilsson et al, 1991), cytoskeleton proteins (Sanchez et al, 1988, Miyata and Yahara, 1991) and also with a 52-59 kDa protein (Renoir et al, 1990, Perdew and Whitelaw, 1991) which has recently been identified.

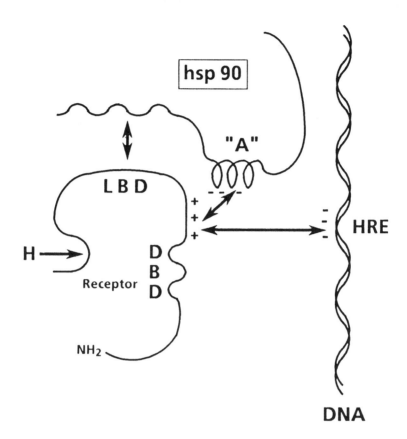

FIGURE 3. Schematical interaction of steroid hormone receptors with hsp 90. H, steroid hormone; HRE: hormone response element. LBD, ligand binding domain; DBD, DNA binding domain.

p59 Association with Steroid Receptors

Much in the same way as the 90K protein was initially observed in our laboratory, Lee Faber, and his colleagues developed a monoclonal antibody from native rabbit progesterone receptor and obtained an antibody against a 59 kDa protein, which was part of the heterooligomers containing the untransformed estrogen, progesterone, glucocorticosteroid or androgen receptors (Tai et al, 1986). Like hsp 90 it did not bind steroids. In fact, we showed that it binds to hsp 90 in 8-9S receptor complexes (Renoir et al, 1990). We prepared calf and rabbit uterus cytosols, aliquots of which were crosslinked with dimethylpimelimidate, and then purified them on an anti-p59 immunoaffinity column. The p59 containing fractions were then subjected to SDS gel electrophoresis and blotted (Fig. 4). Blots A and B contain the same fractions, but blot A was revealed by anti-p59 antibody and blot B by an anti-hsp90 antibody. In lane 1A (uncrosslinked cytosol) there is a large band representing p59, which in lane 2A (crosslinked cytosol) is partly shifted to a high molecular weight form. Blot B shows hsp 90 which is co-immuno-adsorbed with p59. In lane 1B one sees uncrosslinked hsp90, while lane 2B shows that in crosslinked cytosol all the hsp 90 is shifted to a high

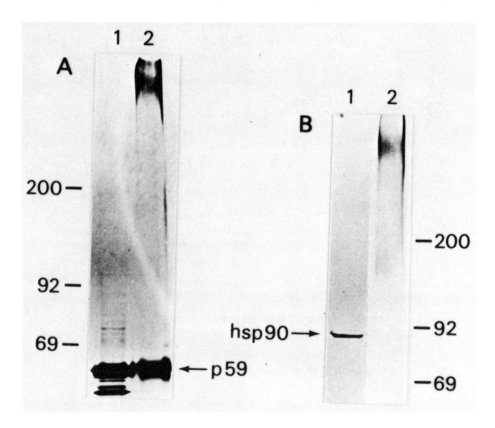

FIGURE 4. Immunoanalysis of p59 and hsp 90 from rabbit uterus cytosol. Cytosol aliquots were crosslinked (2), or not (1), with dimethylpimelimidate and then purified on an anti-p59 immunoaffinity column. The p59-containing fractions were then subjected to SDS polyacrylamide gel electrophoresis and blotted. Blot A was revealed by anti-p59 antibody EC-1, and blot B by anti-hsp 90 antibody AC88.

molecular weight, like p59. This experiment, along with others, where the same type of results were obtained using sucrose gradient analysis, as well as data published by Perdew (Perdew and Whitelaw, 1991) and Gehring (Rehberger et al, 1992), indicate that p59 seems to be associated with hsp 90 and does not interact directly with the receptors, although it cannot be excluded that *in vivo* there are some temporary associations.

Cloning and Sequencing of p59

Lee Faber's monoclonal antibody EC-1 was used to screen an expression library of rabbit liver cDNA, and p59 cDNA was cloned and sequenced (Lebeau et al, 1992). The deduced sequence of 458 amino acids did not correspond to any known protein, however a region of 100 amino acids near the N-terminal presented a 55% homology to peptidyl-prolyl isomerase, and FKBP-12. Using the p59 sequence, Isabelle Callebaut and Jean Paul Mornon applied their Hydrophobic Cluster Analysis, which gives a good idea of the two dimensional structure of the protein (Callebaut et al, 1992). This technique is based on the general principles of protein folding and allows the visualization of hydrophobic clusters that are relevant to the secondary structure and 3D folding of the protein domains. In their model, they suggested that the protein was composed of three relatively similar globular domains + a tail which could bind calmodulin (Fig. 5). The great surprise came from the analogy of these domains with the FK506 Binding protein (or FKBP), a recently described immunosuppressant binding protein. Upon further comparison of the 3D structure of p59 with that of FKBP, Callebaut and Mornon noted that the first domain of p59 could be exactly superimposed on that of FKBP-12, the X-ray structure of which had just been published by Stuart Schreiber's group at Harvard (Michnik et al, 1991). More particularly, in the ligand binding domain (HBI-I), there were no insertions or deletions, and all the structural requirements for immunosuppressant binding were conserved. That p59 could bind immunosuppressant FK506 was confirmed by two reports, one by Deibel's group at Upjohn (Yem et al, 1992) and the other by Lee Faber and co-workers (Tai et al, 1992). These investigators were able to purify p59 on FK506-affinity resins. Since p59 is an immunophilin which can associate with hsp 90, we now call it Heat shock protein Binding Immunophilin (HBI).

p59 - HBI

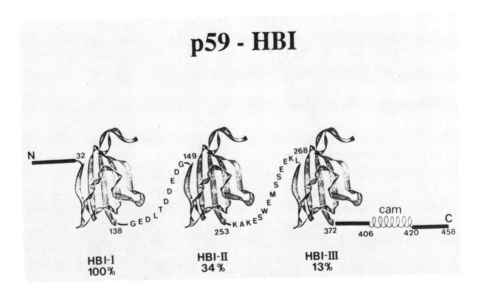

FIGURE 5. Model structure of p59/HBI, showing the domains and the degree of homology of domains II and III relative to domain I (100%). Cam indicates one of the putative calmodulin binding domains.

Immunophilins and Immunosuppressants

The best known protein of the immunophilin family is Cyclophilin (Handschumacher et al, 1984), which is the intracellular receptor of Cyclosporin A. Second generation immunosuppressants such as FK506 and Rapamycin were found initially to bind to FKBP-12, which is a ubiquitous abundant protein, mainly isolated from lymphocyte cell lines (reviewed in Schreiber, 1991). It binds FK 506 and rapamycin with a high affinity (Kd= 0.2-0.4 nM). FKBP-13, which is a membrane protein, found in the endoplasmic reticulum, and FKBP-25 which is a nuclear protein with a binding preference for rapamycin have also been cloned and sequenced (Jin et al, 1991, Hung and Schreiber, 1992, Wiederrecht et al, 1992a). Both proteins contain the binding site for the immunosuppressant, but apart from that are quite different from FKBP-12. And then there is p59, which we cloned in the rabbit, and which has recently been cloned in the human by Harding's group at Vertex (Peattie et al, 1992) and Siekierka's group at Merck (Wiederrecht et al, 1992b), and found to be really a 51-52 kDa protein.

The three main immunosuppressant drugs are shown in Fig 6. These cyclic esters were isolated from soil fungi (for Cyclosporin A), or streptomycetes (for FK506 and rapamycin). While cyclosporin has a different structure, FK506 and rapamycin are closely related. It is strange that cyclosporin and FK506 act in a similar manner involving the Ca^{2+} dependant pathway, while rapamycin has a different mode of action, that is Ca^{2+} independant (reviewed in Schreiber et al, 1991). While the binding domains of FK506 and rapamycin are identical, the other side of their molecules, called "effector domain" (Schreiber, 1991), is very different. This probably is part of the reason why these two drugs can both bind to the same immunophilin with a high affinity, but that the complexes have different targets in the cell. We do not yet know what are the targets of p59/HBI associated with FK506 or rapamycin.

All the immunophilins, including cyclophilin, have peptidyl-prolyl-isomerase (or rotamase) activity (Siekierka et al, 1990). A proline, contained in a peptide, can be in a trans or cis configuration. Rotamase activity catalyzes the cis-trans isomerization of the proline imidic peptide bonds and thereby can change the conformation and folding of a protein (Lang et al, 1987). NMR studies of the Harvard team (Schreiber, 1991) show that part of the FK506 molecule mimicks the Leucine-Proline sequence, which contains a twisted amide bond and is a favoured substrate of peptidyl-prolyl-isomerase. It appears therefore, that via structural analogy, FK506 can inhibit the enzymatic activity by taking the place of the substrate. However, it has been shown by a number of groups using immunosuppressant analogs that bind to immunophilins and inhibit rotamase activity, but are not immunosuppressants, that immunosuppression is not a consequence of rotamase inhibition (Bierer et al, 1990). On the other hand, in the absence of drugs, rotamase activity may play a role in other aspects of cell function, not linked to the immunological response.

Another characteristics of p59 is that it can bind calmodulin (Massol et al, 1992). Since p59/HBI presented the structural requirements for calmodulin binding, it was incubated with calmodulin-Sepharose 4B (Fig. 7). In the presence of Ca^{2+} p59 could be retained and subsequently eluted specifically in the presence of EGTA. The latter observation argued in favour of a specific Ca^{2+}-dependant interaction between p59 and calmodulin. P59/HBI was also treated with calpain, a Ca^{2+}-dependant cysteine protease, which partially proteolyses a limited number of substrates including calmodulin binding proteins. Limited digestion in the presence of Ca^{2+} generated 2 shorter peptides which had lost the C-terminal region of the protein. There is no proteolysis in the absence of Ca^{2+}.

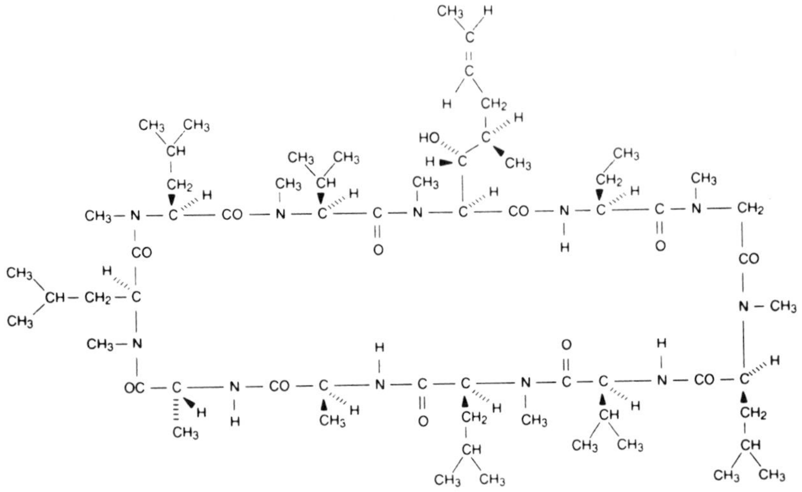

Cyclosporin A

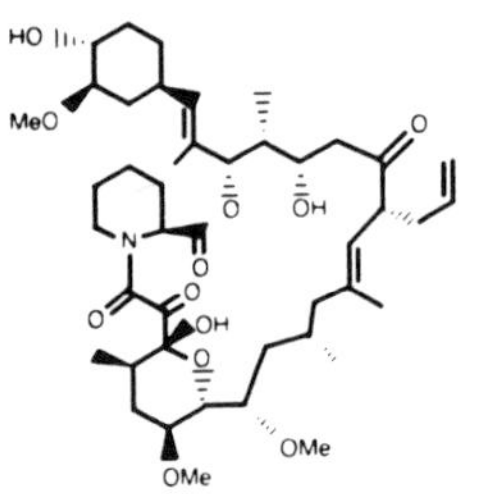

FK506

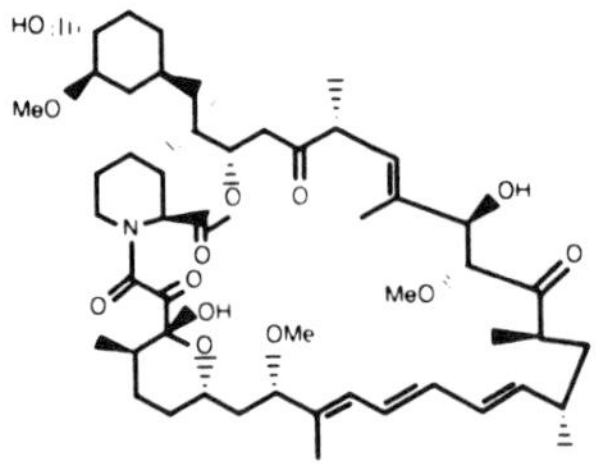

Rapamycin

FIGURE 6. Chemical structures of the three main immunosuppressant drugs currently in use: cyclosporin A (Sandoz), FK506 (Fujisawa) and rapamycin (Wyeth-Ayerst).

Tissue Distribution of p59

We have checked for the presence of p59/HBI mRNA in various tissues by preparing Northern blots which were hybridized with p59 or actin cDNA probes (Fig. 8). P59 mRNA was found in every mammalian tissue we have examined, but in birds hybridization did not resist stringent conditions and the avian equivalent of p59/HBI is probably a different protein (results were similar for Xenopus liver). There was no hybridization of the p59/HBI cDNA probe to a Southern blot of yeast genomic DNA, indicating that p59/HBI is not present in this species either. We will continue however to investigate whether there is an HBI analog in avians, low vertebrates and yeast. We have now produced recombinant p59 in E. coli and Baculovirus and they both bind FK 506 and have rotamase activity.

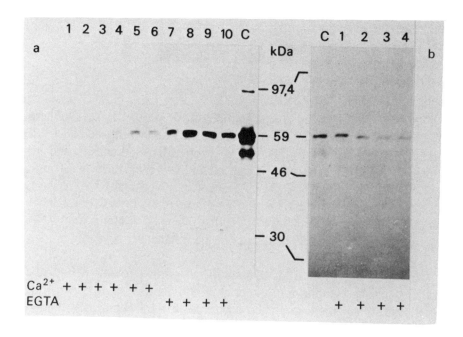

FIGURE 7. Binding of p59 to calmodulin. a) P59/HBI retention on Calmodulin-Sepharose 4B in the presence of Ca^{2+} (lanes 1-6) and elution by EGTA (lanes 8-10); b) Control (C) experiment in the presence of EGTA (lanes 1-4).

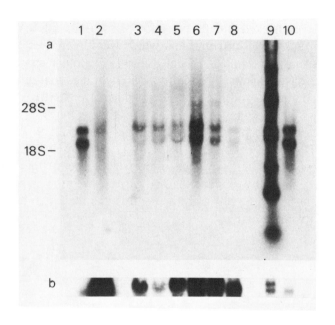

FIGURE 8. Analysis of tissue distribution of p 59/HBI by Northern blots. RNA from various tissues and species was hybridized to a) ^{32}P-rabbit p59 cDNA and b) [32]P- chick actin cDNA. Lanes 1 and 10 represent rabbit liver mRNA (2 µg); lane 2, hen liver mRNA (30 µg); lane 3, rat brain total RNA (34 µg); lane 4, rat liver total RNA (36 µg); lane 5, rat placenta total RNA (36 µg); lane 6, rat thymus total RNA (30µg); lane 7, rat spleen total RNA (30 µg); lane 8, rat uterus total RNA (4 µg); lane 9, ^{32}P labeled lambda Hind III fragments.

In summary, p59/HBI is a mammalian protein found in both the nucleus and the cytoplasm (Gasc et al, 1990, Sanchez et al, 1990, Ruff et al, 1992), and there might be a change in intracellular distribution which depends on the state of the cell. It is bound to hsp 90, but not very tightly and molybdate does not stabilize the association. It is, therefore, necessary to crosslink the two proteins in order to study the hsp 90 and p 59 complex. p59 might be a heat shock protein as has been proposed by Sanchez (Sanchez, 1990). Since the concentration of p 59 in the cell is intermediate between steroid receptors and hsp 90, p59 probably participates in complexes other than receptor heterodigomers.

We, therefore, now know that p59/HBI binds to hsp 90 and to calmodulin, whereby it could play a role in the Ca $^{2+}$-dependant pathway. It binds to the immunosuppressants FK 506 and rapamycin, and it has additional features, such as the capacity to be phosphorylated in the presence of Mn^{2+} or Ca^{2+}/calmodulin (Le Bihan et al, 1993). In the presence of either FK 506 or rapamycin, we have recently observed enhanced binding of the progesterone, estrogen and glucocorticosteroid receptors to their cognate steroid ligands (Renoir et al, 1992). The sedimentation coefficients of the receptors did not change in low salt

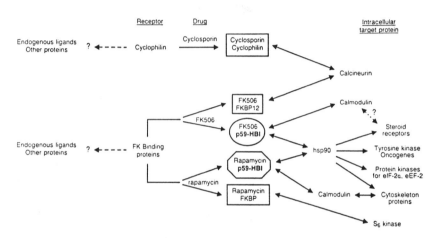

FIGURE 9. Proposed central roles for p59/HBI: immunophilin, heat shock protein and calmodulin binder.

gradients, indicating that there was no dissociation of the heterooligomeric complexes in the presence of the immunosuppressant drugs. A shift of the receptors that was observed in the presence of anti-p59 antibody EC-1 indicated that p59/HBI was still present in the complex. Whether there is a general stabilization of the heterooligomer or a change in the phosphorylation state of the receptors due to the immunosuppressant drugs is not clear at present. P59/HBI appears to be a potentially multifunctional protein which may play a central role in the change of conformation and trafficking of intracellular proteins, including steroid receptors (Fig. 9). It is a molecule situated at the crossroads of endocrinological and immunological processes.

REFERENCES

Baulieu EE (1987): Steroid hormone antagonists at the receptor level: a role for the heat shock protein MW 90,000 (hsp 90). *J Cell Biochem* 35: 161-174.

Bierer BE, Somers PK, Wandless TJ, Burakoff SJ and Schreiber SL (1990): Probing immunosuppressant action with a nonnatural immunophilin ligand. *Science* 250: 556-559.

Binart N, Chambraud B, Dumas B, Rowlands DA, Bigogne C, Levin JM, Garnier J, Baulieu EE and Catelli MG (1989): The cDNA-derived amino acid sequence of chick heat shock protein Mr 90,000 (hsp 90) reveals a "DNA like" structure, potentiel site of interaction with steroid receptors. *Biochem Biophys Res Commun* 159: 140-147.

Brugge JS (1986): Interaction of Rous sarcoma virus protein pp60src with the cellular proteins pp50 and pp90. *Curr Top Microbiol Immunobiol* 123: 1-22.

Cadepond F, Schweizer-Groyer G, Segard-Maurel I, Jibard N, Hollenberg SM, Giguere V, Evans RM and Baulieu EE (1991): Heat shock protein 90 as a critical factor maintaining glucocorticosteroid receptor in a nonfunctional state. *J Biol Chem* 266: 5834-5841.

Callebaut I, Renoir JM, Lebeau M-C, Massol N, Burny A, Baulieu EE and Mornon J-P (1992): An immunophilin that binds Mr 90,000 heat shock protein: main structural features of a mammalian p59 protein. *Proc Natl Acad Sci USA* 89: 6270-6274.

Catelli MG, Binart N, Jung-Testas I, Renoir JM, Baulieu EE, Feramisco JR and Welch WJ (1985): The common 90 kD protein component of non-transformed "8S" steroid receptors is a heat shock protein. *EMBO J* 4: 3131-3135.

Chambraud B, Berry M, Redeuilh G, Chambon P and Baulieu EE (1990): Several regions of human estrogen receptor are involved in the formation of receptor-heat shock protein 90 complexes. *J Biol Chem* 265: 20686-20691.

Denis M and Gustafsson J-A (1989): The Mr=90,000 heat shock protein, an important modulator of ligand and DNA-binding properties of the glucocorticoid receptor. *Cancer Res Suppl* 49: 2275s-2281s.

Denis M, Gustafsson J-A and Wikstrom A-C (1988): Interaction of the Mr=90,000 heat shock protein with the steroid-binding domain of the glucocorticoid receptor. *J Biol Chem* 263: 18520-18523.

Gasc JM, Renoir JM, Faber LE, Delahaye F and Baulieu EE (1990): Nuclear localization of two steroid receptor-associated proteins, hsp 90 and p59. *Exp Cell Res* 186: 362-367.

Gronemeyer H, Turcotte B, Quirin-Stricker C, Bocquel MT, Meyer ME, Krozowski Z, Jeltsch JM, Lerouge T, Garnier JM and Chambon P (1987):The chicken progesterone receptor: sequence, expression and functional analysis. *EMBO J* 6: 3985-3994.

Groyer A, Schweizer-Groyer G, Cadepond F, Mariller M and Baulieu EE (1987): Antiglucocorticosteroid effects suggest why steroid hormone is required for receptors to bind DNA in vivo but not in vitro. *Nature* 328: 624-626.

Handschumacher RE, Harding MW, Rice J, Drugge RJ and Speicher DW (1984): Cyclophilin: a specific cytosolic binding protein for cyclosporin A. *Science* 226: 544-547.

Hollenberg SM, Weinberger C, Ong ES, Cerelli G, Oro A, Lebo R, Thompson EB, Rosenfeld MG and Evans RM (1985): Primary structure and expression of a functional human glucocorticoid receptor cDNA. *Nature* 318: 635-641.

Housley PR, Sanchez ER, Danielsen M, Ringold GM and Pratt WB (1990): Evidence that the conserved region in the steroid binding domain of the glucocorticoid receptor is required for both optimal binding of hsp 90 and protection from proteolytic cleavage. *J Biol Chem* 265: 12778-12781.

Hung DT and Schreiber SL (1992): cDNA cloning of a human 25 kDa FK 506 and rapamycin binding protein. *Biochem Biophys Res Commun* 184: 733-738.

Jin Y-J, Albers MW, Lane WS, Bierer BE, Schreiber SL and Burakoff SJ (1991): Molecular cloning of a membrane-associated FK506- and rapamycin-binding protein FKBP-13. *Proc Natl Acad Sci USA* 88: 6677-6681.

Joab I, Radanyi C, Renoir JM, Buchou T, Catelli MG, Binart N, Mester J and Baulieu EE (1984): Immunological evidence for a common non hormone-binding component in "non-transformed" chick oviduct receptors of four steroid hormones. *Nature* 308: 850-853.

Lang K, Schmid FX and Fischer G (1987): Catalysis of protein folding by prolyl isomerase. *Nature* 329: 268-270.

Lebeau M-C, Massol N, Herrick J, Faber LE, Renoir JM, Radanyi C and Baulieu EE (1992): P59, an hsp 90-binding protein; cloning and sequencing of its cDNA and preparation of a peptide-directed polyclonal antibody. *J Biol Chem* 267: 4281-4284.

Massol N, Lebeau M-C, Renoir JM, Faber LE and Baulieu EE (1992): Rabbit FKBP-59-Heat shock protein Binding Immunophilin (HBI) is a calmodulin binding protein. *Biochem. Biophys. Res. Commun.* 187: 1330-1335.

Matts HS, Robert L and Hurst R (1989). Evidence for the association of the heme-regulated eIF-2 kinase with the 90-kDa heat shock protein in rabbit reticulocyte lysate in situ. *J Biol Chem* 264: 15542-15547.

Michnick SW, Rosen MK, Wandless TJ, Karplus M and Schreiber SL (1991): Solution structure of FKBP, a rotamase enzyme and receptor for FK506 and rapamycin. *Science* 252: 836-839.

Miyata Y and Yahara I (1991): Cytoplasmic 8S glucocorticoid receptor binds to actin filaments through the 90-kDa heat shock protein moiety. *J Biol Chem* 266: 8779-8783.

Nilsson A, Carlberg U and Nygard O (1991): Kinetic characterisation of the enzymatic activity of the eEF-2-specific Ca^{2+}-and calmodulin-dependant protein kinase purified from rabbit reticulocytes. *Eur J Biochem* 195: 377-383.

Peattie DA, Harding MW, Fleming MA, DeCenzo MT, Lippke JA, Livingstone DJ and Benasutti M (1992): Expression and characterization of human FKBP52, an immunophilin that associates with the 90-kDa heat shock protein and is a component of steroid receptor complexes. *Proc Natl Acad Sci USA* 89: 10974-10978.

Perdew GH and Whitelaw ML (1991): Evidence that the 90-kDa heat shock protein (hsp 90) exists in cytosol in heteromeric complexes containing hsp 70 and three other proteins with Mr 63,000, 56,000 and 50,000. *J Biol Chem* 266: 6708-6713.

Picard D and Yamamoto KD (1987): Two signals mediate hormone-dependant nuclear localization of the glucocorticoid receptor. *EMBO J* 6: 3333-3340.

Pratt WB, Jolly DJ, Pratt DV, Hollenberg SM, Giguere V, Cadepond FM, Schweizer-Groyer G, Catelli MG, Evans RM and Baulieu EE (1988): A region in the steroid binding domain determines formation of the non-DNA-binding, 9S glucocorticoid receptor complex. *J Biol Chem* 263: 267-273.

Rafestin-Oblin ME, Couette B, Radanyi C, Lombès M and Baulieu EE (1989): Mineralocorticosteroid receptor of the chick intestine: oligomeric structure and transformation. *J Biol Chem* 264: 9304-9309.

Rehberger P, Rexin M and Gehring U (1992): Heterotetrameric structure of the human progesterone receptor. *Proc Natl Acad Sci USA* 89: 8001-8005.

Renoir JM, Radanyi C, Faber LE and Baulieu EE (1990): The non-DNA-binding heterooligomeric form of mammalian steroid hormone receptors contains a hsp 90-bound 59-kilodalton protein. *J Biol Chem* 265: 10740-10745.

Renoir JM, Radanyi C and Baulieu EE (1992): Un effet des immunosuppresseurs FK506 et rapamycine sur le fonctionnement du récepteur de la progestérone: la protéine p59-HBI, un carrefour de l'immunologie et de l'endocrinologie. *C R Acad Sci Paris* 315: série III, 421-428.

Ruff VA, Yem AW, Munns PL, Adams LD, Reardon IM, Deibel MR Jr and Leach KL (1992): Tissue distribution and cellular localization of hsp 56, an FK506-binding protein. *J Biol Chem* 267: 21285-21288.

Sabbah M, Radanyi C, Redeuilh G and Baulieu EE (1993): The heat shock protein Mr 90,000 (hsp 90) modulates the binding of the estrogen receptor to its cognate DNA (in preparation).

Sanchez ER (1990): Hsp 56: a novel heat shock protein associated with untransformed steroid receptor complexes. *J Biol Chem* 265: 22067-22070.

Sanchez ER, Faber LE, Henzel WJ and Pratt WB (1990): The 56-59-kilodalton protein identified in untransformed steroid recptor complexes is a unique protein that exists in cytosol in a complex with both the 70-and 90-kilodalton heat shock proteins. *Biochemistry* 29: 5145-5152.

Sanchez ER, Redmond T, Scherrer LC, Bresnick EH, Welsh MJ and Pratt WB (1988): Evidence that the 90-kilodalton heat shock protein is associated with tubulin-containing complexes in L cell cytosol and in intact PtK cells. *Molecular Endocrinol* 2: 756-760.

Schreiber SL (1991): Chemistry and biology of the immunophilins and their immunosuppressive ligands. *Science* 251: 283-287.

Schreiber SL, Liu J, Albers MW, Karmacharya R, Koh E, Martin PK, Rosen MK, Standaert RF and Wandless TJ (1991): Immunophilin-ligand complexes as probes of intracellular signaling pathways. *Transplantation Proceed* 23: 2839-2844.

Siekierka JJ, Wiederrecht G, Greulich H, Boulton D, Hung SHY, Cryan J, Hodges PJ and Sigal NH (1990): The cytosolic-binding protein for the immunosuppressant FK-506 is both a ubiquitous and highly conserved peptidyl-prolyl cis-trans isomerase. *J Biol Chem* 265: 21011-21015.

Tai PKK, Maeda Y, Nakao K, Wakim NG, Duhring JL and Faber LE (1986). A 59-kilodalton protein associated with progestin, estrogen, androgen, and glucocorticoid receptors. *Biochemistry* 25: 5269-5275.

Tai PKK, Albers MW, Chang H, Faber LE and Schreiber SL (1992): Association of a 59-kilodalton immunophilin with the glucocorticoid receptor complex. *Science* 256: 1315-1318.

Walter P, Green S, Greene G, Krust A, Bornert JM, Jeltsch JM, Staub A, Jensen E, Scrace G, Waterfield M and Chambon P (1985): Cloning of the human estrogen receptor cDNA. *Proc Natl Acad Sci. USA* 82: 7889-7893.

Wiederrecht G, Martin MM, Sigal NH and Siekierka JJ (1992a): Isolation of a human cDNA encoding a 25 kDa FK-506 and rapamycin binding protein. *Biochem Biophys Res Commun* 185: 298-303.

Wiederrecht G, Hung S, Chan HK, Marcy A, Martin M, Calaycay J, Boulton D, Sigal N, Kincaid RL and Siekierka JJ (1992b): Characterization of high molecular weight FK-506 binding activities reveals a novel FK-506-binding protein as well as a protein complex. *J Biol Chem* 267: 21753-21760.

Yem AW, Tomasselli AG, Heinrikson RL, Zurcher-Neely H, Ruff VA, Johnson RA and Deibel MR Jr. (1992): The hsp56 component of steroid receptor complexes binds to immobilized FK506 and shows homology to FKBP-12 and FKBP-13. *J Biol Chem* 267: 2868-2871.

Ylikomi T, Bocquel MT, Berry M, Gronemeyer H and Chambon P (1992). Cooperation of proto-signals for nuclear accumulation of estrogen and progesterone receptors. *EMBO J* 11: 3681-3694.

Zimiecki A, Catelli MG, Joab I and Moncharmont B (1986): Association of the heat shock protein hsp 90 with steroid hormone receptors and tyrosine kinase oncogene products. *Biochem Biophys Res Comm* 138: 1298-1307.

ESTROGENIC AND DEVELOPMENTAL REGULATION OF 90-KILODALTON HEAT SHOCK PROTEIN GENE EXPRESSION

G. Shyamala
Lawrence Berkeley laboratory
University of California

INTRODUCTION

It is well known that when whole organisms, tissues or cultured cells are exposed to elevated temperature, a wide range of chemicals and other metabolic insults, they synthesize a small number of highly conserved proteins called heat shock proteins (HSPs). The HSPs are also classified as stress proteins or stress-induced proteins. The heat shock response is universal, which has led to the belief that the HSPs synthesized in response to stress ensure the survival of the essential cellular components during the stress period and allow a rapid recovery and resumption of normal cellular functions. Among all the classes of HSPs, proteins with an approximate molecular mass of 90 kDa (HSP-90) and 70 kDa (HSP-70) are the most prominent and also highly conserved among various organisms and tissues (For reviews see 1-5).

Although HSPs were originally identified as proteins made in response to heat shock and other stresses, either HSPs themselves or close relatives of this gene family are present in all organisms at normal temperatures, and are believed to play important roles in normal cellular functions. In support of this, in the last few years, several lines of

STEROID HORMONE RECEPTORS
V. K. Moudgil, Editor
© 1993 Birkhäuser Boston

evidence have established that HSP-90 can complex with steroid receptors *in vitro* and modify their DNA and steroid binding properties. Based on this and other steroid receptor structure-function relationship studies, it is speculated that HSP-90 has a regulatory role in steroid hormone actions (6-11).

A principal characteristic which distinguishes HSP-90 from some other HSPs is that it belongs to a subset whose cellular levels may be subjected to developmental regulation in a wide variety of species. For example, in Drosophila HSP-83, which is analogous to mammalian HSP-90, is expressed at high levels during normal development in the absence of heat shock (12). Sporulating yeast also expresses high levels of HSP-90 in the absence of any stress (13). Similarly, early mouse embryos synthesize at a high rate HSP-90 in the absence of any stress (14). Teratocarcinoma cells constitutively express high levels of HSP-90 which decreases upon *in vitro* differentiation (15). Based on these observations and those which implicate a role for HSP-90 in regulating steroid hormone actions prompts us to postulate that in target tissues for steroids, steroid hormones and cellular HSP-90 may mutually regulate their cellular functions i.e. while HSP-90 may regulate steroid hormone actions, cellular levels of HSP-90 may itself be subjected to regulation by the steroid hormone, especially if the effect of steroid is to regulate growth. Towards testing this hypothesis, we are studying the estrogenic regulation of HSP-90 in female reproductive tissues, in particular the uterine and mammary HSP-90 gene expression and its relationship to development. The present chapter summarizes the results of some of these studies.

Estradiol Regulates Positively the Steady-State Levels of HSP-90

A. *Effect of estrogens on murine uterine HSP-90 protein.* For the last two decades, the rodent uterus has served as an excellent model system for studying estrogen action and estrogen-dependent growth of a normal cell. To this end, initially we examined if estrogen could modulate the steady-state levels of uterine HSP-90. Uterine HSP-90 levels were assayed by immunoblot analysis using a variety of anti-HSP-90 antibodies. The results from these studies, demonstrating a reduction in HSP-90 levels due to ovariectomy and its increase due to estradiol, are shown in Figures 1 and 2, respectively. As shown in Table I, the time course and magnitude of estrogenic stimulation of HSP-90 was similar to that seen with progesterone receptor, a well known marker of estrogen action in the mouse uterus (16). Figure 3 shows that the increase in uterine HSP-90 due to estrogen is steroid specific; as may be seen, in contrast to estradiol and diethylstilbestrol, progestin (R5020) and androgens (testosterone and dihydrotestosterone) had no effect on the basal levels of HSP-90. The target tissue specificity for the effect of estrogen is shown in Figure 4 whereby liver and spleen (in contrast to uterus) do not exhibit an increase in their basal levels of HSP-90 due to estrogen.

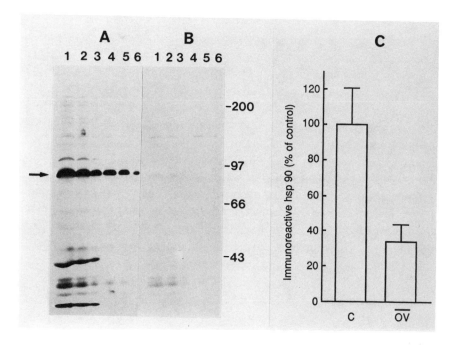

FIGURE 1: *Effect of ovariectomy on uterine HSP-90 level.* Mouse uterine cytosolic extracts corresponding to 6.5 (lanes 3 and 6), 16 (lanes 2 and 5), and 26 (lanes 1 and 4) μl from nonovariectomized (lanes 1-3) and ovariectomized (lanes 4-6) animals were subjected to gel electrophoresis, electroblotted, and probed using a HSP-90 rabbit antiserum. A, With HSP-90 antiserum; B, with normal rabbit serum. The position of HSP-90 is indicated by he arrow. C, Quantitation of immunoreactive HSP-90 level using [^{125}I] protein-A from three sets of nonovariectomized (c) and ovariectomized (ov) animals.

284

B. *Effect of estradiol on murine uterine HSP-90 gene expression.*
In mammalian cells, two forms of HSP-90 encoded by two distinct genes
have been found; in humans, the two forms have been designated as
HSP-89α and HSP-89β (17-20), and their corresponding murine analogs
are HSP-86 and HSP-84 (21, 22). The two forms of HSP-90 are
primarily distinguished by two short segments present in the amino
terminal domain of HSP-86 but absent from HSP-84. Several lines of
evidence suggest that HSP-84 and HSP-86 may have different cellular
functions, particularly as they pertain to proliferation and differentiation.

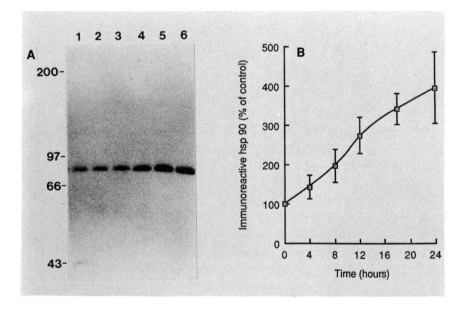

FIGURE 2: *Time course of the effect of estradiol on uterine HSP-90.*
Uterine cytosol was prepared and processed for HSP-90 analysis as
described for Figure 1. A, Autoradiogram of a typical experiment. Lane
1 corresponds to cytosol from saline-injected animals, whereas lanes 2-6
correspond to cytosols from animals injected with 1 μg estradiol for 4,
8, 12, 18 and 24 h, respectively. B, Results from the time-course
experiments. The 100% value represents the content of HSP-90 in the
cytosol of saline-injected (control) animals. The data represents the mean
± SD from three independent experiments. In the absence of HSP-90
antiserum but in the presence of normal rabbit serum no band were
visible in the autoradiogram (data not shown).

Table I. **Effect of estradiol on the uterine concentrations of HSP-90 and progesterone receptor.**

| Treatment | Time (h) | % of control | |
		HSP-90	Progesterone receptor
Saline	24	100	100
Estradiol	4	143 ± 30	144 ± 33
Estradiol	12	275 ± 46	263 ± 17
Estradiol	18	331 ± 43	311 ± 65
Estradiol	24	401 ± 79	367 ± 97

Ovariectomized mice were injected with either saline (control) or 1 μg estradiol for the indicated times before processing the uteri for estimation of HSP-90 and progesterone receptor. The data represent the average ± SD of results from two to five separate experiments.

For example, in embryonic testis, among the various cell types, it is the stem cells with the potential to proliferate which express preferentially the HSP-86 gene as compared to the HSP-84 gene (23). Similarly, teratocarcinoma cells, which constitutively express high levels of HSP-84 and 86 during proliferation when made to differentiate *in vitro*, specifically down regulate HSP-86 and not HSP-84 (15). Therefore, as a first step towards assessing the potential biological importance of estrogen induced HSP-90, we examined the effect of estradiol on HSP-84 versus HSP-86 gene expression.

The data from time course studies (Figure 2) showed that the maximal accumulation of mRNA is observed at approximately four hours after steroid administration. Since the protein steady-state level reaches a maximum at about eighteen hours, after estrogen treatment (Figure 2), these findings suggest that the estrogenic regulation of uterine HSP-90 may be primarily occurring at the level of transcription. As shown in Figure 5, estradiol had a far greater effect (~15 fold) on HSP-86 mRNA than on HSP-84 (~5 fold) mRNA. Northern blot analysis confirmed these results showing that estradiol caused a much greater increase in the 3.1 kilobase transcript corresponding to HSP-86 than in the 2.8 kilobase transcript corresponding to HSP-84 (Figure 6).

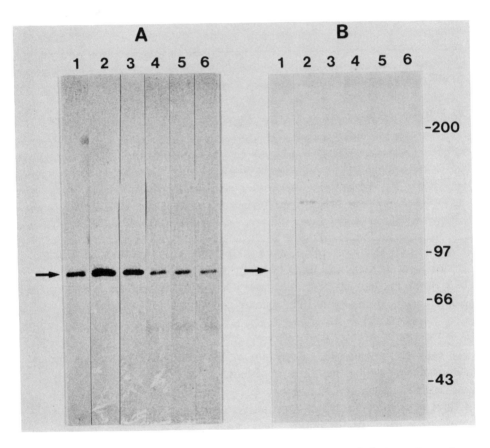

FIGURE 3: *Steroid hormone specificity of the modulation of uterine HSP-90.* Uterine cytosol extracts from ovariectomized mice injected for 18 h with saline (lane 1) or 1 μg each of estradiol (lane 2), diethylstilbestrol (lane 3), progesterone (lane 4), testosterone (lane 5), and dihydrotestosterone (lane 6) 18 h before killing were processed, as described for Figure 1. The blot in A was immunoblotted with the HSP-90 antiserum, whereas the blot in B was incubated with normal rabbit serum. The arrow indicates the position of HSP-90.

Studies on the developmental induction of HSPs have shown that this type of induction is distinguished from that of heat shock by the absence of HSP-70 synthesis (1). Figure 7 shows that, in contrast to HSP-90, uterine HSP-70 is not regulated by estradiol. The glucose regulated protein (GRP-94) is normally induced in cells following glucose deprivation or when exposed to amino acid analogs (24, 25) but is often classified as a member of the HSP-90 family. GRP-94 is distinct from

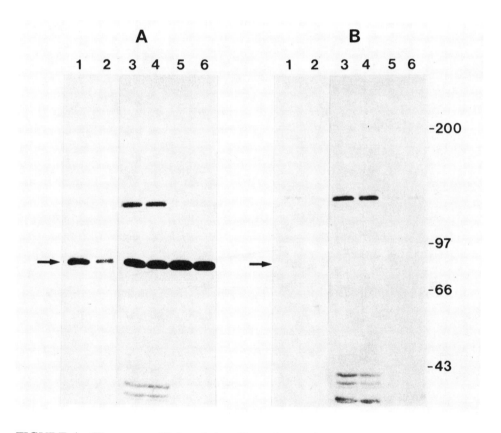

FIGURE 4: *Tissue specificity of the effect of estradiol on HSP-90 level.* Cytosol extracts (16 μl) from uterus (lanes 1 and 2), liver (lanes 3 and 4), and spleen (lanes 5 and 6) from ovariectomized mice injected for 18 h with saline (lanes 2,3, and 5) or 1 mg estradiol (lanes 1, 4, and 6) were processed for HSP-90, analysis as described for Figure 1. A, with HSP-90 antiserum; B, with normal rabbit serum. The arrow indicates the position of HSP-90.

HSP-84 and 86 and is associated with the lumen of the endoplasmic proteins in contrast to HSP-86 and 84 which are predominantly cytoplasmic proteins. GRP-94 has been shown to be regulated by steroid hormones in avian oviduct (26). Figure 7 shows that, in contrast to HSP-70, the levels of GRP-94 mRNA are increased by estrogen but to a much lesser extent than that seen with HSP-86.

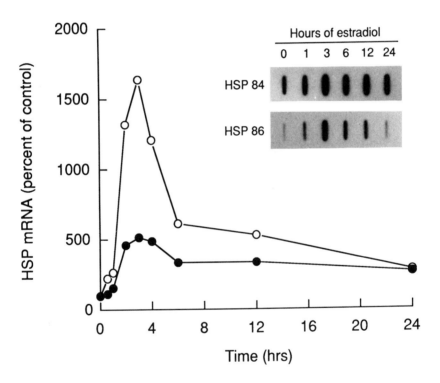

FIGURE 5: *Time course of the effect of estradiol on murine uterine HSP-84 and HSP-86 mRNAs.* Ovariectomized mice were treated with either saline (control) or estradiol for indicated times prior to the processing of the uteri for slot blot analyses of HSP-86 (O) and HSP-84 (●) mRNAs. The data represent the densitometric scan of autoradiographs presented as percentages of control (zero time, saline group) and is the average of two or three separate experiments. Inset, Autoradiograph of a typical experiment.

C. *Immunolocalization of murine uterine HSP-90.* It is well known that when estradiol is administered *in vivo*, there is an influx of foreign cells such as eosinophils into the rodent uterus (27). HSP-90 is an ubiquitous protein, known to be present in almost all cells (1). Therefore, to assess the biological importance of uterine HSP-90 it was necessary to verify that the estrogen dependent increase was intrinsic to the tissue and if so whether there were any differences between the endometrium and the myometrium. Accordingly, an *in situ* analysis for HSP-90 was performed. These studies revealed that the increase in uterine HSP-90

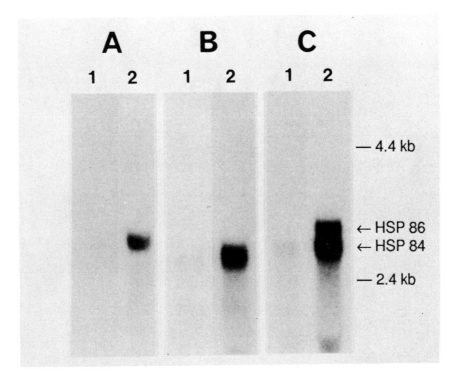

FIGURE 6: *Northern blot analyses of murine uterine HSP-90 mRNAs.* Total cellular RNA (15 μg) isolated from uteri of ovariectomized mice treated with saline (lanes 1) or estradiol (lanes 2) for 3 h was loaded in each lane and hybridized with ^{32}P-labeled cDNA probes of similar specific activity. The blots shown in panels A and B were hybridized with HSP-86 and HSP-84, respectively, while that in panel C was hybridized with both probes. The blots shown in panels B and C were autoradiographed for 7h, while that in panel A was autoradiographed for 17h for better visualization of the basal level of HSP-86.

290

observed in response to estradiol was indeed intrinsic to the tissue and was primarily associated with the endometrium (Figures 8A and B). In addition to the endometrium, HSP-90 was also associated with the luminal epithelium in both control and estrogen treated tissues. In contrast to the endometrium, there was no significant localization of HSP-90 in the myometrium of both control and estrogen treated uteri.

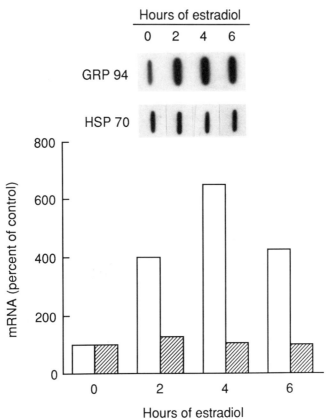

FIGURE 7: *Effect of estradiol on uterine GRP-94 and HSP-70.* Ovariectomized mice were injected with saline or estradiol for indicated times prior to removal of the uteri. Total RNA (15 μg) from each group was immobilized on nitrocellulose membrane and hybridized with ^{32}P-labeled probes specific for either GRP-94 (open bar) or HSP-70 (line bar). The bar graph represents the relative denistometric scanning units presented as percentages of control (zero time, saline group); data are averages of two experiments. The top of the panel shows the autoradiograph from one of these experiments.

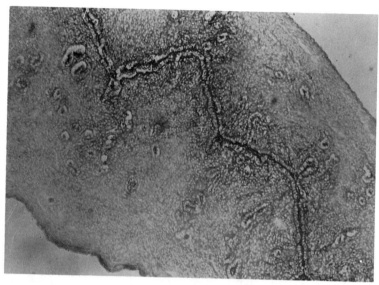

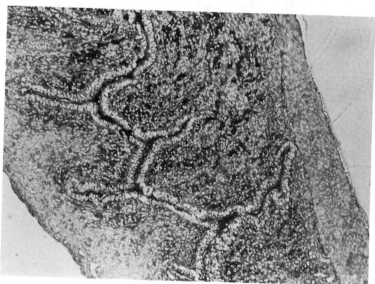

FIGURE 8: *Immunohistochemical analysis of murine uterine HSP-90.*
Cryostat section of fresh, frozen uteri were precessed for
immunohistochemical analysis using rabbit antiserum prepared against
murine HSP-90 (53) and colloidal gold labeling (54). *Panel A* (top)
represents the uterus of an ovariectomized mouse treated with saline for
18 hours *Panel B* (bottom) represents the uterus of an ovariectomized
mouse treated with estradiol for 18 hours.

TABLE II. Relative levels of immunoreactive HSP-90 estrogen receptors and progesterone receptors in normal human cyclic endometrium.

Histologic dating	HSP-90	Estrogen Receptor*		Progesterone receptor*	
	% of control	Epithelium	Stroma	Epithelium	Stroma
Early proliferative	21.5	+ + + +	+ + + +	+ +	+
Mid Proliferative	58.6	+ + + +	+ + + +	+ + +	+ +
Late proliferative	104	+ + + +	+ + + +	+ + + +	+ +
Early secretory	63.8	+ + + +	+ +	+ + + +	+ + +
Mid secretory	75.8	+ +	+ +	+ +	+ + +
Late secretory	72.2	None	+	None	+

For analysis of HSP-90 total cytosolic protein was assayed by Western blot analysis; data are normalized to murine uterine cytosol (100%) analyzed simultaneously.

*Estrogen and progesterone receptor were assayed by immunocytochemistry; data represent the immunohistochemical score corresponding to the sum of both the percentage of positive cells and the staining intensity.

D. *Relative expression of HSP-86 versus HSP-84 and its relationship to estrogen action.* In all target tissues for estrogen examined, estrogen caused a greater increase in HSP-86 mRNA than HSP-84 mRNA. Overall, the relative increase in HSP-86 mRNA due to estrogen co-related well with the intrinsic estrogenic sensitivity of these tissues such that it was uterus > vagina > mammary glands. In the absence of estrogen, in all these tissues of murine origin, HSP-84 mRNA was more abundant; it was approximately two to three fold higher in uterus while in vagina and mammary glands, it was about six to nine fold higher than HSP-86 mRNA. This suggests that if *in vivo*, in the absence

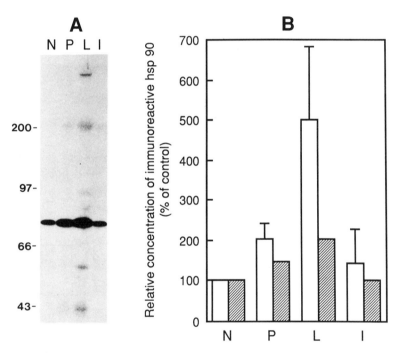

FIGURE 9: *Relative concentrations of HSP-90 in mammary tissues from various development status.* Equal samples of mammary cytosol from the various developmental states (N, nulliparous; P, pregnant, L, lactating; I, involuting) were processed for measurement of HSP-90 by using HSP-90 antiserum and ^{125}I-Protein A (a) Autoradiograph of a typical immunoblot, (b) Amounts of HSP-90 present in equal volumes of the extracts, as measured (open bar), and normalized to total protein concentration (line bar). The data represent the averages of four separate experiments. The 100% control value represents the amount of HSP-90 in mammary cytosol of nulliparous mice.

294

of steroid (estrogen), HSP-90 indeed is responsible for maintaining the receptor (estrogen receptor) in a biologically non-functional state, most likely it is HSP-84 and not HSP-86 which is responsible for mediating this function. This leads us to speculate that the cellular responsiveness to estrogen by its target tissues may be determined by the combined effects of HSP-86 and HSP-84; this in turn may be dictated by the relative magnitude with which each of these genes respond to estrogen and their constitutive synthesis. As such, in female reproductive tissues in which estrogen modulates growth, the relative expression of HSP-86 versus HSP-84 can be important for maintaining cellular replicative homeostasis.

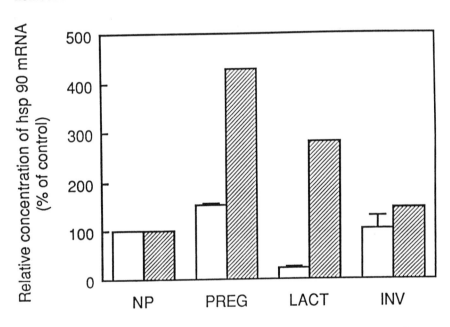

FIGURE 10: *Quantification of the HSP-90 mRNA in mammary tissues from various developmental states.* Mammary tissues from various developmental stages were analyzed for HSP-90 mRNA exactly as described for Figures 5 and 6. For quantification, the autoradiographs were scanned and the data are expressed in terms of relative intensity. The 100% control value represents the scanning intensity observed with RNA from mammary tissues of nulliparous mice: open bar, relative scanning intensity among various tissues normalized to total cellular RNA; line bar, that normalized to total cellular DNA. The data represent means ± S.E.M. of three separate experiments. Abbreviations: N, nulliparous, P, pregnant; L, lactating; I, involuting.

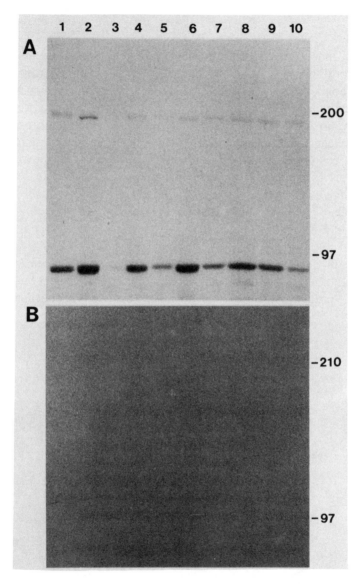

FIGURE 11: *Immunoblot analysis of HSP-90 in human mammary tumors.* Aliquots of cytoplasmic extracts prepared form individual tumors, containing equivalent amounts of protein, were processed for Western blot analysis and probed with either a rabbit antiserum prepared against murine HSP-90 (Panel A) or normal rabbit serum (Panel B). The position of the molecular weight standards is indicated on the right $(x10^{-3})$.

Analysis for HSP-90 in the Human Endometrium During the Menstrual Cycle

In the normal cyclic endometrium, estrogens induce epithelial and stromal proliferation during pre ovulatory phase whereas progesterone induces epithelial secretory differentiation and stromal decidualization during the post ovulatory phase of estrogen primed endometrium (28). Therefore, to assess whether as found with the murine uterus, estrogens can modulate human endometrial HSP-90 and if so, its relationship to epithelial cell proliferation and differentiation, we analyzed for its steady state levels as a function of menstrual cycle. As shown in Table II, during preovulatory phase, there was a progressive increase in the steady-state levels of HSP-90, suggesting the involvement of estrogen. During this phase, a progressive increase in progesterone receptors is also seen (29, 30; Table II). During post ovulatory phase, there was a decline in the levels of HSP-90 which, however, did not return to "baseline" values of early proliferative phase. This suggests that the HSP-90 synthesized during the post-ovulatory phase is most likely estrogen-independent since

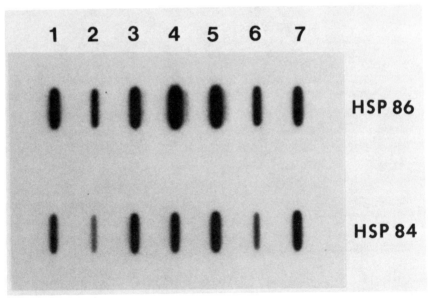

FIGURE 12: *Analysis for HSP-86 (HSP-90-α) and HSP-84 (HSP-84-β) mRNAs in human mammary tumors.* Equal amounts of total cellular RNA from individual tumors were assayed for HSP-86 and HSP-84 mRNA by slot blot analysis using human cDNA probes corresponding to each gene.

this phase is associated with low tissue levels of estrogen receptor and circulating estrogen (31). Interestingly, during this phase, progesterone receptor are present in the stroma at relatively high levels (Table II) and we have previously documented that unlike the progesterone receptors of the epithelium which requires estrogen for their synthesis, stromal progesterone receptors may be synthesized independently of estrogen (32). It is, therefore, likely that the HSP-90 present in the endometrium of the post-ovulatory phase is predominantly associated with the stroma. Based on these observations we propose that, similar to our observations with the rodent tissue, estrogens can augment the steady-state levels of HSP-90 in the human endometrium, the tissue component targeted for estrogen dependent growth.

Developmental Regulation of Mammary HSP-90

Estrogen and progesterone are important for the development and differentiation of normal breasts and their involvement in the growth of human mammary carcinoma was implicated by Beatson more than a century ago. Most of the classic studies on the developmental biology of normal mammary glands have been done with rodents, especially mice, and conclusions derived from the rodents have been found to be frequently applicable to many other species, including humans. These studies have shown that normal mammary growth has two distinct phases: in the adult female, during each menstrual cycle, cyclic proliferative changes and active growth of the ductal tissue occurs, but the full lobuloalveolar development and differentiation of the gland occurs only during pregnancy and lactation (for reviews see 33-36). Both the mammary development occurring during puberty and pregnancy are believed to be regulated by estradiol and progesterone, in support of which estradiol and progesterone have been shown to be important for mouse mammary epithelial cell proliferation (37-40). However, once the mammary epithelial cell achieves a state of full functional differentiation, such as that accompanying lactation, ovarian hormones are no longer required to maintain this physiological state (38). In fact, as documented extensively by our laboratory, lactating mammary glands become refractory to the action of estradiol despite the presence of estrogen receptors (41). The most striking example of this is the estradiol-dependent regulation of progesterone receptor gene expression, whereby during lactation mammary glands totally lack detectable levels of either progesterone receptor mRNA or protein (42). This raised the possibility that mammary HSP-90 may be subjected to both estrogenic and developmental regulation and if so, depending upon the developmental state, HSP-90 may have different biological function.

Analysis for HSP-90 in mammary glands from various developmental states revealed that tissues of pregnant and lactating mice had higher levels of HSP-90 as compared to that present in glands derived from either nulliparous mice or mice undergoing lactational involution (Figure 9). Analysis of HSP-90 mRNA also revealed it to be higher in glands of pregnant mice as compared to that present in tissues from nulliparous mice and mice undergoing lactational involution. However, in contrast to the protein levels, the lactating mammary glands had the lowest level of HSP-90 mRNA (Figure 10). It is well documented that mammary development and differentiation occurring during pregnancy and lactation is accompanied by an increased accumulation of both DNA (due to an increase in epithelial cell proliferation) and total RNA per cell (resulting from an increased synthesis of milk proteins by the epithelial cells). This raised the possibility that the lowest amount of HSP-90 mRNA found in lactating tissue might have resulted from a dilution of this mRNA and not necessarily due to a lesser amount per cell. Indeed, when the relative level of HSP-90 mRNA were normalized to total DNA, similar to the data obtained with protein, lactating mammary glands were found to contain higher levels of HSP-90 mRNA as compared to tissue from either nulliparous mice or mice undergoing lactational involution (Figure 10).

The developmental regulation of HSP-90 appeared to be an estrogen independent phenomenon, such that in lactating mammary glands, ovariectomy and administration of estrogen had no effect on the steady state levels of either HSP-84 or HSP-86 mRNA. In the mammary glands of ovariectomized nulliparous mice, however, estradiol did cause an increase in the steady state levels of both HSP-84 and HSP-86 mRNA (by 1.5 and 2.5 fold respectively).

Relevance of HSP-90 to Clinical Management of Carcinomas

The importance of cellular HSP-90 in cellular proliferation is suggested by observations that altered expression of HSP-90 may be to a common characteristic of transformed cells (43-46). Indeed, in a preliminary survey of randomly chosen human breast tumors, we have found a huge variation in their steady-state levels of HSP-90 both at the level of protein (Figure 11) and mRNA (Figure 12). At the mRNA level, as shown in Figure 12, this was reflected both in terms of the total amount present in each tumor and the relative expression of HSP-86 versus HSP-84 genes.

SUMMARY AND CONCLUSIONS

Studies from several laboratories during the past eight to ten years have established that HSP-90 can complex with steroid receptors and maintain them in an inactive state. Interestingly, cellular level of HSP-90 itself appears to be under the regulation of steroid hormone, as documented by our studies on the estrogenic regulation of HSP-90. Based on these observations we postulate that in target tissues for steroids, steroid hormones and cellular HSP-90 may mutually regulate their cellular functions i.e. while HSP-90 may regulate steroid hormone actions, cellular levels of HSP-90 may itself be subjected to regulation by the steroid hormone. HSP-90 belongs to a subset of HSPs which are developmentally regulated in many species and alterations in its expressions are commonly found in transformed cells. As such, in female reproductive tissues in which estrogen modulates growth, this mutual regulatory phenomenon by HSP-90 and estrogen can be important for maintaining cellular replicative homeostasis.

Our studies also clearly reveal that among the two genes (HSP-84 and HSP-86) coding for HSP-90, the response to estrogen is much greater for HSP-86. Therefore, if HSP-90 is indeed involved in promoting estrogen dependent growth, most likely it is HSP-86 which is responsible for mediating this function. At present the biological function of HSP-90 is speculated to be that of a "molecular chaperone", critical for the correct assembly and transport of key regulatory molecules (47-49). This leads us to propose that HSP-86 may be involved in regulating estrogen dependent growth by acting as a "molecular chaperone" for cellular components critical for cell replication. In all murine target tissues for estrogen, which we have tested so far, HSP-84 is more abundant than HSP-86 in the absence of estrogen. This suggests that if *in vivo*, in the absence of steroid (estrogen), HSP-90 indeed is responsible for maintaining the receptor (estrogen receptor) in a biologically non-functional state, most likely it is HSP-84 and not HSP-86 which is responsible for mediating this function. Thus it is interesting to note that during mammary development and differentiation the steady-state levels of HSP-90 reaches a maximum during lactation, when the tissue becomes refractory to the action of estrogen despite the presence of estrogen receptor. The synthesis of HSP-90 in lactating mammary glands occurs independent of estrogen and HSP-90 is present predominantly as HSP-84. This leads us to speculate that the net cellular responsiveness to estrogen by its target tissues may be determined by the combined effects of HSP-86 and HSP-84; this in turn may be dictated by the relative magnitude with which each of these genes respond to estrogen and their constitutive synthesis.

Alterations in the expression of HSP-90 are commonly found in transformed cells and we have evidence that this occurs with human breast tumors. Furthermore, a preliminary study analyzing simultaneously for cellular HSP-90, estrogen receptor and progesterone receptor suggests that there may be a positive relationship in the steady-state levels of these three proteins. We believe that within the context of tumor cells, alterations in the cellular regulation of HSP-90 (regardless of their origin), may have particularly important implications for the clinical management of carcinomas, especially that of the breast and the endometrium, with regard to the outcome of certain treatment modalities such as radiation, hyperthermia and chemotherapy. This is because to the extent almost any treatment modality designed to kill tumor cells constitutes a metabolic insult, it can evoke the synthesis of HSPs; this can counteract the effectiveness of a particular treatment which in turn will depend on the intrinsic ability of the tumor to synthesize HSPs and its rate of decay. And in the case of certain ovarian hormone dependent cancers, e.g. the ER positive human mammary and endometrial cancers, this situation may be further compounded by the possibility that ovarian hormones may themselves regulate the synthesis of HSP-90; this in turn may modify the response of these tumors to subsequent exposure to hyperthermia, radiation and chemotherapy. This speculation is strengthened by observations that if tamoxifen, an estrogen antagonist which in selected instances may behave as an agonist, is included in chemotherapeutic regimen of melphalan and fluorouracil, it adversely affects the survival of several breast cancer patient subsets (50). In fact, it has been shown that pretreatment of human breast cancer cells with tamoxifen results in greater tumor cell survival upon subsequent exposure to melphalan as compared to cells treated with melphalan alone (51). Similarly, it has been reported that the radiosensitivity of MCF-7 human breast cancer cells can be modified by estradiol and tamoxifen (52). Based on these observations we propose that future studies designed specifically to examine the relative importance of HSP-90 in mediating growth of normal and neoplastic tissues of the female reproductive tract including the breast will offer important insights with regard to the functionality of HSP-90. In addition, the results from these studies may also have utility in the clinical management of endometrial and breast carcinomas.

ACKNOWLEDGMENTS

These studies were supported by NIH grant CA 54828 and a grant from the National Cancer Institute of Canada. We thank Dr. A. Ferenczy for providing the human endometrial tissues, Dr. M. Willingham for his assistance with the immunocytochemical analysis, Dr. E. Hickey for providing the human HSP-90 cDNA probes and Ms. J. Knox for preparing the manuscript.

REFERENCES

Baez M, Sargan DR, Elbrecht A, Kulomaa MS, Zaruki-Schultz T, Tsai MS and O'Malley BW (1987): Steroid hormone regulation of the gene encoding the chicken heat shock protein HSp 108. *J Biol Chem* 262: 6582-6588.

Banerjee MR (1976): Responses of mammary cells to hormones. *Int Res Cytol* 47: 1-97.

Barnier JV, Bensuade O, Morange M and Babinet E (1987): Mouse 89 KD heat shock protein. Two polypeptides with distinct developmental regulation. *Exp Cell Res* 170: 186-194.

Baulieu EE (1981): Steroid hormone antagonists at the receptor level. A role for the heat-shock protein mw 90,000 (hsp 90). *J Cell Biochem* 35: 161-174.

Bensuade O and Morange M (1983): Spontaneous high expression of heat shock proteins in mouse embryonal carcinoma cells and ectoderm from day 8 mouse embryo. *EMBO J* 2: 173-177.

Bergeron C, Ferenczy A, Toft DO, Schneider W and Shyamala G (1988): Immunocytochemical study of progesterone receptors in the human endometrium during the menstrual cycle. *Lab Investigation* 59, 862-869.

Bresciani F (1965): Effect of ovarian hormones on duration of DNA synthesis in cells of the C3H mouse mammary gland. *Expt Cell Res* 38: 13-32.

Bresciani F (1968): Topography of DNA synthesis in the mammary gland of the C_3H mouse and its control by ovarian hormones: an autoradiographic study. *Cell Tiss Kinetics* 2: 51-63.

Bresnick EH, Dalman F, Sanchez EC and Pratt WB (1989): Evidence that the 90-kDa heat shock protein is necessary for the steroid binding conformation of the L cell glucocorticoid receptor. *J Biol Chem* 264: 4992-4997.

Burdon RA (1986): Heat shock and the heat shock proteins. *Biochem J* 240: 313-324.

Clarke CL, Zaino RJ, Feil PD, Miller JV, Steck ME, Ohlsson-Wilhelm BM and Satyaswaroop PG (1987): Monoclonal antibodies to human progesterone receptor: characterization by biochemical and immunohistochemical techniques. *Endocrinology* 121: 1123-1132.

Ellis RJ and Hemmingsen SM (1989): Molecular chaperons: Proteins essential for the biogenesis of some macromolecular structures. *Trends Biochem Sci* 14: 339-342.

Feil PD, Glasser SR, Toft DO and O'Malley BW (1972): Progesterone binding in the mouse and rat uterus. *Endocrinology* 91: 738-746.

Fisher B, Redmond C and Brown A (1986): Adjuvant chemotherapy with and without tamoxifen in the treatment of primary breast cancer: 5-year results from the National Surgical Adjuvant Breast and Bowel Project. *J Clin Oncology* 4: 459-471.

Haagensen CD (1986): In *Diseases of the Breast*, W.B. Saunders Company, Philadelphia, PA.

Haslam SZ and Shyamala G (1981): Relative distribution of estrogen and progesterone receptors among the epithelial, adipose, and connective tissue components of the normal mammary gland. *Endocrinology* 108: 825-830.

Hickey E, Brandon SE, Sadis S, Smale G and Weber LA (1986): Molecular cloning of sequences encoding the human heat-shock proteins and their expression during hypothermia. *Gene* 43: 147-154.

Hickey E, Brandon SE, Smale G, Lloyd D and Weber LA (1989): Sequence and regulation of a gene encoding a human 89-kilodalton heat shock protein. *Mol Cell Biol* 9: 2615-2626.

Hightower LE (1991): Heat shock, stress proteins, chaperons, and proteotoxicity. *Cell* 66: 191-197.

Hollman KH (1974): In *Lactation: A comprehensive treatise* eds. Larson, B.L., Smith V.R.

Howard KJ and Distelhorst CW (1988): Evidence for intracellular association of the glucocorticoid receptor with the 90-kDa heat shock protein. *J Biol Chem.* 263: 3474-3481.

Howard KJ, Holley SJ, Yamamoto KR and Distelhorst CW (1990): Mapping the hsp90 binding region of the glucocorticoid receptor. *J Biol Chem* 265: 11928-11935.

King WJ, Allen TC and DeSombre ER (1981): Localization of uterine peroxidase activity in estrogen-treated rats. *Biol Reprod* 25: 859-870.

Kurtz S and Lindquist S (1984): Changing patterns of gene expression during sporulation in yeast (*in vitro* translation/heat shock protein/nitrogen starvation/development. *Proc Natl Acad Sci USA* 81: 7323-7327.

Lanks KW (1986): Modulators of the eukaryotic heat shock response. *Exp Cell Res* 165: 1-10.

La Thangue MB and Latchman DS (1988): A cellular protein related to heat shock protein 90 accumulates during Herpes Simplex virus infection and is over expressed in transformed cells. *Exp Cell Res* 1: 169-179.

Lee SJ (1990): Expression of HSP86 in male germ cells. *Mol Cell Biol* 10: 3239-3242.

Lindquist S (1986): The heat shock response. *Ann Rev Biochem* 55: 1151-1191.

Lindquist S and Craig EA (1988): The heat shock proteins. *Ann Rev Genet* 22: 631-677.

Lyons WR, Li CH and Johnson RE (1958): The hormonal control of mammary growth and lactation. *Rec Progr Horm Res* 14: 219-254.

Mazzarella RA and Green M (1987): ERp99, an abundant, conserved glucoprotein of the endoplasmic reticulum, is homologous to the 90-kDa heat shock protein (hsp 90) and the 94-kDa glucose regulated protein (GRP 94). *J Biol Chem* 262: 8875- .

McNab JC, Orr M and La Thangue NB (1985): Cellular proteins expressed in herpes simplex virus transformed cells also accumulate on herpes simplex virus infection. *EMBO J* 4: 3223-3228.

Moore SK, Kozak C, Robinson EA, Ullrich SJ and Appella E (1987): Cloning and nucleotide sequence of the murine hsp 84 cDNA and chromosome assignment of related sequences. *Gene* 56: 29-40.

Moore SK, Kozak C, Robinson EA, Ullrich SJ and Appella E (1989): Murine 86- and 84-kDa heat shock proteins, cDNA sequences, chromosome assignments, and evolutionary origins. *J Biol Chem* 264: 5343-5351.

Nandi S (1958): Endocrine control of mammary gland development and function in the $C_3H/HeCr_g1$ mouse. *J Natl Cancer Inst* 21: 1039-1063.

Omar RA and Lanks KW (1984): Heat shock protein synthesis and cell survival in clones of normal and simian virus 40 - transformed mouse embryo cells. *Cancer Res* 44: 3976-3982.

Osborne CK, Kitten L and Arteaga CL (1989): Antagonism of chemotherapy induced cytotoxicity for human breast cancer cells by antiestrogens. *J Clin Oncology* 7: 710-717.

Picard D and Yamamoto KR (1987): Two signals mediate hormone-dependent nuclear localization of the glucocorticoid receptor. *EMBO J* 6: 3333-3340.

Pratt WB (1987): Transformation of glucocorticoid and progesterone receptors to the DNA-binding state. *J Cell Biochem* 31: 51-68.

Rebbe NF, Wave J, Bertina RM, Modrich P and Stafford DW (1987): Nucleotide sequence of a cDNA for a member of the human 90-kDa heat-shock protein family. *Gene* 53: 235-245.

Rebbe NF, Hickman WS, Ley TS, Stafford DW and Hickman S (1989): Nucleotide sequence and regulation of a human 90-kDa heat shock protein gene. *J Biol Chem* 264: 15006-15011.

Ross GT, Vandwicke RL and Frantz AG (1981): The ovaries and the breasts. In: *Textbook of Endocrinology* ed. R.H. Williams, W.B. Saunders Company, Philadelphia

Rothman JE (1989): Polypeptide chain binding proteins: catalysts of protein folding and related processes in cells. *Cell* 59: 591-601.

Schlesinger MJ (1990): Heat shock proteins. *J Biol Chem* 265: 12111-12114.

Schmidt-Gollwitzer M, Genz T, Schmidt-Gollwitzer K, Pollow B and Pollow K: In *Endometrial Cancer*. Ed Brush MG, King RJB and Taylor RW, Bailliere Tindall, London.

Shyamala G (1985): Regulation of mammary responsiveness to estrogen: An analysis of differences between mammary gland and uterus. In *Molecular Mechanisms of Steroid Hormone Action*, ed. Moudgil, V.K., Walter de Gruyter Press, pp. 413-435.

Shyamala G, Schneider W and Schott D (1990): Developmental regulation of murine mammary progesterone receptor gene expression. *Endocrinology* 126: 2882-2889.

Sorger PK and Pelham HRB (1987): The glucose regulated protein grp 94 is related to heat shock protein hsp 90. *J Mol Biol* 194: 341-344.

Topper YS and Freeman CS (1980): Multiple hormonal interactions in the developmental biology of the mammary gland. *Physiol Res* 60: 1049-1106.

Ullrich SJ, Robinson EA, Law LW, Willingham M and Appella E (1986): A mouse tumor-specific transplantation antigen is a heat-shock related protein. *Proc Natl Acad Sci* USA 83: 3121-3125.

Wazer DE, Tercilla OF, Lin P-S and Schmidt-Ullrich R (1989): Modulation in the radiosensitivity of MCF-7 human breast carcinoma cells by 17β estradiol and tamoxifen. *J Radiol* 62: 1079-1083.

Willingham MC (1990): Immunocytochemical methods: useful and informative tools for screening hybridomas and evaluating antigen expression. *Focus* 12: 62-67.

Yahara I, Ishida H and Koyasu S (1986): A heat shock variant of chinese hamster cell line constitutively expressing heat shock proteins of mr 90,000 at high level. *Cell Struct Funct* 11: 65-73.

Zimmerman JL, Petri W and Maselson M (1983): Accumulation of a specific subset of D-melanogaster heat shock mRNAs in normal development without heat shock. *Cell* 32: 1161-1170.

RECEPTOR MODIFICATION AND HORMONE/ANTIHORMONE INTERACTIONS

PHOSPHORYLATION AND PROGESTERONE RECEPTOR FUNCTION

Nancy L. Weigel
Department of Cell Biology
Baylor College of Medicine

Angelo Poletti
Istituto di Endocrinologia
Milano, Italy

Candace A. Beck
Dean P. Edwards
Department of Pathology
University of Colorado Health Science Center

Timothy H. Carter
Department of Biological Sciences
St John's University

Larry A. Denner
Texas Biotechnology Corp.
Houston, Texas

STEROID HORMONE RECEPTORS
V. K. Moudgil, Editor
© 1993 Birkhäuser Boston

INTRODUCTION

The progesterone receptor is a member of a superfamily of ligand activated transcription factors which includes the steroid receptors, thyroid hormone, retinoic acid and vitamin D receptors as well as a large number of proteins which are termed orphans because their ligands and/or their functions are unknown (Evans,1988; Wang et al, 1989). A variety of studies have shown that phosphorylation/dephosphorylation plays a major role in the regulation of the activity of numerous transcription factors. Phosphorylation can alter DNA binding, either by enhancing or inhibiting the activity, by changing DNA binding specificity or by altering interactions with other transcription factors (Englander et al, 1991; Prywes et al, 1988; Raychaudri et al, 1989; Boyle et al, 1991). In addition, phosphorylation is often important for transcriptional activation of the factors (Binetruy et al, 1991; Jackson et al, 1991). Moreover, transcription factors are frequently multiply phosphorylated and these phosphorylations may play different roles (Barber and Verma, 1987; Boyle et al, 1991). In some cases, both specific phosphorylations as well as specific dephosphorylations are required for full activity (Boyle et al, 1991).

Most, if not all, of the members of the steroid receptor superfamily are phosphorylated and recent evidence suggests that phosphorylation plays a key role in regulating the activity of these transcription factors. Progesterone receptors are phosphorylated in the absence of hormone and show enhanced phosphorylation in response to hormone administration (Beck et al, 1992; Sheridan et al, 1989; Logeat et al, 1985; Denner et al, 1989; Sullivan et al, 1988). Both estrogen receptors and glucocorticoid receptors also show enhanced phosphorylation in response to hormone administration (Hoeck et al, 1989; Orti et al, 1989; Washburn et al, 1991), but the phosphorylation of the androgen receptor does not appear to be increased, rather the stability of the protein itself is enhanced (Kemppainen et al, 1992).

Moreover, hormone dependent phosphorylation of the rabbit, human, and chicken progesterone receptors alters the mobility of receptors on SDS gels, indicating that the increase in phosphorylation is due at least in part to phosphorylation of additional sites. Thus, there is abundant evidence that the progesterone receptors are phosphoproteins which show enhanced phosphorylation concomitant with activation suggesting that these phosphorylations are likely to play an important role in the regulation of receptor function.

Phosphorylation of the chicken progesterone receptor

Figure 1 shows the structure of the chicken progesterone receptor and the locations of the phosphorylation sites which have been identified. The chicken progesterone receptor is expressed as two forms, PR_A (Mr=72,000) and PR_B (Mr=86,000), which are produced from the same gene either from separate messages (Jeltsch et al, 1990) or by alternate initiation of translation from the same message (Conneely et al, 1987). PR_A lacks the first 128 amino acids of PR_B, but is otherwise identical. Both are transcriptionally active, although their relative activities with different promoters differs (Tora et al, 1988). Both proteins contain all four of the phosphorylation sites thus far identified. Sites 1-3 were identified by isolating [^{32}P] labeled receptor from oviduct tissue minces, isolating tryptic peptides, and identifying both the peptides and the phosphorylated serines by protein sequencing (Denner et al, 1990a). Sites 1 and 2, Ser211 and Ser260 respectively, are phosphorylated in the absence of hormone, but show enhanced phosphorylation in response to hormone (Denner et al, 1990a) whereas Site 3, Ser530, is phosphorylated only in response to hormone administration.

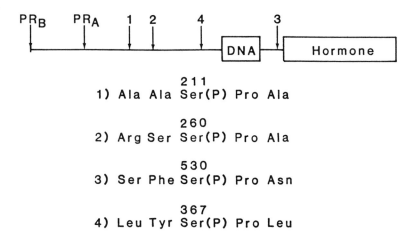

FIGURE 1. Location of the Phosphorylation Sites in the Chicken Progesterone Receptor. PR_B and PR_A indicate the start sites for PR_B and PR_A respectively. DNA, the DNA binding domain; Hormone, the hormone binding domain. 1-4 indicate the location of the phosphorylation sites and the sequences surrounding the phosphorylation sites are given in the lower portion of the figure.

These sites were identified in receptor isolated from cytosol of chicken oviduct slices (Denner et al, 1990a) which had been labeled only briefly with [^{32}P]phosphate, because of the lability of the oviduct slices. Thus, there may be additional phosphorylation sites which were not detected in these studies. For example, constitutive sites which turn over slowly or sites which are phosphorylated only in the nuclear fraction might not have been sufficiently phosphorylated to be detected.

Phosphorylation of progesterone receptor expressed in yeast (Saccharomyces cerevisiae)

In order to look for additional sites which might be labelled slowly as well as to determine whether the yeast expression system is a suitable model for the study of the role of receptor phosphorylation in receptor function, we characterized the phosphorylation of chicken progesterone receptor expressed in yeast. We have developed a yeast expression system for the synthesis of chicken progesterone receptor (Poletti et al, 1992) which produces intact, biologically active, hormone binding progesterone receptor. The progesterone receptor is synthesized as a ubiquitin fusion protein (Mak et al, 1989; Poletti et al, 1992) and the endogenous ubiquitinase rapidly cleaves the protein *in vivo* so that only authentic receptor of the correct size is detected. The progesterone receptor expression plasmid contains two estrogen response elements upstream of the coding region for the chicken progesterone receptor. Yeast cells co-transformed with an estrogen receptor expression plasmid as well as the chicken progesterone receptor expression plasmid will rapidly and efficiently produce progesterone receptor in response to estrogen administration (Poletti et al, 1992). This provides an ideal system, therefore, to look at receptor phosphorylation under conditions where the receptor can be uniformly labeled with [^{32}P]phosphate. Receptor was immunopurified from [^{32}P]phosphate labelled yeast expressing progesterone receptor and peptide maps were prepared using reversed phase HPLC as previously described (Denner et al, 1990a). In the absence of progesterone, only Ser211 and Ser260 were phosphorylated, just as has been observed in chicken oviduct slices. In the presence of progesterone, PR$_B$ was phosphorylated at Ser211, Ser260, and Ser530. In addition, another phosphopeptide was detected which had not been detected in the previous studies of receptor isolated from chicken oviduct cytosol. This raised the question of whether the receptor expressed in yeast was anomalously phosphorylated, or whether the site had simply not been detected in the chicken experiments. One difference between the yeast and the chicken studies is that whole cell extracts were analyzed in the yeast experiments whereas only the cytosol fraction was analyzed in the initial chicken slice experiments (Denner et al, 1990a). Therefore, any phosphorylation which occurs only in the nuclear compartment might have been missed. Analysis of whole cell extracts of [^{32}P]phosphate labelled oviduct slices subsequently showed that the same phosphopeptide was present in oviduct. The site has now been identified as Ser367 (Poletti and Weigel, 1993) and is designated as site 4 (see figure 1).

Early studies had suggested that phosphorylation might be necessary for hormone binding of the glucocorticoid receptor (Munck and Brinck-Johnson, 1968; Mendel et al, 1986) and several papers have been published by Auricchio and colleagues (Migliaccio et al, 1984; Migliaccio et al, 1991) suggesting that estrogen receptor must be phosphorylated on an as yet unidentified tyrosine in order to produce the hormone binding form of the receptor. In contrast, other investigators (Washburn et al, 1991) have not detected phosphotyrosine in the estrogen receptor. The studies of chicken progesterone receptor expressed in yeast show that only Ser211 and Ser260 are phosphorylated in the absence of hormone administration, yet the receptor binds hormone with normal affinity (Mak et al, 1989; Poletti et al, 1992). Since Ser211 and Ser260 are both in regions which can be deleted without altering the hormone binding of receptor expressed either in mammalian cells (Dobson et al, 1989) or in yeast (Mak et al, 1989), the chicken progesterone receptor does not require phosphorylation for hormone binding.

Figure 1 shows the sequences and the locations of the phosphorylation sites in the chicken progesterone receptor. It is interesting that there is a hormone-dependent phosphorylation in each of the regions shown to be important for transcriptional activation of the receptor. Bocquel et al, (1989) have shown that the two activation domains assume different relative levels of importance in activating the receptor depending on the promoter and the cell type tested. Thus the relative importance of the two phosphorylation sites may depend both on the cell and the target gene activated. None of these sites are good consensus Protein Kinase A (PKA) or Protein Kinase C (PKC) phosphorylation sites. The single common motif is that of a consensus Ser-Pro sequence. There are four Ser-Pro sequences in the chicken progesterone receptor and all four of these are phosphorylated. Many of the phosphorylation sites identified in the glucocorticoid receptor also contain the consensus sequence Ser-Pro (Bodwell et al, 1991). This sequence is found in the recognition sequences for a number of cell-cycle

314

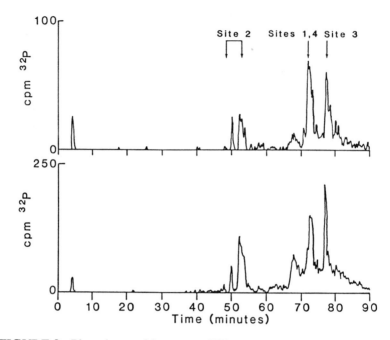

FIGURE 2. Phosphopeptide maps of PR$_B$. Upper panel: Strain CMY478 was transformed with a PR$_B$ expression plasmid whose expression is induced by copper (Mak et al, 1989). The receptor synthesis was induced by treatment with copper (Mak et al, 1989) and the yeast were treated with [^{32}P] phosphate and 2 uM progesterone at the non-permissive temperature (37°C) for 4 hours. The cells were harvested and receptor was immunopurified, digested with trypsin and the peptides analyzed by reversed phase HPLC as previously described (Denner et al, 1990a). Sites indicate the elution position of the phosphotryptic peptides which were further characterized by redigestion with additional proteases as previously described (Denner et al, 1990a). Lower panel: Control Phosphopeptide Map of PR$_B$ in strain BJ3505 (Poletti et al, in press). PR$_B$ whose synthesis was under the control of the estrogen receptor as previously described for PR$_A$ (Poletti et al, 1992) was induced as described previously (Poletti et al, 1992) and labeled with ^{32}P in the presence of progesterone at 30°. Samples were prepared and analyzed as described in the upper panel. Both the receptor from the strain containing the cdc28 mutant (upper panel) as well as the control receptor (lower panel) contained all 4 known phosphorylation sites. The decreased incorporation in the cdc28 strain (note Y axis) is due to decreased recovery of intact receptor due to proteolysis in this strain, rather than to decreased incorporation/mole of receptor.

and/or growth regulated kinases such as cdc2 (Mulkhopadhyay et al, 1992), CDK2 (Elledge et al, 1992), and the ERK or MAP kinases (Mulkhopadhyay et al, 1992; Boulton et al, 1991). The kinase or kinases responsible for receptor phosphorylation are not yet known. Since two of the sites are phosphorylated constitutively and two are phosphorylated in response to hormone, either there are multiple kinases which phosphorylate the receptor, or the hormone dependent sites are occluded in the hsp90-receptor complex. In order to determine whether cdc2 is responsible for phosphorylating any of the sites in the chicken progesterone receptor, phosphopeptide mapping analysis of [^{32}P] labelled receptor isolated from a yeast strain (CMY478) (Elledge et al, 1991) containing a temperature sensitive CDC28 (an *S. cerevisiae* cdc2 homolog) was done. Figure 2 shows that the receptor was efficiently phosphorylated at the non-permissive temperature (upper panel)and was phosphorylated on the same sites as the receptor isolated from cells containing wild type kinase (lower panel). This suggests, then, that cdc2 is not required for receptor phosphorylation.

Phosphorylation of chicken progesterone receptor during *in vitro* transcription assays

Klein-Hitpass et al, (1990) demonstrated specific progesterone receptor dependent *in vitro* transcription using receptor purified from control non-hormone treated oviducts and HeLa nuclear extract as the source of general transcription factors. The fact that the receptor was active suggested that either the hormone-dependent phosphorylation was not required for receptor activation providing that the heat shock proteins had been removed, or that the HeLa nuclear extract contains kinases which phosphorylate the receptor. Figure 3 shows that the receptor is, in fact, phosphorylated during the *in vitro* transcription assays. The upper panel shows that ^{32}P is incorporated into the receptor when the reaction is supplemented with [^{32}P] ATP, but the extent of the phosphorylation cannot be determined because of the ATP generating system present in the incubations (Weigel et al, 1992). However, when receptor is analyzed for changes in mobility on SDS gels (lower panel), it is clear that the receptor is rapidly modified producing two forms with reduced mobility which persist throughout the assay. Thus, the *in vitro* assay measures the activity of both the original as well as the modified forms of the receptor.

Characterization of this activity showed that the activity is dependent upon added DNA (Weigel et al, 1992) and that the DNA must be double stranded. Two groups have described a double-stranded DNA-dependent

316

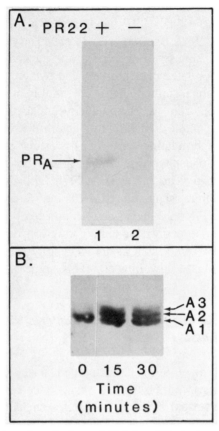

FIGURE 3. Phosphorylation of Chicken PR during *in vitro* Transcription Assays. Panel **A**: Each sample of PR_A (0.6 ug) was preincubated as described previously (Weigel et al, 1992) with 18 ul HeLa nuclear extract, 3 ug herring sperm DNA, and 0.3 ug test template PRE 2+pLovTATA in a final volume of 69 ul. After the preincubation, 21 ul of mix, including 15 uCi [^{32}P]ATP were added and incubated. Samples were immunopurified as previously described (Weigel et al, 1992) and analyzed by SDS gel electrophoresis and [^{32}P] was detected by autoradiography. Lane 1, Complete reaction; lane 2, no PR22. Panel **B**, Effect of phosphorylation of PR during the *in vitro* transcription assay on receptor mobility. PR_A was incubated with assay components exactly as described in A, except that [^{32}P]ATP was omitted. At the indicated times after the preincubation, aliquots containing 0.12 ug PR_A were removed and samples were analyzed by SDS gel electrophoresis and immunoblotting as previously described (Weigel et al, 1992) Figure reproduced from Weigel et al, (1992) *Mol Endocrinol* 6: 8-14. Reprinted with permission.

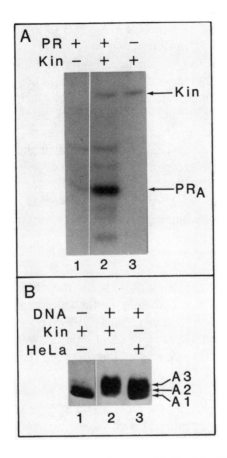

FIGURE 4. Phosphorylation of PR$_A$ with Purified DNA-PK.
A, Receptor (0.6 ug) was incubated for 15 min at 37°C in 30 ul as previously described (Weigel et al, 1992) with 1.2 ug double stranded herring sperm DNA, 0.12 ug PRE 2+pLov and 20 uCi [^{32}P]ATP. The samples were run on SDS-gels and ^{32}P was detected by wet gel autoradiography. Lane 1, PR$_A$ alone; lane 2, PR$_A$ plus 2 ul purified DNA-PK; lane 3, 2 ul DNA-PK alone. **B**, Effect of phosphorylation by DNA-PK on PR mobility. Receptor (0.06 ug) was incubated for 15 min at 37°C with the indicated amounts of DNA and kinase. Samples were run on SDS-gels, and receptor was detected by immunoblot. Lane 1, 4 ul of DNA-PK with no DNA; lane 2, 1.6 ug PRE 2+pLov DNA and 4 ul DNA-PK; lane 3, 1.6 ug PRE 2+pLov DNA and 3 ul HeLa extract. Reproduced from Weigel et al, (1992) *Mol Endocrinol* 6: 8-14. Reprinted with permission.

kinase isolated from HeLa cells (Carter et al, 1990; Lees-Miller et al, 1990). Panel A of Figure 4 shows that the purified DNA-dependent protein kinase (DNA-PK) isolated by Carter et al, (1990) phosphorylates the receptor *in vitro*. The immunoblot in Panel B of Figure 4 shows that the phosphorylation is DNA-dependent, as expected, and the altered mobility reproduces the pattern produced by the HeLa nuclear extract. Preliminary peptide mapping studies as well as studies using receptor isolated from progesterone treated oviduct cytosol (Weigel et al, 1992) indicate that the DNA-dependent phosphorylations are in addition to the Ser530 phosphorylation produced by treatment of oviducts with progesterone. Although evidence for DNA-dependent phosphorylation of the chicken progesterone receptor *in vivo* has not yet been published, Takimoto et al, (1992) have shown that at least one of the phosphorylations of the human progesterone receptor in T47D breast cancer cells requires DNA binding of the receptor.

Phosphorylation and progesterone receptor function

Numerous roles have been suggested for the phosphorylations in steroid receptors including regulating hormone binding as discussed previously, DNA binding (Denner et al, 1989; Beck et al, 1992), transcriptional activation (Denner et al, 1990b; Power et al, 1991a; Power et al, 1991b) and receptor shuttling between the cytoplasm and nuclei (DeFranco et al, 1991).

Since progesterone receptor undergoes a hormone-dependent phosphorylation which changes the mobility on SDS gels, this change in mobility can be used to determine at what stage in receptor activation this phosphorylation occurs and to gain information about the role of this phosphorylation in receptor function.

Identification of the phosphorylation site responsible for altered mobility on SDS gels

Previous studies have shown that Ser530 is phosphorylated in response to hormone administration and that the phosphorylated receptor displays altered mobility on SDS gels (Denner et al, 1990a). Recent studies have shown that there is a second hormone dependent phosphorylation site, Ser367 (Poletti and Weigel, 1993), but several lines of evidence suggest that it is phosphorylation of Ser530 which is responsible for the altered mobility. First, the upshift is observed in the cytosol fraction which correlates better with observed phosphorylation of

Ser530 than of Ser 367 (Denner et al, 1990a). Second, analysis of a receptor mutant expressed in yeast showed that the mutant which was phosphorylated on Ser530, but not on Ser367 still showed reduced mobility on SDS gels (Poletti et al, in press). Moreover, phosphorylation of progesterone receptor *in vitro* with Protein Kinase A results in altered mobility of receptor on SDS gels and phosphorylation of Ser528. Incubation of PR$_A$ with an antibody against this region blocks phosphorylation of Ser528 and prevents the change in mobility on SDS gels, whereas incubation with a control antibody, PR22, directed against a distal region allows phosphorylation and produces receptor with altered mobility. Thus, it appears that phosphorylation of the hinge region of the chicken progesterone receptor causes the altered mobility on SDS gels observed either in receptor isolated from cytosol of oviducts treated with progesterone or receptor phosphorylated *in vitro* with Protein Kinase A.

Identification of 8S complexes as substrates for hormone dependent phosphorylation

In the absence of hormone, the chicken progesterone receptor is found in chicken oviduct cytosol and is associated with a number of heat shock proteins which form a complex which migrates at 8S on a sucrose gradient (Denner et al, 1989). We have shown that the region containing Ser530 is occluded in the hsp90-receptor complex since an antibody raised to a synthetic peptide containing this sequence cannot interact with the hsp90-receptor complex whereas it does react with the salt dissociated receptor (Weigel et al, 1989). Treatment, *in vivo*, with progesterone results in the production of a 4S complex which initially is in the cytosol (Denner et al, 1989) and subsequently binds tightly to the nuclear fraction. Our previous studies have shown that the receptor containing the hormone dependent phosphorylation at Ser530 can be isolated from the chicken oviduct cytosol indicating that this phosphorylation occurs prior to the tight nuclear binding step (Denner et al, 1990a). In order to determine whether this phosphorylation occurs prior to 8S receptor complex dissociation or after dissociation, 8S and 4S receptors were isolated from oviduct cytosol of progesterone treated oviducts using phosphocellulose chromatography as previously described (Denner et al, 1989). The receptors were immunopurified as previously described (Denner et al, 1989) and analyzed on SDS polyacrylamide gels and receptor detected by silver staining. Figure 5 shows the PR$_A$ regions of silver stained gels. On the left is the PR$_A$ isolated from cytosol of control and progesterone treated oviducts. Whereas the control receptor consists of a single band, the progesterone treated sample also shows a band of reduced mobility. On the right is shown the receptor isolated from the 8S and 4S complexes. The 8S complex shows a substantial amount of receptor with reduced mobility showing that the phosphorylation does occur in the 8S complex. The 4S receptor is almost exclusively receptor with reduced mobility. This study suggests that hormone binding alters the conformation of the 8S complex which allows the phosphorylation at Ser530. This produces a further conformational change which is important for the dissociation of the 8S complex. The observation that the 4S pool also contains some receptor with control mobility suggests that hormone may be sufficient to dissociate the 8S complex at a slower rate in the absence of phosphorylation or, alternatively, that phosphatase activity has not been completely inhibited during purification.

Smith (Smith et al, 1991; Smith and Toft, 1992) has found that the hsp90-receptor complexes can be reconstituted using immunopurified chicken progesterone receptor and rabbit reticulocyte lysates which contain heat shock proteins. The reconstitution requires ATP and is only successful in the absence of hormone. This suggests that phosphorylated hormone bound receptor does not form stable complexes with hsp90 and the other components of the receptor complex.

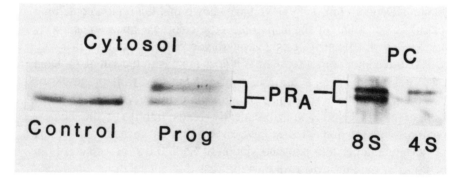

FIGURE 5. Analysis of PR_A for Mobility on SDS Gels. Left: receptor was immunopurified from cytosol of either control or progesterone treated chickens as previously described (Denner et al, 1990a) and analyzed by SDS gel electrophoresis. The receptor was detected by silver staining. Right: 8S and 4S receptors were prepared from cytosol of progesterone treated chickens using phosphocellulose to separate the fractions as previously described (Denner et al, 1989), immunopurified, analyzed by SDS gel electrophoresis and detected by silver staining.

The hinge regions of steroid receptors are not highly conserved regions. Despite this, all of the steroid receptors contain a Ser/Thr-Pro consensus sequence in their hinge regions (Denner et al, 1990a). In contrast, a number of the other members of this family such as the Vitamin D receptor do not contain the consensus sequence in the hinge region. That these members of the family also do not bind to heat shock proteins suggests that the phosphorylation of a hinge region site may be a common feature in dissociation of heat shock protein receptor complexes during receptor activation.

Phosphorylation and DNA binding

None of the phosphorylation sites identified in either the chicken progesterone receptor or the glucocorticoid receptor are in the conserved region containing the two zinc fingers which is the minimal sequence required for DNA binding. However, there is evidence that some of the phosphorylations may alter DNA binding either directly by changing the affinity of the receptor for the DNA or indirectly by enhancing receptor dimerization. Since receptors bind to DNA as dimers (Tsai et al, 1988), any modification which strengthens receptor dimerization may also enhance DNA binding.

The studies of Denner et al, (1989) demonstrated that receptor isolated from hormone treated oviducts showed higher DNA binding affinity and capacity than receptor isolated from control oviducts. Furthermore when 8S and 4S fractions were isolated from progesterone treated oviducts using phosphocellulose, followed by activation of all receptor to the 4S form using salt and hormone *in vitro*, the receptor from the 4S pool showed a substantially higher affinity for DNA than that from the pool which had contained 8S receptor (Denner et al, 1989). The correlation of the 4S pool containing the predominantly phosphorylated receptor (figure 5) with the enhanced DNA binding, suggests that the phosphorylation is enhancing the DNA binding.

Studies of the human progesterone receptor isolated from T47D cells also suggest that phosphorylation enhances DNA binding. The human progesterone receptor is also expressed as two forms, PR-B and PR-A, which are somewhat larger than the corresponding chicken forms. The phosphorylation of the human progesterone receptor is much more complex than that of the chicken progesterone receptor (Sheridan et al, 1989). There are PR-B specific phosphorylations which reduce mobility of the control receptor. In the absence of hormone, PR-B appears as a triplet on SDS gels whereas the PR-A is a single band. Treatment with hormone ultimately results in decreased mobility of all of these forms. Panel A of Figure 6 shows the phosphorylation of human progesterone receptor in T47D cells as a function of the time of hormone treatment as described previously (Beck et al,1992). There is a rapid increase in phosphorylation in the first 5-10 minutes of R5020 treatment and virtually no net change in phosphorylation beyond this time. However, the phosphorylation which results in a change in mobility occurs much more slowly (Panel B) and takes 2 or more hours to reach completion. Takimoto et al, (1992) have reported that this slow phosphorylation which alters the mobility of the receptors is a DNA-dependent phosphorylation.

In contrast to the chicken progesterone receptor (Rodriguez et al, 1989), the human progesterone receptor does not bind to DNA in the absence of hormone even when treated with salt to dissociate the receptor complex (Beck et al, 1992). Panel C of Figure 6 shows the DNA binding activity as measured by gel retardation of extracts from control cells and from cells treated with R5020 for the indicated periods of time. In the absence of added R5020, there is essentially no DNA binding.

FIGURE 6. Time Course of Progestin Stimulation of PR Phosphorylation and DNA-Binding Activity. **A,** T47D cells were labeled to steady state with [^{32}P]orthophosphate, and R5020 (80 nM) was added for the times indicated as previously described (Beck et al, 1992). Cells were then harvested rapidly and cooled (5 min), whole cell extracts were prepared by cell lysis in 50 mM potassium phosphate (pH 7.4), 50 mM sodium fluoride, 1 mM EDTA, 1 mM EGTA and 12 mM monothio-glycerol containing protease inhibitors, plus 0.5 M NaCl and progesterone receptor was immune-isolated with receptor-specific MAb, AB-52. Isolated receptors were submitted to SDS-PAGE, and proteins were transferred to nitrocellulose and exposed to x-ray film. Inset, progesterone receptors were processed in the same manner in a separate experiment after treatment of T47D cells for 0,1, and 120 min with R5020. **B,** the same nitrocellulose filter in A was rehydrated and Western blotted with AB-52 using an ELISA detection method. **C,** Time course of R5020 activation of PR-DNA binding activity *in vivo*. T47D cells were incubated without or with R5020 (80 nM) for the times indicated. Cells were harvested and quickly cooled in ice, and whole cell extracts were prepared as described in A. Aliquots of whole cell extracts (50 fmol PR) were incubated with a [^{32}P]PRE oligonucleotide and submitted to a gel mobility shift assay. *In vitro* hormone addition to the indicated whole cell extracts was performed for 3 h at 0 C with 80 nM R5020. Reproduced from Beck et al (1992) *Mol Endocrinol* 6: 607-620.

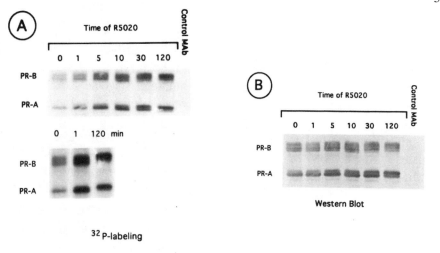

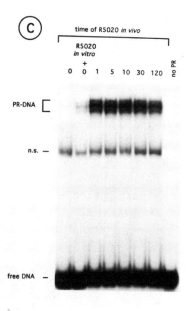

FIGURE 6. Time Course of Progestin Stimulation of PR Phosphorylation and DNA-Binding Activity.

Incubation *in vitro* with R5020 yields the characteristic triplet of bands which is the result of B-B, B-A, and A-A dimers interacting with DNA. As little as 1 minute of treatment with hormone *in vivo* rather than *in vitro* produces a dramatic increase in DNA binding. Since all samples are whole cells extracts, the only known difference between the 0 and 1 minute samples is the degree of phosphorylation (see Panel A). Thus, for the human progesterone receptor, enhanced phosphorylation strongly correlates with enhanced DNA binding. Whether this enhancement is a result of stronger binding to the DNA directly or to a strengthening of the affinity of receptor monomers to produce dimers is not yet known.

Phosphorylation and transcriptional activation

Although the effect of mutation of phosphorylation sites on receptor activation has not yet been reported, a number of investigators have examined the effect of modulating kinase and/or phosphatase activities on the overall transcriptional activity of steroid receptors. The transcriptional activity of the human progesterone receptor has been characterized in T47D cells stably transfected with MMTVCAT (Clone B11)(Beck et al, 1992). The activity is strictly dependent upon addition of R5020. However, addition of 8-Br cAMP which activates Protein Kinase A or of okadaic acid which inhibits phosphatases 1 and 2A stimulates hormone dependent activity 2-4 fold (Beck et al, 1992). The basal activity in the absence of hormone and the activity of a control reporter pSV_2CAT is unaffected by these treatments (Beck et al, 1992). Thus either the activity of the receptor itself, or a factor which is important for receptor-dependent, but not general transcription is being enhanced. Subsequently Rangarajan et al (1992) reported that the activity of the glucocorticoid receptor can be enhanced by these treatments. These data suggest that phosphorylation plays a role in transcriptional activation of steroid receptors.

Even stronger evidence for phosphorylation affecting transcriptional activation of steroid receptors is the finding of Denner et al, (1990b) that the chicken progesterone receptor co-transfected into CV1 cells with a reporter gene, PREtkCAT, can be activated by 8-Br cAMP or by okadaic acid in the absence of hormone. The activity was strictly dependent upon the presence of receptor and could be stimulated to levels similar to progesterone alone. Addition of progesterone and either 8-Br cAMP or okadaic acid resulted in still higher activity similar to the enhanced activity observed with human progesterone receptor (Beck et al, 1992) and with glucocorticoid receptor (Rangarajan et al, 1992). Subsequent studies by Power et al, have shown that other receptors such as the COUP-TF (Power et al, 1991a) and human estrogen receptor (Power et al, 1991b) can be activated by both okadaic acid and by dopamine. Thus, it appears that some, but perhaps not all, of the members of this family are capable of ligand-independent activation.

The known phosphorylation sites in chicken progesterone receptor are not consensus Protein Kinase A sites (Denner et al, 1990a). This suggests that if the activation is through phosphorylation of receptor, it is either through an alternate site such as Ser528 or that Protein Kinase A acts through a cascade to ultimately alter receptor phosphorylation. A third possibility is that the ability of Protein Kinase A to inactivate Phosphatase 1 in some cells, may functionally increase the activity of the kinase which is responsible for receptor phosphorylation.

One surprising finding was that the chicken progesterone receptor can be activated in the absence of ligand (Denner et al, 1990b), but apparently the human progesterone receptor cannot (Beck et al, 1992). Although different cells were used for the two sets of studies, it is likely that the different properties of the two receptors, rather than the different cell types are responsible for the differences. The chicken progesterone receptor can be salt activated *in vitro* and purified in the absence of hormone. This receptor binds to DNA (Rodriguez et al, 1989), is active in *in vitro* transcription assays (Klein-Hitpass et al, 1990) and can undergo DNA-dependent phosphorylation in the absence of hormone; addition of hormone has little effect. By contrast, the partially purified human progesterone receptor will not bind to DNA (Bagchi et al, 1988, Beck et al, 1992), activate transcription *in vitro* (Bagchi et al, 1991), or undergo DNA-dependent phosphorylation (Bagchi et al, 1992) in the absence of added hormone.

The hormone-dependent transcriptional activity of both the human (Beck et al, 1992) and chicken (Denner et al, 1990b) progesterone receptors is inhibited by H8, an inhibitor of Protein Kinase A further indicating that phosphorylation of receptors is important for hormone-dependent transcriptional activation.

A proposed model for the role of phosphorylation in receptor function is shown in Figure 7. In the absence of hormone the receptor is phosphorylated only on constitutive sites and is associated with a number of non-steroid binding proteins including hsp90 and hsp70. Treatment with hormone results in a conformational change which allows a second round of phosphorylation which for the chicken progesterone receptor includes Ser530, dissociation of the receptor complexes, receptor dimerization and binding to DNA. We would then propose that there is

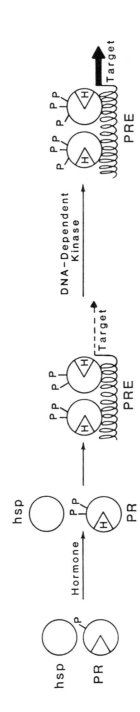

FIGURE 7. A Model for Transcriptional Activation of Progesterone Receptor by Phosphorylation. hsp, Heat shock protein-90 and other nonsteroid-binding proteins. Each P represents a class of phosphorylation sites rather than an individual site. Reproduced from Weigel et al, (1992) Mol. Endocrinol. 6, 8-14. Reprinted with permission.

an additional DNA-dependent round of phosphorylation which is necessary to produce the fully active form of the receptor. The stimulation of receptor activity through modulation of kinase and/or phosphatase activity would occur through enhanced phosphorylation either of the sites normally phosphorylated in the hormone dependent pathway or through phosphorylation of novel sites which also contribute to activation of the receptors.

Recent studies by several groups have implicated kinases as playing a role in either ligand-independent or ligand-dependent activation of sterid receptors. Aronica and Katzenellenbogen (1991) have shown that cAMP as well as other kinase activators can activate the estrogen receptor. Ignar-Trowbridge et al, (1992) have shown that EGF can cause nuclear localization and the change in mobility on SDS gels characteristic of hormone dependent phosphorylation of the estrogen receptor. Thus, it appears that kinases and cell signaling pathways play an integral role in the regulation of the activity of steroid receptors.

REFERENCES

Aronica SM and Katzenellenbogen BS (1991): Progesterone receptor regulation in uterine cells: stimulation by estrogen, cyclic adenosine 3'5'-monophosphate, and insulin-like growth factor I and suppression by antiestrogens and protein kinase inhibitors. *Endocrinology* 128: 2045-2052.

Bagchi MK, Elliston JF, Tsai SY, Edwards DP, Tsai MJ and O'Malley BW (1988): Steroid hormone-dependent interaction of human progesterone receptor with its target enhancer element. *Mol Endocrinol* 2: 1221-1229.

Bagchi MK, Tsai SY, Tsai M-J and O'Malley BW (1992): Ligand and DNA-dependent phosphorylation of human progesterone receptor in vitro. *Proc Natl Acad Sci USA* 89: 2664-2668.

Barber JR and Verma IM (1987): Modification of fos proteins: phosphorylation of c-fos, but not v-fos, is stimulated by 12-tetradecanoyl-phorbol-13-acetate and serum. *Mol Cell Biol* 7: 2201-2211.

Beck CA, Weigel NL and Edwards DP (1992): Effects of hormone and cellular modulators of protein phosphorylation on transcriptional activity, DNA binding, and phosphorylation of human progesterone receptors. *Mol Endocrinol* 6: 607-620.

Binetruy B, Smeal T and Karin M (1991): Ha-Ras augments c-Jun activity and stimulates phosphorylation of its activation domain. *Nature* 351: 122-127.

Bocquel MT, Kumar V, Stricker C, Chambon P and Gronemeyer H (1989): The contribution of the N- and C-terminal regions of steroid receptors to activation of transcription is both receptor and cell-specific. *Nucleic Acids Res* 17: 2581-2595.

Bodwell JE, Orti E, Coull JM, Pappin DJ, Swift F and Smith LI (1991): Identification of the phosphorylated sites in the mouse glucocorticoid receptor. *J Biol Chem* 266: 7549-7555.

Boulton TG, Nye SH, Robbins DJ, Ip NY, Radziejewska E, Morgenbesser SD, DePinho RA, Panayotatos N, Cobb MH and Yancopolous GD (1991): ERKS: a family of protein-serine/threonine kinases that are activated and tyrosine phosphorylated in response to insulin and NGF. *Cell* 65: 663-675.

Boyle WJ, Smeal T, Defize LHK, Angel P, Woodgett JR, Karin M and Hunter T (1991): Activation of protein kinase C decreases phosphorylation of c-Jun at sites that negatively regulate its DNA-binding activity. *Cell* 64: 573-584.

Carter T, Vancurova I, Sun I, Lou W and DeLeon S (1990): A DNA-activated protein kinase from HeLa cell nuclei. *Mol Cell Biol* 10: 6460-6471.

Conneely OM, Maxwell BL, Toft DO, Schrader WT and O'Malley BW (1987): The A and B forms of the chicken progesterone receptor arise by alternate initiation of translation of a unique mRNA. *Biochem Biophys Res Commun* 149: 493-501.

DeFranco DB, Qi M, Borror KC, Garabedian MJ and Brautigan DL (1991): Protein Phosphatases Types 1 and/or 2A Regulate Nucleocytoplasmic Shuttling of Glucocorticoid Receptors. *Mol Endocrinol* 5: 1215-1228.

Denner LA, Bingman WE III, Greene GL and Weigel NL (1987): Phosphorylation of the chicken progesterone receptor. *J Steroid Biochem* 27: 235-243.

Denner LA, Weigel NL, Schrader WT and O'Malley BW (1989): Hormone-dependent regulation of chicken progesterone receptor deoxyribonucleic acid binding and phosphorylation. *Endocrinology* 125: 3051-3058.

Denner LA, Schrader WT, O'Malley BW and Weigel NL (1990a): Hormonal regulation and identification of chicken progesterone receptor phosphorylation sites. *J Biol Chem* 265: 16548-16555.

Denner LA, Weigel NL, Maxwell BL, Schrader WT, and O'Malley BW (1990b): Regulation of progesterone receptor-mediated transcription by phosphorylation. *Science* 250: 1740-1743.

Dobson ADW, Conneely OM, Beattie W, Maxwell BL, Mak P, Tsai M-J, Schrader WT and O'Malley BW (1989): Mutational Analysis of the chicken progesterone receptor. *J Biol Chem* 264: 4207-4211.

Elledge SJ, Mulligan JT, Ramer SW, Spottswood M and David RW (1991): YES: A multifunctional cDNA expression vector for the isolation of genes by complementation of yeast and *Escherichia coli* mutations. *Proc Natl Acad Sci USA* 88: 1731-1735.

Elledge SJ, Richman R, Hall FL, Williams RT, Lodgson N and Harper JW (1992): CDK2 encodes a 33-kDa cyclin A-associated protein kinase and is expressed before CDC2 in the cell cycle. *Proc Natl Acad Sci USA* 89: 2907-2911.

Englander EW, Widen SG and Wilson SH (1991): Mammalian beta-polymerase promoter: phosphorylation of ATF/CRE-binding protein and regulation of DNA binding. *Nucl Acids Res* 19: 3369-3375.

Evans R (1988): The steroid and thyroid hormone receptor superfamily. *Science* 240: 889-895.

Hoeck W, Rusconi S and Groner B (1989): Down-regulation and phosphorylation of glucocorticoid receptors in cultured cells. *J Biol Chem* 264: 14396-14402.

Ignar-Trowbridge D, Nelson KG, Bidwell MC, Curtis SW, Washburn TF, MacLachlan JA and Korach KS (1992): Coupling of dual signalling pathways: epidermal growth factor action involves the estrogen receptor. *Proc Natl Acad Sci* 89: 4658-4662.

Jackson SP, MacDonald JJ, Lees-Miller S and Tjian R (1990): GC box binding induces phosphorylation of Sp1 by a DNA-dependent protein kinase. *Cell* 63: 155-165.

Jeltsch J-M, Turcotte B, Garnier J-M, Lerouge T, Krozowsloski Z, Gronemeyer H and Chambon P (1990): Characterization of multiple mRNAs originating from the chicken progesterone receptor gene. *J Biol Chem* 265: 3967-3974.

Kemppainen JA, Lane MV, Sar M and Wilson EM (1992): Androgen receptor phosphorylation, turnover, nuclear transport and transcriptional activation. Specificity for steroid and antihormone. *J Biol Chem* 267: 968-974.

Klein-Hitpass L, Tsai SY, Weigel NL, Allan GF, Riley D, Rodriguez R, Schrader WT, Tsai M-J and O'Malley BW (1990): The progesterone receptor stimulates cell free transcription by enhancing the formation of a stable preinitiation complex. *Cell* 260: 247-257.

Lees-Miller SP, Chen YR and Anderson CW (1990): Human cells contain a DNA-activated protein kinase that phosphorylates simian virus 40 T antigen, p53, and other regulatory proteins. *Mol Cell Biol* 10: 6472-6481.

Logeat F, Le Cunff M, Pamphile R and Milgrom E (1985): The nuclear-bound form of the progesterone receptor is generated through a hormone-dependent phosphorylation. *Biochem Biophys Res Commun* 131: 421-427.

Mak P, McDonnell, DP, Weigel NL, Schrader WT and O'Malley BW (1989): Expression of functional chicken oviduct progesterone receptors in yeast (*Saccharomyces cerevisiae*). 260 *J Biol Chem* 264: 21613-21618.

Mendel D, Bodwell JE and Munck A (1986): Glucocorticoid receptors lacking hormone-binding activity are bound in nuclei of ATP-depleted cells. *Nature* 324: 478-485.

Migliaccio A, Rotondi A and Auricchio F (1984): Calmodulin-stimulated phosphorylation of 17-beta-estradiol receptor on tyrosine. *Proc Natl Acad Sci USA* 81: 5921-5925.

Migliaccio A, Castoria G, deFalco A, Di-Domenico M, Galdiero M, Nola E, Chambon P and Auricchio F (1991): In vitro phosphorylation and hormone binding activation of the synthetic wild type human estradiol receptor. *J Steroid Biochem Mol Biol* 38: 407-413.

Mukhopadhyay NK, Price DJ, Kyriakis JM, Pelech S, Sanghera J and Avruch J (1992): An array of insulin-activated, proline-directed serine/threonine protein kinases phosphorylate the p70 S6 kinase. *J Biol Chem* 267: 3325-3335.

Munck A and Brinck-Johnson T (1968): Specific and non-specific physicochemical interactions of glucocorticoids and related steroids with rat thymus cells *in vitro*. *J Biol Chem* 243: 5556-5565.

Orti E, Mendel DB, Smith LI and Munck A (1989): Agonist-dependent phosphorylation and nuclear dephosphorylation of glucocorticoid receptors in intact cells. *J Biol Chem* 264: 9728-9731.

Poletti A, Weigel NL, McDonnell DP, Schrader WT, O'Malley BW and Conneely OM (1992): A novel, highly regulated, rapidly inducible system for the expression of chicken progesterone receptor, cPR$_A$ in *Saccharomyces cerevisiae*. *Gene* 114: 51-58.

Power RF, Lydon JP, Conneely OM and O'Malley BW (1991): Dopamine activation of an orphan of the steroid receptor superfamily. *Science* 252: 1546-1548.

Power RF, Mani SK, Codina J, Conneely OM and O'Malley BW (1991): Dopaminergic and ligand-independent activation of steroid hormone receptors. *Science* 254: 1636-1639.

Prywes R, Dutta A, Cromlish JA and Roeder RG (1988): Phosphorylation of serum response factor, a factor that binds to the serum response element of the c-fos enhancer. *Proc Natl Acad Sci USA* 85: 7206-7210.

Raychaudhuri P, Bagchi S and Nevins JR (1989): DNA binding activity of the adenovirus-induced E4F transcription factor is regulated by phosphorylation. *Genes and Develop* 3: 620-627.

Rodriguez R, Carson MA, Weigel NL, O'Malley BW and Schrader WT (1989): Hormone induced changes in the in vitro DNA-binding activity of the chicken progesterone receptor. *Mol Endocrinol* 3: 356-362.

Sheridan PL, Evans RM and Horwitz KB (1989): Phosphotryptic peptide analysis of human progesterone receptors. *J Biol Chem* 264: 6520-6528.

Smith DF, Stensgard PA, Welch WJ and Toft DO (1991): Assembly of progesterone receptor with heat shock proteins and receptor activation are ATP mediated events. *J Biol Chem* 267: 1350-1356.

Smith DF and Toft DO (1992): Composition, assembly and activation of the avian progesterone receptor. *J Ster Biochem Mol Biol* 41: 201-207.

Sullivan WP, Smith DF, Beito TG, Krco CJ and Toft DO (1988): Hormone-dependent processing of the avian progesterone receptor. *J Cellular Biochem* 36: 103-119.

Takimoto G, Tasset DM, Eppert AC and Horwitz KB (1992): Hormone-induced progesterone receptor phosphorylation consists of sequential DNA-independent and DNA-dependent stages--analysis with zinc finger mutants and the progesterone antagonist ZK98299. *Proc Natl Acad Sci USA* 89: 3050-3054.

Tora L, Gronemeyer H, Turcotte B, Gaub M-P and Chambon P (1988): The N-terminal region of the chicken progesterone receptor specifies target gene activation. *Nature* 333: 185-188.

Tsai SY, Carlstedt-Duke J, Weigel NL, Dahlman K, Gustafsson J-A, Tsai MJ and O'Malley BW (1988): Molecular interactions of steroid hormone receptor with its enhancer element:evidence for receptor dimer formation. *Cell* 55: 361-369.

Wang L-H, Tsai SY, Cook RG, Beattie WG, Tsai M-J and O'Malley BW (1989): COUP transcription factor is a member of the steroid receptor superfamily. *Nature* 340: 163-166.

332

Washburn T, Hocutt A, Brautigan DL and Korach KS (1991): Uterine estrogen receptor in vivo: phosphorylation of nuclear specific forms on serine residues. *Mol Endocrinol* 5: 235-242.

Weigel NL, Schrader WT and O'Malley BW (1989): Antibodies to chicken progesterone receptor peptide 523-536 recognize a site exposed in receptor-deoxyribonucleic acid complexes but not in receptor-heat shock protein-90 complexes. *Endocrinology* 125: 2494-2501.

Weigel NL, Carter TH, Schrader WT and O'Malley BW (1992): Chicken progesterone receptor is phosphorylated by a DNA-dependent protein kinase during *in vitro* transcription assays. *Mol Endocrinol* 6: 8-14.

References added in proof:

Poletti A, and Weigel NL (1993). Identification of a hormone dependent phosphorylation site adjacent to the DNA-binding domain of the chicken progesterone receptor. *Mol Endocrinol* 7: 241-246.

Poletti A, Conneely OM, McDonnell DP, Schrader WT, O'Malley BW and Weigel NL (1993) chicken progesterone receptor expressed in (oven) Saccharo mycis cerevisroe is correctly phosphorylated at all Four Ser-Pro phosphorylation sites. *Biochemistry* (in press).

ESTROGEN RECEPTOR ACTIVATION BY LIGAND-DEPENDENT AND LIGAND-INDEPENDENT PATHWAYS

Carolyn L. Smith
Orla M. Conneely
Bert W. O'Malley
Department of Cell Biology
Baylor College of Medicine

INTRODUCTION

Estrogens are a class of female sex steroid hormones important for the growth and differentiation of mammary and reproductive tissues and brain. The biological activities of these hormones are mediated by the estrogen receptor (ER), an intracellular phosphoprotein which binds a comparatively small molecular weight steroidal ligand and transduces its signal to the nuclear genetic apparatus. This protein belongs to a gene superfamily of ligand-activated transcription factors whose members are distinguished by the presence of three regions of homology, named C1, C2 and C3 (Carson-Jurica et al, 1990). The region of greatest conservation, C1, is composed of a sequence motif of 66-68 amino acids proposed to form two type II zinc fingers (Evans, 1988; Carson-Jurica et al, 1990; Green and Chambon, 1991 and references therein). Each finger contains 4 invariant cysteine residues thought to coordinate a single zinc atom and together these fingers in the context of the surrounding amino acid sequence determine the specificity of receptor binding to

STEROID HORMONE RECEPTORS
V. K. Moudgil, Editor
© 1993 Birkhäuser Boston

DNA (reviewed by Freedman, 1992). The characteristic homology of this region has been exploited to allow the rapid identification of new gene superfamily members in species as diverse as sea urchin (Chan et al, 1992), *Drosophila* (Mlodzik et al, 1990) and man (Wang et al, 1989), and although the final size of this superfamily is presently unknown, estimates suggest it may eventually encompass as many as 50 members (O'Malley and Conneely, 1992). Of the members identified to date, there are two major subgroups: those for which a ligand has been identified such as the steroid/thyroid/vitamin receptors, and those for which a putative ligand, if any, is currently unknown (O'Malley, 1990). These latter proteins, referred to as 'orphan' receptors, are the subject of ongoing investigations to determine their physiological ligand and/or other factors that may regulate their transcriptional activity.

The isolation and sequence analysis of steroid receptor cDNAs has greatly facilitated our understanding of their structure. The first ER cDNA was isolated from a human breast cancer cell line (MCF-7) library and encoded a protein of 595 amino acids with a molecular weight of approximately 66 kiloDaltons (Green et al, 1986; Greene et al, 1986). Subsequently, the cDNAs of chicken (Krust et al, 1986), rat (Koike et al, 1987), mouse (White et al, 1987) rainbow trout (Pakdel et al, 1989) and *Xenopus* (Weiler et al, 1987) ER also have been characterized and the deduced primary structures indicated six regions (designated A - F) of differing homology. Mutagenesis of several of these cDNAs combined with functional analysis has allowed specific functions, such as transcriptional activation, DNA binding, receptor dimerization and hormone binding to be assigned to distinct portions (*e.g.* region E is required for ligand binding) of the receptor molecule (Kumar et al, 1986; Fawell et al, 1990) and has refined our understanding of the ability of ER to modulate the transcription of estrogen-responsive genes.

Our current understanding of the process of ligand-regulated, steroid receptor-mediated biological activity is derived from studies of several steroid receptor superfamily members, including the ER. Briefly, that portion of steroid in plasma not bound by protein is considered available for diffusion across the plasma membrane of target cells whereupon receptor proteins are able to bind their cognate ligand with high affinity. Receptors subsequently undergo a conformational change and dissociate from an inhibitory heteroligomeric complex of receptor and heat shock and other proteins (Pratt, 1987; Denis et al, 1988; Bagchi et al, 1990; Allan et al, 1992). Receptors then dimerize and bind to a specific DNA sequence, termed a steroid response element (Payvar et al, 1981; Kumar and Chambon, 1988; Tsai et al, 1988). These elements, such as the consensus estrogen-response element, GGTCAnnnTGACC, are typically found within the regulatory regions of steroid-sensitive genes and function as enhancers since they confer steroid-responsiveness regardless of orientation and position with respect to their target gene promoters (Chandler et al, 1983; Klein-Hitpass et al, 1986; Green and Chambon, 1991). Subsequent modifications such as phosphorylation by DNA-dependent and other kinases occur and interactions with the transcriptional apparatus enable receptor to alter gene transcription (Klein-Hitpass et al, 1990; Bagchi et al, 1992; Weigel et al, 1992).

LIGAND-INDEPENDENT ACTIVATION OF STEROID RECEPTOR SUPERFAMILY MEMBERS

Reversible, post-translational, phosphorylation/dephosphorylation reactions long have been recognized to be important regulators of intracellular enzymatic activities (*e.g.* glycogen phosphorylase; Krebs and Beavo, 1979), and it was first suggested by Munck et al (1972) that phosphorylation of steroid receptors could influence their activities. Indeed, all of the steroid receptors studied to date, including the ER, are phosphoproteins (reviewed by Ortí et al, 1992) and increases in overall receptor phosphorylation (Ortí et al, 1989; Denner et al, 1990a; Washburn et al, 1991; Denton et al, 1992; Ali et al, 1993), as well as increased phosphorylation of specific amino acid residues (Denner et al, 1990a; Poletti and Weigel, 1993) have been demonstrated following exposure of cells to the receptor's cognate ligand. Recent mutational analysis of amino acid 118 of the human ER from a serine to an alanine reduced the overall phosphorylation of this receptor as well as its ability to mediate the transcription of various reporter constructs and this suggests phosphorylation may directly influence gene expression (Ali et al, 1993). Current models of ligand-dependent steroid receptor activation include multiple phosphorylation steps (Bagchi et al, 1992; Ortí et al, 1992; Weigel et al, 1992) and all instances of ligand-independent activation of steroid receptor superfamily members require the use of compounds, either synthetic or naturally occurring, capable of altering cellular phosphorylation pathways. The mechanism by which steroid receptors are activated in the apparent absence of ligand are not currently understood, but it is presumed that their phosphorylation and/or phosphorylation of other rate limiting co-factors with which receptors must interact to facilitate transcription are intimately involved in enhancing target gene expression.

It is important to note in the discussion below that our definition of ligand-independent activation relates to receptor-dependent target gene transcription in the absence of exogenous ligand in cells which are not known to produce detectable amounts of ligand. It should be appreciated, particularly in the case of orphan receptors, that certain cells may produce a substance(s) that receptors bind with low affinity, and thus we are unable to discount the presence of very weak agonists in experimental systems.

Ligand-Independent Activation of Steroid Receptors Other Than ER

The membrane-permeable agents, 8-bromo-cyclic adenosine monophosphate (8-Br-cAMP; an activator of cAMP-dependent protein kinase) and okadaic acid (an inhibitor of protein phosphatases 1 and 2A) activated chicken progesterone receptor-dependent target gene transcription in the absence of progesterone (Denner et al, 1990b). The human retinoic acid receptor-α is also ligand-independently activated in cells transfected with the catalytic subunit of protein kinase-A (Huggenvik et al, 1993). Although, the human progesterone receptor, in the absence of ligand, is not activated by either 8-Br-cAMP or okadaic acid both of these compounds were able to augment ligand-dependent, progesterone receptor-mediated transcription in the presence of the synthetic progestin R5020 (Beck et al, 1992). In the absence of the synthetic glucocorticoid dexamethasone, the addition of 8-Br-cAMP to cells transiently expressing the human glucocorticoid receptor or co-expression of the cAMP-dependent protein kinase catalytic subunit did not induce glucocorticoid receptor-dependent gene transcription, but their addition in the presence of ligand resulted in the synergistic activation of glucocorticoid receptor-mediated gene expression (Rangarajan et al, 1992). Similarly, the protein kinase C activating phorbol ester, 12-O-tetradecanoylphorbol-13-acetate (TPA) was unable to activate glucocorticoid receptor in the absence of ligand, but concurrent treatment of several T cell lines with dexamethasone and TPA synergistically increased glucocorticoid receptor-dependent transcription (Maroder et al, 1993). However, in NIH 3T3 fibroblasts, TPA treatment antagonizes dexamethasone-induced glucocorticoid receptor-mediated transcription (Vacca et al, 1989; Maroder et al, 1993) and the ability of signaling pathways to modulate receptor activity therefore appears to vary by cell type and could potentially even be gene specific. Moreover, not all steroid receptors can be activated in the absence of ligand and the explanation for this remains to be determined.

In addition to pharmacological modulators of cellular phosphorylation levels, the catecholaminergic neurotransmitter, dopamine, appears to ligand-independently activate the chicken progesterone receptor and human vitamin D_3 receptor in CV_1 (green monkey kidney epithelial) cells via a signal transduction pathway initiated from dopamine receptors located on the cell surface (Power et al, 1991b). There are two classes of dopaminergic receptors: those of the D_1 subtype which stimulate adenylyl cyclase and phospholipase C activities (Kebabian and Caine, 1979; Felder et al, 1989) and the D_2 subtype receptors which inhibit the production of cAMP by adenylyl cyclase (Kebabian and Caine, 1979). A dose-dependent increase in cAMP levels has been detected in CV_1 cells exposed *in vitro* to increasing dopamine concentrations, indicating dopamine receptors of the D_1 subtype are expressed in this cell line (Power et al, 1991b)

Ligand-Independent Activation of Orphan Receptors

Orphan receptors may be categorized into three subclasses: receptors which appear to be constitutively activated; those which may be activated in a ligand-independent fashion, and those that may be activated by a 'classical', but as yet unknown ligand binding event (Lydon et al, 1992). Regulation of the constitutively active orphan receptor's ability to mediate gene transcription may reside in their synthesis as immediate early gene products, as appears to be the case for nur77 (Hazel et al, 1988; Davis et al, 1991), or other unknown mechanisms as for the orphans, estrogen receptor-related 1 (ERR1) and ERR2 (Giguère et al, 1988; Lydon et al, 1992). In either case, it is difficult to ascertain the absence of endogenous ligands in the experimental systems utilized, and it has been postulated that nutritionally or metabolically derived molecules may act as ligands in autocrine or intracrine pathways to activate certain orphan receptors (O'Malley, 1988;1990).

Several orphan receptors have been activated in a ligand-independent fashion. In CV_1 cells, a chimera of the chicken ovalbumin upstream promoter transcription factor (COUP-TF) in which the chicken progesterone receptor DNA binding domain was substituted for the COUP-TF DNA binding domain, has been activated by the neurotransmitter, dopamine and the dopamine receptor agonist α-ergocryptine to induce target gene transcription (Power et al, 1991a). Neither of these compounds were bound with high affinity by COUP-TF and a deletion within the carboxyl-terminal region of the chimeric receptor rendered it unresponsive to dopaminergic activation.

Additionally, the COUP-TF chimera could also be activated by 8-Br-cAMP and okadaic acid (Power et al, 1991a). Using a similar strategy as that employed for COUP-TF, Lydon et al (1992) also have demonstrated that a chimera of the orphan receptor TR2 (Chang and Kokontis, 1988) was activated by dopamine in the apparent absence of its cognate ligand.

Ligand-Independent Activation of Estrogen Receptor

A number of investigators have reported measurable transcription of reporter genes containing estrogen response elements in the absence of exogenous estrogens (Tora et al, 1989a; Berry et al, 1990), and the ability of antiestrogens to block this apparent ligand-independent transcription suggests this gene expression is ER-dependent (Tora et al, 1989a). It therefore has been suggested that ER enhances transcription in the absence of estrogen, perhaps through its hormone-independent transcription activating function (TAF-1), or that medium used for the cultured cells contained trace levels of estrogen as might be expected for media supplemented with charcoal-treated serum (Tora et al, 1989a; Reese and Katzenellenbogen, 1992).

Several laboratories have examined the ability of ER to be activated in a ligand-independent fashion. Cho and Katzenellenbogen (1993) used activators of protein kinase A [cholera toxin (ADP-ribosylates G_s allowing it to persistently activate adenylate cyclase) and 3-isobutyl-1-methylxanthine (IBMX; a phosphodiesterase inhibitor)] and protein kinase C (TPA) to treat MCF-7 and CHO cells, and in neither case was an increase in ER-dependent reporter gene expression observed in the absence of estradiol. The human ER, however, has been ligand-independently activated by okadaic acid in CV_1 cells (Power et al, 1991b) and the apparent inability of the protein kinase activators mentioned above to modulate ER activity in the absence of estradiol may reflect promoter or cell-type specific differences. Indeed, Cho and Katzenellenbogen (1993) reported that the ability of TPA to synergize with estradiol in the activation of ER-dependent transcription varies with these two parameters.

Similar to observations on the chicken progesterone receptor, dopamine, at a concentration similar to that previously utilized to stimulate adenylyl cyclase in cultured cells (Dearry et al, 1990), stimulates ER-dependent gene expression in a dose-dependent manner in the absence of estradiol (Power et al, 1991b). These initial studies utilized CV_1 cells, and the ability of ER to be activated by dopamine in

HeLa (human epithelioid carcinoma) cells suggests that ligand-independent activation of ER and other steroid receptor superfamily members (see above) is not restricted to a single cell-type (Figure 1 and Smith et al, 1993). As for CV_1 cells (Power et al, 1991b), a dose-dependent augmentation of cAMP production was observed in HeLa cells exposed *in vitro* to increasing dopamine concentrations (Smith et al, 1993) indicating that HeLa cells express dopamine receptors of the D_1 subtype and suggesting that a similar activation of ER may occur in other cell types which express functional dopamine receptors of this general subtype. Nevertheless, it is probable that dopamine-stimulated ER transcriptional activity will vary with cell type and the promoter context of the target gene as has been shown previously for basal (Reese and Katzenellenbogen, 1992) and ligand-stimulated ER-dependent transcription (Berry et al, 1990).

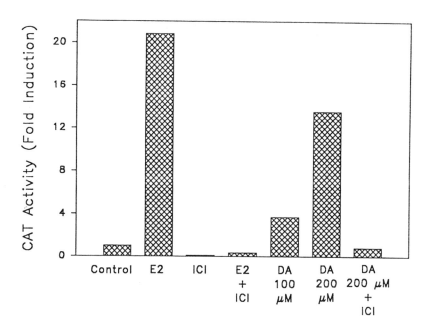

FIGURE 1. The wild type human ER is activated by 17β-estradiol and dopamine. HeLa cells were cotransfected with 2 μg of a wild type ER expression vector and 5 μg of the ER-responsive reporter plasmid, ERE-E1bCAT by the polybrene method (Denner et al, 1990b). Hormones (E2, 1 nM estradiol; ICI, 100 nM ICI 164,384; DA, dopamine) were added 24 h after transfection and 24 h thereafter cells were harvested for CAT activity measurements by phase extraction assay (Seed and Sheen, 1988).

The first human ER cDNA isolated (Green et al, 1986) contained within its ligand binding domain, a point mutation relative to wild type ER which resulted in a glycine to valine substitution at residue 400. The binding affinity of this ER mutant (ER-Val[400]) for estradiol is reduced in comparison to ER-Gly[400] as is its stability in cytoplasmic extracts incubated *in vitro* (Tora et al, 1989a) and its ability to bind DNA *in vivo* (Reese and Katzenellenbogen, 1992). Although ER-Val[400] is activated by estradiol (Kumar et al, 1987; Tora et al, 1989a; Tora et al, 1989b; Berry et al, 1990), dopamine alone could not induce ER-Val[400]-dependent transcription (Smith et al, 1993). This inability was unlikely to be due to the relative instability of mutant receptor expressed in dopamine-treated cells since the steady-state levels of wild type and mutant receptor were similar after treatment with either estradiol or dopamine in comparison to receptor levels present in control cells (Smith et al, 1993). This further suggested that a specific functional property intrinsic to the ER is required for the receptor to mediate ligand-independent activation and that at least a portion of the ligand binding domain is required to mediate this effect. This is consistent with data obtained for the chicken progesterone receptor which demonstrated that the carboxyl-terminal half of the receptor was required for activation by dopamine and okadaic acid (Power et al, 1991b). Furthermore, deletion of the A/B (TAF-1) domain of the human ER did not interfere with the ability of cholera toxin and IBMX to modulate ER-dependent gene transcription (Cho and Katzenellenbogen, 1993).

When ER-Val[400] has been expressed transiently in HeLa cells, the basal transcription of ER-dependent reporter genes was low to non-detectable in the absence of exogenous ligand while the transcriptional activity of cells expressing ER-Gly[400] was relatively high. Although this basal activity has been attributed to residual estrogen contamination of media supplemented with charcoal-stripped serum (Tora et al, 1989a; Berry et al, 1990) our experiments suggest that the relative level of receptor-dependent transcription in the apparent absence of cognate ligand is correlated with the ability of receptor to respond to cellular signaling pathways. The appreciable level of basal ER-Gly[400]-dependent transcription therefore may represent receptor activation by an extracellular factor or factors present in stripped serum or serum substitutes that are able to initiate a second messenger pathway(s). In support of this, under culture conditions apparently free of progesterone, basal transcription of a progesterone response element-containing transgene is undetectable in cells transiently transfected with a chicken progesterone receptor variant which contains a serine[628] to threonine[628] mutation and is not activated by dopamine, while unstimulated transcription in cells expressing wild type progesterone receptor is comparatively high (Power et al, 1991b).

Most studies to date have relied upon pharmacological activators of intracellular signaling pathways to examine the ligand-independent activation of steroid receptor superfamily members, and it is possible that the dopaminergic activation of ER in HeLa and CV_1 cells may simply represent the preferential activation of a signaling pathway able to communicate with this receptor. In consideration of this, it is possible that other extracellular molecules which transduce their signal by initiating signaling pathways that activate protein kinases or inhibit phosphatases may activate members of the steroid receptor superfamily, such as the ER. Indeed, it has been known for several years that epidermal growth factor (EGF), and insulin-like growth factor-I, in addition to the phorbol ester, TPA, and elevated intracellular cAMP levels increase progesterone receptor synthesis in fetal uterine and MCF-7 breast cancer cells (Sumida et al, 1988; Sumida and Pasqualini, 1989; Katzenellenbogen and Norman, 1990; Sumida and Pasqualini, 1990; Aronica and Katzenellenbogen, 1991). Increased progesterone receptor synthesis is classically thought to be an estrogen-regulated process and the ability of antiestrogens to block these effects supports the suggestion that ER may be activated in a ligand-independent fashion (Sumida and Pasqualini, 1989; Katzenellenbogen and Norman, 1990; Sumida and Pasqualini, 1990; Aronica and Katzenellenbogen, 1991). Furthermore, epidermal growth factor has been shown to mimic the estrogen-induced production of lactoferrin (a secreted iron-binding glycoprotein) as well as promoting growth in mouse uterus (Nelson et al, 1991). Epidermal growth factor also induces biochemical changes in mouse uterine ER *in vivo*, such as increased DNA binding and production of heterogeneous forms of nuclear receptor (Ignar-Trowbridge et al, 1992) indicating that a peptide modulator of cellular phosphorylation pathways, EGF, is able to influence ER in living animals.

In view of these reports we undertook an examination of whether EGF could induce ER-dependent gene transcription in a *trans*-activation assay in EGF receptor-positive (Berkers et al, 1991) HeLa cells transiently transfected with a human ER expression vector (supplied by G. Greene, University of Chicago) and the ERE-E1bCAT reporter construct which consists of a fragment of the vitellogenin A2 gene promoter (-331 to -87) upstream of the adenovirus E1b TATA box linked to a bacterial chloramphenicol acetyl transferase (CAT) gene (supplied by J.A. Cidlowski and V.E. Allgood, University of North Carolina).

Utilizing a serum-free cell culture medium, as Katzenellenbogen and Norman (1990) did for their studies on the induction of progesterone receptor synthesis by IGF-1, human recombinant EGF stimulated human ER-dependent transcription of the estrogen-responsive reporter construct, ERE-E1bCAT (Figure 2A). Stimulation of these cells by EGF concentrations similar to the binding affinity of this growth factor for its membrane receptor (Fitzpatrick et al, 1984; Mukku and Stancel, 1985) resulted in a dose-dependent increase in reporter gene expression; transcription was dependent on the presence of ER but was independent of added estrogen. Furthermore, EGF-treatment of cells expressing the human progesterone receptor from the same mammalian expression vector that directed ER production was unable to increase target gene transcription. Our results indicate that ER is required to mediate the EGF-induced target gene expression and that this increase is not the result of an overall enhancement in cellular transcriptional activity.

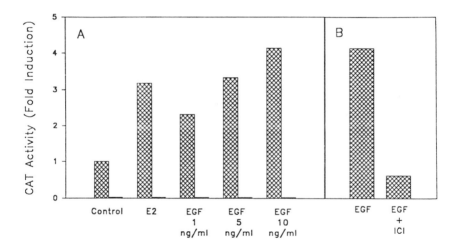

FIGURE 2. Wild type human estrogen receptor-dependent gene transcription is stimulated in HeLa cells by epidermal growth factor. Cells were cotransfected with 2 μg of ER (cross-hatch bars) or human PR$_B$ (solid bars) expression vector and 5 μg of the estrogen-responsive reporter plasmid, ERE-E1bCAT as described in Figure 1. (**A**) One nM estradiol (E2) or various concentrations of epidermal growth factor (EGF) were added to cells 24 h after transfection. Twenty-four hours after hormone treatment, cells were harvested for CAT assay. (**B**) The antiestrogen ICI 164,384 (ICI; 100 nM) blocked EGF(10 ng/ml)-induced transcription of the ERE-E1bCAT reporter plasmid.

MODULATION OF THE LIGAND-INDEPENDENT ACTIVATION OF ER BY HORMONE AND ANTIHORMONE

Experimental systems utilizing transient transfection techniques have provided useful information concerning how individual factors may influence ER transcriptional activity. However, cells *in vivo* are likely to be simultaneously exposed to a wide variety of complex signals, including those that would initiate intracellular signaling pathways as well as receptor ligands. To determine how different modes of signal transduction may impinge on ER, we and others have examined the consequences of simultaneously treating various cell types with ligands and activators of protein kinases and/or intracellular signaling pathways on receptor-mediated transcription.

As noted above, Cho and Katzenellenbogen (1993) found that neither TPA nor cholera toxin and IBMX alone were able to increase ER-dependent transcription. However, the addition of either of these compounds in the presence of a broad range of estradiol concentrations resulted in a synergistic increase in reporter gene transcription, suggesting that either of these two signaling pathways may converge on the liganded ER or a cofactor with which this receptor directly interacts.

We also have examined the impact of simultaneously activating either human wild type ER (ER-Gly^{400}) or the dopamine-unresponsive ER mutant (ER-Val^{400}) by ligand-dependent and ligand-independent mechanisms (Smith et al, 1993). Concurrent treatment with estradiol and dopamine activated wild type ER such that ER-dependent gene expression was greater than that induced by either compound alone. This effect was not, however, synergistic, and this suggests that the estrogenic and dopaminergic signaling pathways do not produce a "superactive" wild type ER. However, similar treatment of cells expressing ER-Val^{400} did lead to the synergistic induction of reporter gene expression, suggesting that ligand is able to convert mutant ER-Val^{400} to a dopamine-activatable form, whereas the aporeceptor is unresponsive to the catecholamine. Secondary structure predictions of the region surrounding amino acid^{400} of wild type ER suggest it is located in a β-turn between an α-helix and a β-strand; it is possible that substitution of a valine for the more compact glycine residue may destabilize this turn and the surrounding local conformation (Tora et al, 1989a). It is likely that the interaction between estradiol and the ER-Val^{400} mutant induces a conformational change and/or stabilization of the receptor that allows it to respond to a dopamine-initiated, signal transduction pathway(s).

In addition to agonistic ligands, synthetic compounds have been developed which are bound by ER with high affinity and act as antihormones. Although these antagonists increase receptor binding to an estrogen response element *in vitro* and *in vivo* (Kumar and Chambon, 1988; Reese and Katzenellenbogen, 1992), they are unable to effectively activate the receptor to stimulate gene transcription (Berry et al, 1990; Reese and Katzenellenbogen, 1992). We have examined the impact of two antiestrogens, *trans*-4-hydroxytamoxifen (4HT) and ICI 164,384 on the ability of dopamine to stimulate ER-dependent gene transcription. The triphenylethylene-derived antagonist 4HT, acts as a partial agonist/antagonist of ER activity (Jordon, 1984) and binds to ER with a binding affinity of 0.36 relative to estradiol (1.00; Jordan et al, 1977) while the antiestrogen, ICI 164,384, which has a relative binding affinity of 0.23 for ER (Arbuckle et al, 1992) exerts no agonist activity and is thus classified as a pure antiestrogen (Wakeling and Bowler, 1988; Wakeling, 1990).

Although, 4HT blocked the estrogenic activation of ER-Gly^{400} and ER-Val^{400}, it was unable to attenuate the ligand-independent activation of the wild type ER; it exhibited agonist activity in dopamine-treated cells expressing either form of receptor (Smith et al, 1993). Indeed, 4HT appeared to have stabilized ER-Val^{400} in much the same fashion as estradiol and enabled dopamine to further increase transgene transcription.

The ability of 4HT to block ligand-dependent but not ligand-independent ER activation suggests these two events differ mechanistically. The ligand-activated transcriptional activation function (TAF-2), located within the hormone binding domain (region E) of the ER, is thought to be blocked by 4HT, leaving the amino-terminal transactivation function (TAF-1) able to initiate gene transcription in a promoter context-dependent and cell-specific fashion (Webster et al, 1988; Berry et al, 1990). Expression of CAT activity in ER-Gly^{400}-expressing cells treated with 4HT and dopamine should therefore represent the combined activity of TAF-1 and that induced by dopamine. Because amino acid 400 of the ER influences the induction of reporter gene expression by dopamine, it seems reasonable to suggest that this agent may act primarily, though not necessarily exclusively, through TAF-2. This is consistent with previous studies using the avian progesterone receptor where dopamine activation required the carboxyl-terminal region and not the amino-terminal domain of this receptor. (Lydon et al, 1992).

The pure antihormone ICI 164,384 has been shown to reduce steady-state levels of transiently expressed mouse ER in COS-1 cells (Dauvois et al, 1992) and endogenous receptor in mouse uterus (Gibson et al, 1991) and several cell lines (Reese and Katzenellenbogen, 1992); several mechanisms by which ICI 164,384 blocks ER-dependent gene transcription have been offered (Gibson et al, 1991; Reese and Katzenellenbogen, 1991; Arbuckle et al, 1992; Reese and Katzenellenbogen, 1992). We also observed an ICI 164,384 induced reduction of ER levels (15% of untreated controls) and this likely contributed to the very low to non-detectable ER-dependent gene expression measured in HeLa cells treated with this antiestrogen and either dopamine or estradiol (Smith et al, 1993). At a concentration of ICI 164,384 sufficient to block estradiol-induced transcription, this pure antiestrogen is also able to block EGF-induced ER-mediated transcriptional enhancement (Figure 2B). Additionally, ICI 164,384 effectively blocked ER-dependent reporter gene expression in MCF-7 cells stimulated by estradiol and cholera toxin/IBMX or estradiol and TPA (Cho and Katzenellenbogen, 1993) indicating that in all cases examined, ICI 164,384 is an effective blocker of both ligand-dependent and ligand-independent activation of ER-mediated gene transcription.

FUTURE DIRECTIONS

In the last several years it has become increasingly apparent that the transcriptional activity of certain steroid/thyroid/vitamin/orphan receptor superfamily members, such as the ER, may be regulated in the apparent absence of ligand by agents capable of modulating cellular signaling pathways as well as by the receptor's cognate ligand. The initial studies of ligand-independent receptor activation were performed with pharmacological activators of cellular signaling pathways (*i.e.* phorbol esters, protein phosphatase inhibitors and agents which elevate cAMP) and suggested that extracellular hormonal signals able to stimulate similar pathways may communicate with these transcription factors and cause their activation. The ability of Neurotransmitters (*e.g.* dopamine) and growth factor (*e.g.* EGF) to activate ER-dependent transcription supports this contention and suggests that steroid receptor superfamily members may facilitate the transduction of multiple, yet unidentified, extracellular signals to the genome. In view of this, we propose that ligand-independent activation be incorporated into the classical model of steroid receptor action.

346

Significance in Brain

Estrogen receptor mRNA is located throughout the brain of both males and females, and its expression appears to be particularly high in the hypothalamus (Simerly et al, 1990). Although ligand-modulated ER activity in the ventromedial nucleus of the hypothalamus is associated with female mating behaviors (lordosis) in rats, the function of this steroid receptor in male brain, where estradiol levels are virtually undetectable remains a matter of debate. It has been observed, however, that intracerebral administration of dibutyryl cAMP can substitute for progesterone in facilitating female mating behaviors in estrogen-primed rats (Kubli-Garfias and Whalen, 1977) as can apomorphine, a dopamine receptor stimulant, and SKF 38393, a D_1 subtype receptor agonist (Mani et al, 1993). Such studies suggest the possibility that steroid receptors in

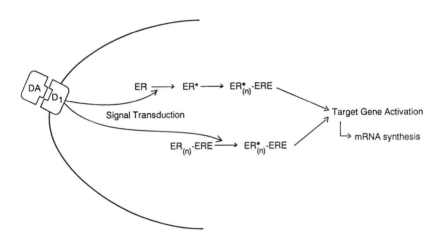

FIGURE 3. Hypothetical model of estrogen receptor activation in the brain. Dopamine (DA) binds to its plasma membrane receptor (D_1) and initiates a signal transduction pathway able to transform transcriptionally-inactive estrogen receptor (ER) to an activated form (ER*) and thereby stimulate ER-dependent gene expression. Dopaminergic activation of unliganded receptor may occur prior to receptor binding to an estrogen response element (ERE; upper pathway). Alternatively, target gene activation may arise from post-translational modification of nuclear (n) ER bound to an estrogen response element or another co-factor necessary for efficient transcription of ER-dependent genes (lower pathway).

the brain may be ligand-independently activated, and in view of the ability of dopamine to activate ER in the absence of exogenous estrogen, physiological, estrogen-independent activation of this receptor, as illustrated in Figure 3, may be possible in this organ.

Significance in Breast and Uterus

In females, ER is concentrated in reproductive organs and the growth and differentiation of these tissues is influenced by estradiol. This steroid increases the mRNA and/or protein levels of a variety of genes in these tissues, such as pS2, cathepsin-D and lactoferrin (Cavailles et al, 1988; Nunez et al, 1989; Nelson et al, 1991), and estrogen-response elements have been located in the promoter regions of several of these genes (Nunez et al, 1989; Liu and Teng, 1992). Interestingly, estrogen-mediated growth of both uterine and breast tissues has been associated with the transcription and synthesis of growth factors (such as EGF, insulin-like growth factor-1, transforming growth factor-α) and their receptors (Mukku and Stancel, 1985; Murphy et al, 1987; DiAugustine et al, 1988; Berthois et al, 1989; Clarke et al, 1991; Nelson et al, 1992).

Our data indicate ER activity may be influenced by both ligand and non-steroidal signaling pathways and suggests a model (Figure 4) in which ER may be transcriptionally activated by either steroid hormone or growth factor. It is very likely that these two signals acting in concert increase the overall ER activity by simultaneously promoting derepression and activation of receptor or the ability of receptor to interact with transcriptional co-factors. Furthermore, since estradiol increases EGF and EGF-receptor levels, estrogen-dependent gene transcription may be amplified in turn by the products of its target genes via a positive loop feedback circuit. Whether or not estradiol and EGF could synergistically activate ER remains to be determined, as does the question of target gene specificity of ligand activated or growth-factor activated receptor.

Taken together, these data support our hypothesis (Power et al, 1991a) that steroid receptor function in target tissues may be regulated via pathway "cross-talk" from membrane receptors. The events by which signals are transduced from these cell surface receptors to steroid receptor superfamily members are presently unknown but are likely to be complex and involve multiple downstream kinases and/or phosphatases. We believe it probable that ligand dependent and independent pathways may be utilized under many physiological situations and the relative

348

contributions of these pathways to the activation of steroid receptor superfamily members should therefore be considered when evaluating steroid receptor function *in vivo*. In fact it is quite possible under certain conditions that the dominant activation mechanism for selected receptors

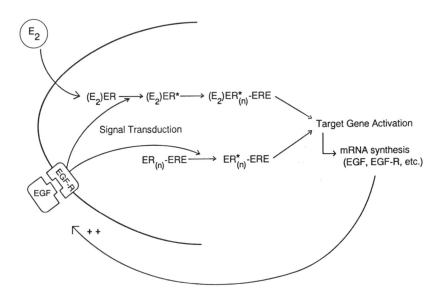

FIGURE 4. Model for the ligand-dependent and ligand-independent activation of estrogen receptor in female reproductive tissues. In the classical model of steroid hormone action, inactive estrogen receptor (ER) binds its cognate ligand, estradiol (E_2), is then activated (ER*), binds to an estrogen response element (ERE) and through interaction with the general transcriptional machinery, expression of target genes such as epidermal growth factor (EGF) and its receptor (EGF-R) is activated. In the ligand-independent model, unliganded ER may be activated by an EGF-initiated signaling pathway and thereby increase ER-dependent gene expression; receptor may be nuclear (n) and already bound to an ERE (lower pathway) or not yet associated with DNA. It is also possible that estradiol and EGF may simultaneously activate the ER to increase the overall level of target gene activation (upper pathway) and in either case, increased gene expression may also arise through modification of transcription co-factors able to interact with ER. These pathways may also allow estradiol-stimulated gene transcription to be amplified since estradiol increases EGF and EGF-R levels and these gene products may in turn further activate ER-dependent gene transcription.

may be initiated by a membrane-receptor transduced signal. In addition to normal physiological situations, the implication of ligand dependent and independent activation of the ER for the initiation, progression and ultimately the management of cancers such as breast, ovarian and endometrial could be important; of special interest would be estrogen-independent or tamoxifen-resistant tumors. Future consideration to both ER activation mechanisms may be prudent in planning the treatment of these diseases.

REFERENCES

Ali S, Metzger D, Jean-Marc B and Chambon P (1993): Modulation of transcriptional activation by ligand-dependent phosphorylation of the human oestrogen receptor A/B region. *EMBO J* 12: 1153-1160.

Allan GF, Leng X, Tsai SY, Weigel NL, Edwards DP, Tsai M-J and O'Malley BW (1992): Hormone and antihormone induce distinct conformational changes which are central to steroid receptor activation. *J Biol Chem* 267: 19513-19520.

Arbuckle ND, Dauvois S and Parker MG (1992): Effects of antioestrogens on the DNA binding activity of oestrogen receptors *in vitro*. *Nuc Acids Res* 20: 3839-3844.

Aronica SM and Katzenellenbogen BS (1991): Progesterone receptor regulation in uterine cells: stimulation by estrogen, cyclic adenosine 3',5'-monophosphate, and insulin-like growth factor I and suppression by antiestrogens and protein kinase inhibitors. *Endocrinology* 128: 2045-2052.

Bagchi MK, Tsai SY, Tsai M-J and O'Malley BW (1990): Identification of a functional intermediate in receptor activation in progesterone-dependent cell free transcription. *Nature* 345: 547-550.

Bagchi MK, Tsai SY, Tsai M-J and O'Malley BW (1992): Ligand and DNA-dependent phosphorylation of human progesterone receptor *in vitro*. *Proc Natl Acad Sci USA* 89: 2664-2668.

Beck CA, Weigel NC and Edwards DP (1992): Effects of hormone and cellular modulators of protein phosphorylation on transcriptional activity, DNA binding, and phosphorylation of human progesterone receptors. *Mol Endocrinol* 6: 607-620.

Berkers JAM, van Bergen en Henegouwen PMP and Boonstra J (1991): Three classes of epidermal growth factor receptors on HeLa cells. *J Biol Chem* 266: 922-927.

Berry M, Metzger D and Chambon P (1990): Role of the two activating domains of the oestrogen receptor in the cell-type and promoter-context dependent agonistic activity of the anti-oestrogen 4-hydroxytamoxifen. *EMBO J* 9: 2811-2818.

Berthois Y, Dong XF and Martin PM (1989): Regulation of epidermal growth factor-receptor by estrogen and antiestrogen in the human breast cancer cell line MCF-7. *Biochem Biophys Res Commun* 159: 126-131.

Carson-Jurica MA, Schrader WT and O'Malley BW (1990): Steroid receptor family: structure and functions. *Endocrine Rev* 11: 201-220.

Cavailles V, Augereau P, Garcia M and Rochefort H (1988): Estrogens and growth factors induce the mRNA of the 52K-pro-cathepsin-D secreted by breast cancer cells. *Nuc Acids Res* 16: 1903-1919.

Chan SM, Xu N, Niemeyer CC, Bone JR and Flytzanis CN (1992): SpCOUP-TF: a sea urchin member of the steroid/thyroid hormone receptor superfamily. *Proc Natl Acad Sci USA* 89: 10568-10572.

Chandler VL, Maler BA and Yamamoto KR (1983): DNA sequences bound specifically by glucocorticoid receptor *in vitro* render a heterologous promoter hormone responsive *in vivo*. *Cell* 33: 489-499.

Chang C and Kokontis J (1988): Identification of a new member of the steroid receptor super-family by cloning and sequence analysis. *Biochem Biophys Res Commun* 155: 971-977.

Cho H and Katzenellenbogen BS (1993): Synergistic activation of estrogen receptor-mediated transcription by estradiol and protein kinase activators. *Mol Endocrinol* 7: 441-452.

Clarke R, Dickson RB and Lippman ME (1991): The role of steroid hormones and growth factors in the control of normal and malignant breast. In: *Nuclear Hormone Receptors,* Parker, M.G., ed. London: Academic Press.

Dauvois S, Danielian PS, White R and Parker MG (1992): Antiestrogen ICI 164,384 reduces cellular estrogen receptor content by increasing its turnover. *Proc Natl Acad Sci USA* 89: 4037-4041.

Davis IJ, Hazel TG and Lau LF (1991): Transcriptional activation by nur77, a growth factor-inducible member of the steroid hormone receptor superfamily. *Mol Endocrinol* 5: 854-859.

Dearry A, Gingrich JA, Falardeau P, Fremeau Jr RT, Bates MD and Caron MG (1990): Molecular cloning and expression of the gene for a human D1 dopamine receptor. *Nature* 347: 72-76.

Denis M, Poellinger L, Wikstrom A-C and Gustafsson J-Å (1988): Requirement of hormone for thermal conversion of the glucocorticoid receptor to a DNA-binding state. *Nature* 333: 686-688.

Denner LA, Schrader WT, O'Malley BW and Weigel NL (1990a): Hormonal regulation and identification of chicken progesterone receptor phosphorylation sties. *J Biol Chem* 265: 16548-16555.

Denner LA, Weigel NL, Maxwell BL, Schrader WT and O'Malley BW (1990b): Regulation of progesterone receptor-mediated transcription by phosphorylation. *Science* 250: 1740-1743.

Denton RR, Koszewski NJ and Notides AC (1992): Estrogen receptor phosphorylation: hormonal dependence and consequence on specific DNA binding. *J Biol Chem* 267: 7263-7268.

DiAugustine RP, Petrusz P, Bell GI, Brown CF, Korach KS, McLachlan JA and Teng CT (1988): Influence of estrogens on mouse uterine epidermal growth factor precursor protein and messenger ribonucleic acid. *Endocrinology* 122: 2355-2363.

Evans RM (1988): The steroid and thyroid hormone receptor superfamily. *Science* 240: 889-895.

Fawell SE, Lees JA, White R, Parker MG (1990): Characterization and localization of steroid binding and dimerization activities in the mouse estrogen receptor. *Cell* 60: 953-962.

Felder CC, Blecher M and Jose PA (1989): Dopamine-1-mediated stimulation of phospholipase C activity in rat renal cortical membranes. *J Biol Chem* 264: 8739-8745.

Fitzpatrick SL, LaChance MP and Schultz GS (1984): Characterization of epidermal growth factor receptor and action on human breast cancer cells in culture. *Cancer Res* 44: 3442-3447.

Freedman LP (1992): Anatomy of the steroid receptor zinc finger region. *Endocrine Rev* 13: 129-145.

Gibson MK, Nemmers LA, Beckman Jr WC, Davis VL, Curtis SW and Korach KS (1991): The mechanism of ICI 164,384 antiestrogenicity involves rapid loss of estrogen receptor in uterine tissue. *Endocrinology* 129: 2000-2010.

Giguère V, Yang N, Segui P and Evans RM (1988): Identification of a new class of steroid hormone receptors. *Nature* 331: 91-94.

Green S and Chambon P (1991): The oestrogen receptor: from perception to mechanism. In *Nuclear Hormone Receptors,* Parker, M.G., ed. London: Academic Press.

Green S, Walter P, Kumar V, Krust A, Bornert J-M, Argos P and Chambon P (1986): Human oestrogen receptor cDNA: sequence, expression and homology to v-*erb-A*. *Nature* 320: 134-139.

Greene GL, Gilna P, Waterfield M, Baker A, Hort Y and Shine J (1986): Sequence and expression of human estrogen receptor complementary DNA. *Science* 231: 1150-1154.

Hazel TG, Nathans D, Lau LF (1988): A gene inducible by serum growth factors encodes a member of the steroid and thyroid hormone receptor superfamily. *Proc Natl Acad Sci USA* 85: 8444-8448.

Huggenvik JI, Collard MW, Kim Y-W and Sharma RP (1993): Modification of the retinoic acid signaling pathway by the catalytic subunit of protein kinase-A. *Mol Endocrinol* 7: 543-550.

Ignar-Trowbridge DM, Nelson KG, Bidwell MC, Curtis SW, Washburn TF, McLachlan JA, Korach KS (1992): Coupling of dual signaling pathways: epidermal growth factor action involves the estrogen receptor. *Proc Natl Acad Sci USA* 89: 4658-4662.

Jordon VC (1984): Biochemical pharmacology of antiestrogen action. *Pharm Rev* 36: 245-276.

Jordan VC, Collins MM, Rowsby L and Prestwich G (1977): A monohyroxylated metabolite of tamoxifen with potent antioestrogenic activity. *J Endocrinol* 75: 305-316.

Katzenellenbogen BS and Norman MJ (1990): Multihormonal regulation of the progesterone receptor in MCF-7 human breast cancer cells: interrelationships among insulin/insulin-like growth factor-1, serum, and estrogen. *Endocrinology* 126: 891-898.

Kebabian JW and Caine DB (1979): Multiple receptors for dopamine. *Nature* 277: 93-96.

Klein-Hitpass L, Schorpp M, Wagner U, Ryffel GU (1986): An estrogen-responsive element derived from the 5' flanking region of the *Xenopus* vitellogenin A2 gene functions in transfected human cells. *Cell* 46: 1053-1061.

Klein-Hitpass L, Tsai SY, Weigel NL, Allan GF, Riley D, Rodriguez R, Schrader WT, Tsai M-J and O'Malley BW (1990): The progesterone receptor stimulates cell-free transcription by enhancing the formation of a stable preinitiation complex. *Cell* 60: 247-257.

Koike S, Sakai M, Muramatsu M (1987): Molecular cloning and characterization of rat estrogen receptor cDNA. *Nuc Acids Res* 15: 2499-2513.

Krebs EG and Beavo JA (1979): Phosphorylation-dephosphorylation of enzymes. *Ann Rev Biochem* 48: 923-959.

Krust A, Green S, Argos P, Kumar V, Walter P, Bornet J-M and Chambon P (1986): The chicken oestrogen receptor sequence: homology with v-*erbA* and the human oestrogen and glucocorticoid receptors. *EMBO J* 5: 891-897.

Kubli-Garfias C and Whalen RE (1977): Induction of lordosis behavior in female rats by intravenous administration of progestins. *Horm Behav* 9: 380-386.

Kumar V and Chambon P (1988): The estrogen receptor binds tightly to its responsive element as a ligand-induced homodimer. *Cell* 55: 145-156.

Kumar V, Green S, Stack G, Berry M, Jin J-R and Chambon P (1987): Functional domains of the human estrogen receptor. *Cell* 51: 941-951.

Kumar V, Green S, Staub A and Chambon P (1986): Localization of the estradiol-binding and putative DNA-binding domains of the human oestrogen receptor. *EMBO J* 5: 2231-2236.

Liu Y and Teng CT (1992): Estrogen response module of the mouse lactoferrin gene contains overlapping chicken ovalbumin upstream promoter transcription factor and estrogen receptor-binding elements. *Mol Endocrinol* 6: 355-364.

Lydon JP, Power RF and Conneely OM (1992): Differential modes of activation define orphan subclasses within the steroid/thyroid receptor superfamily. *Gene Express* 2: 273-283.

Mani SK, Allen JMC, Clark JH and O'Malley BW (1993): Convergent molecular pathways for progesterone receptor activation in the control of sexual behavior in female rats. 75th Annual meeting of the Endocrine Society, Abstract #1559.

Maroder M, Farina AR, Vacca A, Felli MP, Meco D, Screpanti I, Frati L and Gulino A (1993): Cell-specific bifunctional role of jun oncogene family members on glucocorticoid receptor-dependent transcription. *Mol Endocrinol* 7: 570-584.

Mlodzik M, Hiromi Y, Weber U, Goodman CS and Rubin GM (1990): The *Drosophila* seven-up gene, a member of the steroid receptor superfamily, controls photoreceptor cell fates. *Cell* 60: 211-224.

Mukku VR and Stancel GM (1985): Regulation of epidermal growth factor receptor by estrogen. *J Biol Chem* 260: 9820-9824.

Mukku VR and Stancel GM (1985): Receptors of epidermal growth factor in the rat uterus. *Endocrinology* 117: 149-154.

Munck A, Wira C, Young DA, Mosher KM, Hallahan C and Bell PA (1972): Glucocorticoid-receptor complexes and the earliest steps in the action of glucocorticoids on thymus cells. *J Steroid Biochem* 3: 567-578.

Murphy LJ, Murphy LC and Friesen HG (1987): Estrogen induces insulin-like growth factor-1 expression in the rat uterus. *Mol Endocrinol* 1: 445-450.

Nelson KG, Takahashi T, Bossert NL, Walmer DK and McLachlan JA (1991): Epidermal growth factor replaces estrogen in the stimulation of female genital-tract growth and differentiation. *Proc Natl Acad Sci USA* 88: 21-25

Nelson KG, Takahashi T, Lee DC, Luetteke NC, Bossert NL, Ross K, Eitzman BE and McLachlan JA (1992): Transforming growth factor-α is a potential mediator of estrogen action in the mouse uterus. *Endocrinology* 131: 1657-1664.

Nunez A-M, Berry M, Imler J-L and Chambon P (1989): The 5' flanking region of the pS2 gene contains a complex enhancer region responsive to oestrogens, epidermal growth factor, a tumor promoter (TPA), the c-Ha-*ras* oncoprotein and the c-*jun* protein. *EMBO J* 8: 823-829.

O'Malley BW (1988): Editorial: did eucaryotic steroid receptors evolve from intracrine gene regulators? *Endocrinology* 125: 1119-1120

O'Malley BW (1990): The steroid receptor superfamily: more excitement predicted for the future. *Mol Endocrinol* 4: 363-369.

O'Malley BW and Conneely OM (1992): Orphan receptors: in search of a unifying hypothesis for activation. *Mol Endocrinol* 6: 1359-1361.

Ortí E, Bodwell JE and Munck A (1992): Phosphorylation of steroid hormone receptors. *Endo Rev* 13: 105-128.

Ortí E, Mendel DB, Smith LI and Munck A (1989): Agonist-dependent phosphorylation and nuclear dephosphorylation of glucocorticoid receptors in intact cells. *J Biol Chem* 264: 9728-9731.

Pakdel F, Le Guellec C, Vaillant C, Le Roux MG and Valtaire Y (1989): Identification and estrogen induction of two estrogen receptors (ER) messenger ribonucleic acids in the rainbow trout liver: sequence homology with other ERs. *Mol Endocrinol* 3: 44-51.

Payvar F, Wrange Ö, Carlstedt-Duke J, Okret S, Gustafsson J-Å and Yamamoto KR (1981): Purified glucocorticoid receptors bind selectively *in vitro* to a cloned DNA fragment whose transcription is regulated by glucocorticoids *in vivo*. *Proc Natl Acad Sci USA* 78: 6628-6632.

Poletti A and Weigel NL (1993): Identification of a hormone-dependent phosphorylation site adjacent to the DNA-binding domain of the chicken progesterone receptor. *Mol Endocrinol* 7: 241-246.

Power RF, Lydon JP, Conneely OM and O'Malley BW (1991a): Dopamine activation of an orphan of the steroid receptor superfamily. *Science* 252: 1546-1548.

Power RF, Mani SK, Codina J, Conneely OM and O'Malley BW (1991b): Dopaminergic and ligand-independent activation of steroid hormone receptors. *Science* 254: 1636-1639.

Pratt WB (1987): Transformation of glucocorticoid and progesterone receptors to the DNA-binding state. *J Cell Biochem* 35: 51-68.

Rangarajan PN, Umesono K and Evans RM (1992): Modulation of glucocorticoid receptor function by protein kinase A. *Mol Endocrinol* 6: 1451-1457.

Reese JC and Katzenellenbogen BS (1991): Differential DNA-binding abilities of estrogen receptor occupied with two classes of antiestrogens: studies using human estrogen receptor overexpressed in mammalian cells. *Nucleic Acids Res* 19: 6595-6602.

Reese JC and Katzenellenbogen BS (1992): Examination of the DNA-binding ability of estrogen receptor in whole cells: implications for hormone-independent transactivation and the actions of antiestrogens. *Mol Cell Biol* 12: 4531-4538.

Seed B and Sheen J-Y (1988): A simple phase-extraction assay for chloramphenicol acyltransferase activity. *Gene* 67: 271-277.

Simerly RB, Chang C, Muramatsu M and Swanson IW (1990): Distribution of androgen and estrogen receptor mRNA-containing cells in the rat brain: an *in situ* hybridization study. *J Comp Neurol* 294: 76-95.

Smith CL, Conneely OM and O'Malley BW (1993): Modulation of the ligand-independent activation of the human estrogen receptor by hormone and antihormone. *Proc Natl Acad Sci USA* 90: 6120-6124

Sumida C, Lecerf F and Pasqualini JR (1988): Control of progesterone receptors in fetal uterine cells in culture: effects of estradiol, progestins, antiestrogens, and growth factors. *Endocrinology* 122: 3-11.

Sumida C and Pasqualini JR (1989): Antiestrogens antagonize the stimulatory effect of epidermal growth factor on the induction of progesterone receptor in fetal uterine cells in culture. *Endocrinology* 124: 591-597.

Sumida C and Pasqualini JR (1990): Stimulation of progesterone receptors by phorbol ester and cyclic AMP in fetal uterine cells in culture. *Mol Cell Endocrinol* 69: 207-215.

Tora L, Mullick A, Metzger D, Ponglikitmongkol M, Park I and Chambon P (1989a): The cloned human oestrogen receptor contains a mutation which alters its hormone binding properties. *EMBO J* 8: 1981-1986.

Tora L, White J, Brou C, Tasset D, Webster N, Scheer E and Chambon P (1989b): The human estrogen receptor has two independent nonacidic transcriptional activation functions. *Cell* 59: 477-487.

Tsai SY, Carlstedt-Duke J, Weigel NL, Dahlman K, Gustafsson J-Å, Tsai M-J and O'Malley BW (1988): Molecular interactions of steroid hormone receptor with its enhancer element: evidence for receptor dimer formation. *Cell* 55: 361-369.

Vacca A, Screpanti I, Maroder M, Petrangeli E, Frati L and Guilino A (1989): Tumor-promoting phorbol ester and ras oncogene expression inhibit the glucocorticoid-dependent transcription from the mouse mammary tumor virus long terminal repeat. *Mol Endocrinol* 3: 1659-1665.

Wakeling AE (1990): Novel pure antiestrogens: mode of action and therapeutic prospects. *Ann NY Acad Sci* 595: 348-356.

Wakeling AE and Bowler J (1988): Novel antioestrogens without partial agonist activity. *J Steroid Biochem* 31: 645-653.

Wang L-H, Tsai SY, Cook RG, Beattie WG, Tsai M-J and O'Malley BW (1989): COUP transcription factor is a member of the steroid receptor superfamily. *Nature* 340: 163-166.

Washburn T, Hocutt A, Brautigen DL and Korach KS (1991): Uterine estrogen receptor in vivo: phosphorylation of nuclear specific forms on serine residues. *Mol Endocrinol* 5: 235-242.

Webster NJG, Green S, Jin JR and Chambon P (1988): The hormone-binding domains of the estrogen and glucocorticoid receptors contain an inducible transcription activation function. *Cell* 54: 199-207.

Weigel NL, Carter TH, Schrader WT and O'Malley BW (1992): Chicken progesterone receptor is phosphorylated by a DNA-dependent protein kinase during *in vitro* transcription assays. *Mol Endocrinol* 6: 8-14.

Weiler IJ, Lew D and Shapiro DJ (1987): The *Xenopus laevis* estrogen receptor: sequence homology with human and avian receptors and identification of multiple estrogen receptor messenger ribonucleic acids. *Mol Endocrinol* 1: 355-362.

White R, Lees JA, Needham M, Ham J and Parker M (1987): Structural organization and expression of the mouse estrogen receptor. *Mol Endocrinol* 1: 735-744.

WHY ARE STEROID RECEPTOR ANTAGONISTS SOMETIMES AGONISTS?

Kathryn B. Horwitz
Kimberly K. Leslie
Department of Medicine
University of Colorado Health Sciences Center

INTRODUCTION

Faithful expression of genetic information is lost in tumor cells due to the formation of spontaneous cell variants. In breast cancer, this evolution is marked by progression of tumors from hormone-dependent, through hormone-responsive, to hormone-resistant states. Many resistant tumors no longer express estrogen (ER) and progesterone receptors (PR) and this may be the basis for their hormone resistance. However, half of all advanced breast cancers are receptor-positive, yet they too fail to respond to antiestrogen therapy. Both the cellular heterogeneity that mark progression of the disease, and the hormone-resistance that characterize the end-stages of the disease, have been long-standing clinical problems that are slowly yielding to basic research focused both on solid tumors taken directly from patients, and on breast cancer cell lines derived from such tumors. Studies addressing issues of molecular and cellular heterogeneity of receptors and tumor cells and their relationship to progression and resistance of breast cancer are reviewed below.

STEROID HORMONE RECEPTORS
V. K. Moudgil, Editor
© 1993 Birkhäuser Boston

Estrogen receptors

The molecular biology of estrogen receptors has been extensively explored in recent years. Their cDNA was independently cloned and sequenced from MCF-7 breast cancer cells by Green et al (1986) and Greene et al (1986), and the ER gene was cloned and analyzed two years later (Ponglikitmongkol et al, 1988). The protein is comprised of 595 amino acids within which Kumar et al (1988) distinguished six functional domains identified by the letters A through F. The A/B domains contain regions that regulate the transcriptional function of the proteins. The C domain contains two DNA-binding zinc fingers and is the region of the protein that binds to the estrogen response element (ERE). Mutations in this portion of the protein change its affinity for DNA, resulting in sub-optimal, or complete loss of DNA binding. The hormone binding properties of the receptors reside in region E, as shown by mutagenesis analysis. Since these two functions, DNA binding and hormone binding, are carried out by separate parts of the protein, they are to some extent independent. Thus, it is possible to have variant receptors that can bind to DNA with limited affinity without first binding hormone, and vice versa (Kumar et al, 1986, 1988). Three additional important specialized regions of steroid receptors have also been identified: a nuclear localization signal, a heat shock protein (hsp 90) binding region, and a dimerization domain. The nuclear localization signal, located downstream of the DNA binding domain, is a region of the protein that must be present for the receptor to remain within the nucleus in the absence of ligand (Guiochon-Mantel et al, 1988). It has been identified in progesterone receptors (PR) and is presumed to be similar in ER. The hsp 90 appears to bind to regions in the hormone binding domain of some steroid receptors when ligand is absent, and its binding is believed to prevent receptor dimerization and DNA binding (Baulieu E-E, 1987). Ligand activation leads to hsp 90 dissociation, and monomer dimerization in solution (Kumar et al, 1988). The dimerization domain that mediates this interaction between two ER molecules has been localized to the carboxy-terminal end of the hormone binding domain (Fawell et al, 1990). A weak dimerization domain may also be present in the second zinc finger of the DNA-binding domain (Kumar et al, 1988). Additional sites for heterologous protein-protein interactions may also be located in the hormone binding domain (Adler et al, 1988) and covalent modifications by phosphorylation (Washburn et al, 1991) further enhance the complexity of this protein molecule.

Molecular Heterogeneity: Estrogen Receptors

Several reports of naturally occurring mutant or variant ER forms have recently appeared (Murphy LC, 1990). In addition, polymorphic forms of the ER gene have been described (Hill et al, 1989; Castagnoli et al, 1987; Parl et al, 1989). The majority of these genetic changes are found in introns, which do not directly encode the mRNA, or in turn, the protein. Of interest is the recent report by Keaveney et al (1991) identifying an alternative estrogen receptor mRNA which appears to be the primary transcript present in the human uterus, as opposed to the breast cancer line MCF-7. This newly identified transcript is alternatively spliced in the 5'-untranslated region, and has an additional exon with two small open reading frames upstream of the alternative splice site. Although the receptor proteins encoded by these two types of messages are identical, the nucleotide sequences which flank the translated regions are different, and are likely to lead to differential regulation of the protein depending upon which type of message predominates in the tissue in question. Equally interesting is a truncated ER message specific to pituitary cells (Shupnik et al, 1989). This deletion involves the translated region and presumably encodes a variant receptor, although expression of the protein has not yet been documented. Thus, in normal cells, the regulation of ER gene transcription, and even ER protein structure, may be tissue-specific.

1. *Mutant Estrogen Receptors in Solid Tumors*

Turning to malignant cells, there is now mounting evidence to show that in addition to silent mutations and regulatory heterogeneity, mutations in ER exons exist that would influence protein structure and protein function. Garcia et al (1988, 1989) identified a polymorphic variant in the B region of ER mRNA in some human breast cancer biopsies. This variant has since been correlated with lower than normal levels of hormone binding activity, and preliminary evidence suggests that women who are heterozygous for this variant have a higher proportion of spontaneous abortions than those who are homozygous at the same locus (Lehrer et al, 1990).

Wild-type ER mRNAs from several normal and malignant tissues and species are reported to be approximately 6.2 kb in size. However, Murphy and Dotzlaw (1989) have identified truncated ER-like mRNAs in human breast cancer biopsy samples by Northern blotting. These messages appear to lack significant portions of the 3' region including the hormone binding domain. By polymerase chain reaction amplification of mRNA from breast tumor specimens, Fuqua et al (1991) have also identified mutant forms of ER missing part of the hormone binding domain due to deletion of exon 5 and exon 7. These mutants are an alternatively spliced form, capable of constitutively activating transcription of an ER-dependent gene, or of dominantly inhibiting the activity of wild-type ER. PCR amplification was also used to identify a mutation in the D domain of ER mRNA expressed in a murine transformed Leydig cell line, B-1 F (Hirose et al, 1991). The functional significance of these mutations has yet to be fully explored but they clearly suggest mechanisms by which mutant receptor forms can subvert the activity of wild-type forms, when both are co-expressed in the same tumor cell.

The weakness in all these analyses is the assumption that message variants reflect protein variants. While this may indeed be the case, until recently, given the immunologic tools currently available, no mutant proteins had been detected. This may have been rectified by two studies in which gel shift assays were used to examine the ability of tumor ERs to bind an ERE (Foster et al 1991; Scott et al, 1991). The studies show that some tumors containing abundant immunoreactive ER failed to demonstrate DNA-binding ER, or the DNA-binding ER forms appeared to be truncated, or they were immunologically ER-negative but positive by the mobility shift assay. Based on these preliminary data, the prevalence of non-DNA binding ER forms or of truncated ER forms among ER-positive or PR-positive tumors may exceed 50%; a significant number whose structural analysis may become a critically important prognostic tool.

2. *Mutant Estrogen Receptors in Breast Cancer Cell Lines*

Estrogen receptors play a critical role in the development, progression and hormone-responsiveness of breast cancers. Their structural analysis, by methods like those described above, can be used to generate functional predictions. Alternatively, a product of ER action can be monitored, and PR have served this role for many years (Horwitz et al, 1975). In all estrogen/progesterone target tissues, estradiol is required for PR induction. This relationship holds true for breast cancers (Horwitz and McGuire 1978) and led us to propose that the presence of PR could be used as a tool to predict the hormone dependence of human breast tumors. Thus, a tumor that contains PR would, of necessity, have a functional ER. This idea has in general been borne out by studies which show that ER-positive tumors that also have PR are much more likely (75%) to respond to hormone treatment than tumors that are ER-positive but PR-negative (35%) (Horwitz et al, 1975). These studies also identify a small group of puzzling tumors that are ER-negative but PR-positive and have a higher response rate than is usually expected of ER-negative tumors. They are puzzling because according to dogma such tumors should not exist. Thus, either PR synthesis in these tumors is entirely independent of ER action in defiance of a long accepted principle, or a variant or other unmeasured form of ER is stimulating PR synthesis.

In 1978, while measuring the steroid receptor content of a series of cultured human breast cancer cells, we found one cell line, T47D, that had no soluble ER by sucrose density gradient analysis, yet had the highest PR levels of any cell line surveyed (Horwitz et al, 1978; Keydar et al, 1979). These cells seemed to be ideally suited to study this ER-negative but PR-positive paradox.

We subsequently found that a subline, which we called T47Dco, did have ER, but they were in a permanently activated state in the nucleus. The ER were not sensitive to the action of estrogens, suggesting that the estrogen regulatory mechanism was defective at a step beyond the initial interaction of the steroid-receptor complex with DNA. The PR were also insensitive to estradiol or to antiestrogens but were synthesized in extraordinary amounts and were functional. Additional studies suggested that the PRs retained characteristics of inducible proteins. Thus, we suggested that persistent nuclear ER were constitutively stimulating PR, even in the absence of exogenous estradiol (Horwitz, 1981; Horwitz et al, 1982). Recently, the tools became available to test this conjecture. Two cDNA libraries were constructed from T47Dco cells, that have yielded mutant ER cDNA clones encoding the putative ER proteins (Graham et al, 1990). Library I, made from total poly(A)+ mRNA, yielded two clones having wild-type ER sequences and one clone that encodes part of ER exon 2 and all of exon 3. At the precise junction between exons 3 and 4, the ER sequence homology ends and the adjacent sequence is not found in ER cDNA, in the reported ER intron sequences, or in the DNA data base. Compared to wild-type ER, the region immediately upstream of the exon 3/intron 3 junction has two inserted T-residues. This leads to a disruption of the protein-reading frame by generating a TGA termination codon. There are also two point mutations just distal to the insertions. Together, these mutations may be responsible for the abnormal splicing reaction observed. This cDNA would encode an ER truncated at aa 250 in the DNA binding domain, just beyond the last cysteine (aa 245) of the second DNA-binding finger. The putative mutant protein would lack the nuclear localization signal and hormone-binding domains of ER, and lacks epitopes for most anti-ER antibodies.

Library I also yielded an ER clone that appears to be an RNA-processing intermediate or splicing error and contains ~1 kb of intron 5 linked upstream of exon 6, and three clones with an insertion in exon 5. The insert in the three clones contains at least two blocks of direct repeats of ~130 nucleotides terminating in A residues that are 70-85% homologous to the human alu family.

To ensure that the first library did not have an overrepresentation of unprocessed nuclear mRNAs, a second library was made from cytoplasmic mRNA. In addition to clones consistent with wild-type ER, library II yielded mutant cDNAs that would encode proteins with potentially important biological activity. One clone has a point deletion in the hormone-binding domain just upstream of the end of exon 5. This leads to a frame-shift and a translation termination seven codons later. This mutant cDNA would encode an ER truncated in the middle of the hormone binding domain at aa 417, with a unique 7 aa COOH-terminal end. The putative mutant protein would be unable to bind hormone or the anti-ER antibody H222, but could be constitutively active.

Library II also yielded two independent clones having an identical in-frame deletion. These clones contain a wild-type upstream sequence from the 5' untranslated region to the first four codons of exon 4. There follows a loss of 460 nucleotides, including most of exons 4 and 5. The last codon of exon 5 is preserved, as are downstream exon 6 sequences. Interestingly, the nucleotide sequences at the borders of the deletion are identical (GAC), and could represent either nucleotides 1004-1006 (aa 258) of exon 4 or nucleotides 1463-1465 (aa 411) of exon 5. This cDNA would encode a mutant ER of 442 aa instead of the normal 595 aa, having a 153 aa deletion from the end of the DNA-binding domain C, through the hinge region D, to the mid-hormone-binding domain E. The deletion originates in the sequence encoding the putative nuclear localization signal (aa 256-263; R-K-D-R-R-G-G-R). However, the aa sequence encoded by the deletion mutant (R-K-D-R-N-Q-G-K) preserves four of the five basic aa residues of the wild-type sequence.

We do not know whether the abnormal proteins are expressed. Gel mobility shift analyses of T47Dco nuclear extracts show considerable amounts of specific ERE-binding proteins which neither comigrate with wild-type receptors, nor are supershifted by antibodies H222 or H226. The identity of these proteins is still under investigation. However, based on deletion mutagenesis analyses (Kumar and Chambon, 1988), we can begin to predict the consequence to the cells of such mutant ERs. Especially in T47Dco sublines with HT subpopulations (see further below) which contain 4-5 alleles of the ER gene (Graham et al, 1989b), cells having a mixture of wild-type and mutant receptors could co-exist. Heterodimers of the wild-type and mutant monomers, having dominant positive or dominant negative activity (Forman et al, 1989), could override the estrogen requirement of the wild-type receptors. This would result in ER-positive but estrogen-resistant cells, a phenotype that describes 50% of hormone-resistant breast cancers.

Consequences of Mutant Estrogen Receptors: Cellular Heterogeneity?

The consequences of this newly discovered molecular diversity in ER, may reach beyond issues of hormone-dependence, to the broader problems of tumor progression and cellular heterogeneity that also characterize advanced breast cancer. Cellular heterogeneity has usually been assumed to exist within tumors, but has been difficult to demonstrate. The concept is, however, important, since it means that in practice, the clinician must treat not just one tumor, but a variety of possibly heterogenous, subtumors. Is it possible that heterogeneity of ER among cells can lead to heterogeneity of cells among tumors? While the analyses of ER described above have led to the discovery of variant receptor forms, the methods cannot answer a fundamental question. Do all, or only some of the cells carry the variants? Moreover, wild-type ER are always present together with the variants. Are wild-type ER present alone in some cells of the tumor or are they always co-expressed with the variants in any one cell? We postulated that the genetic diversity of ER would be reflected in heterogeneity of other molecular markers and set out to develop an assay that could simultaneously measure DNA content and PR heterogeneity in subpopulations of tumor cells (Graham et al, 1989 a,b). We have used this immunologic, dual-parameter flow cytometry (FCM)-based assay to demonstrate and quantitate a remarkable heterogeneity in PR content, DNA ploidy, and mitotic indices among subpopulations of breast cancer cells (Graham et al, 1992).

1. Cellular Heterogeneity of Progesterone Receptor Levels

Progesterone receptor heterogeneity is illustrated in figure 1 by three cell lines derived from T47Dco, in which only the first, T47Dv (panel A), has the PR phenotype that most current receptor measurement methods

364

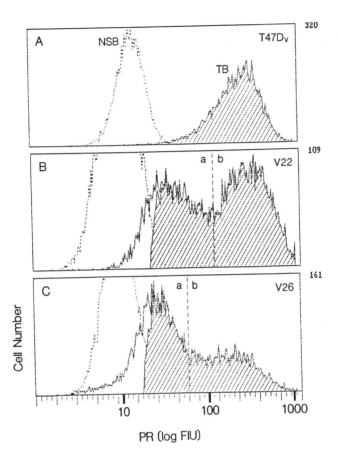

FIGURE 1. Heterogeneity of PR levels in three T47D cell lines.
Total (TB) and nonspecific (NSB) PR were measured by dual parameter
flow cytometry in T47Dv cells and two clonal sublines, V22 and V26.
Fluorescence intensity unit (FIU) levels in 10,000 cells of each set were
analyzed and plotted.   Specific binding (SB) of PR-positive cells
indicated by the shaded areas, was calculated by curve subtraction. Cells
with TB levels falling under the NSB curve are considered to be
PR-negative. Using the 1-Par program, subpopulations with low ("a")
and high ("b") PR levels were gated and analyzed as described in Table
1 and the text. The vertical scale of each panel is different as shown by
the number on the right. PR levels are plotted on a log scale. In panels
B and C the peak of the NSB curve has been cut off in order to expand
the SB curves. Originally published in Graham et al (1992). Reproduced
with permission.

assume, namely that cells are positive at a level greater than measured background. However, even T47Dv cells have PR levels which range extensively as shown by the width of the receptor peak on the log scale. Panels B and C illustrate entirely different PR-positive patterns; two cell lines (V22 and V26) that have more than one PR-positive population despite the fact that they were derived as single cell clones from T47Dv.

To quantitate PR in the subpopulations we have developed a computer program entitled 1-par. Calculations using this software show that 12.8% of cells in V22 and 23.2% of cells in V26 are PR-negative, and that in addition, each cell line also contains two distinctly different PR-positive subpopulations. Starting with a cell-line having a PR-negative subpopulation and cloning by limiting dilution plus FCM analysis, we have generated new T47D cell lines, in which 100% of the cells are PR-negative by FCM and by enhanced chemiluminescence immunoblotting (unpublished).

Does a two-population model adequately describe cells such as those depicted in panels B and C of Figure 2? Probably not. Bimodality of a single variable like PR hints at still greater numbers of subpopulations when a second variable is analyzed simultaneously. The simultaneous analysis of PR and DNA indices shows that V26 is a mixture of 47.2% hyperdiploid (HD) cells and 52.8% hypertetraploid (HT) cells. The HD cells, with 24.3% of cells in S and G2M, grow slightly faster than the HT cells which have 17.0% of cells in the proliferating fraction (Graham et al, 1992). Combining the PR and DNA data shows there are two distinct HD subpopulations: one contains cells with low PR levels and the other cells with high PR levels. In addition, the HT cells also contain subpopulations with low and high PR levels. Thus, there are at least four subpopulations in this cell line, each having a different combination of PR, DNA content, and mitotic indices. V22 cells are similarly heterogenous.

2. *Tumor Cell "Remodeling" by Tamoxifen*

The practical consequence of this PR heterogeneity in breast cancer cells is illustrated by an experiment in which the T47Dv cell line was treated for eight weeks with or without 1 µM tamoxifen. Tamoxifen at 1 µM generally suppresses growth and PR in estrogen target tissues (Horwitz and McGuire, 1978) that carry a normal ER, and is the major endocrine therapeutic drug used in breast cancer (Lerner and Jordan, 1990). But, what is the effect of tamoxifen in cells that carry not only normal but also variant ER?

Figure 2 shows a univariate PR analysis of control cells (panel A), 1 μM tamoxifen-treated cells (panel B), and a comparison of the two (panel C) to demonstrate the shifts in PR patterns after eight weeks under the influence of the drug. Cell growth was suppressed by 40% (not shown), and there was a marked shift in the PR pattern--mostly to the left, reflecting a complete loss or decrease in PR, as shown in panel B, and by the large shadowed area in panel C. However, there was an unexpected small subpopulation shifted to the right, in which PR levels have apparently been induced by tamoxifen. This subpopulation represents 5.2% of the cells in this experiment, and contains an average PR of 571.6 fluorescence intensity units (FIU), or greater than one million PR molecules/cell--levels that none of the untreated cells attain. Thus, tamoxifen, while decreasing PR levels in a majority of cells, appears paradoxically to increase PR levels in a selected subset of cells. The ominous consequence of tumor cell populations that may be stimulated by tamoxifen requires little comment.

In addition, analysis of the DNA indices (Graham et al, 1992) demonstrates that tamoxifen has a dual effect on proliferation. First, for the same number of cells, fewer tamoxifen-treated cells are in mitosis, and second, the populations that are in mitosis under tamoxifen differ from the controls. Thus, while the overall growth of the tamoxifen-treated cells lags behind that of the control cells, the DNA data show what we term the remodeling influence of the drug; the growth and emergence of at least two new subpopulations of cells that are not present in controls: a PR-negative or low-PR, HD subset; and an ultra-high PR, HT subset. If the biologic behavior of this cell line mimics the pattern seen in patients with metastatic breast cancer who have an initial growth inhibitory response to tamoxifen but then relapse, it may be these emerging subpopulations that lead to later tumor progression and our present impression of recurrent breast cancer as an incurable disease.

A variety of mechanisms have been proposed for development of the acquired resistance to tamoxifen that arises in animal model systems (Osborne et al, 1987; Gottardis and Jordan, 1988) and in virtually all patients (Lippman, 1984; McGuire et al, 1987) undergoing hormone therapy. Genetic mechanisms include the variant and mutant forms of ER described above which may exert dominant controls over estrogen and antiestrogen-regulated growth. Additionally, heterogeneity of ERs and mutant ERs, may in part explain the extreme PR heterogeneity documented here. Epigenetic mechanisms center on pharmacokinetic issues related to drug absorption, distribution and metabolism. While some of the metabolites of tamoxifen are more potent antiestrogens than

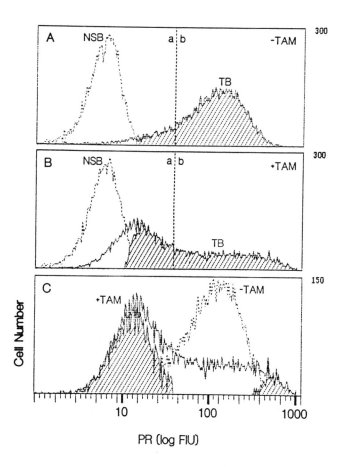

FIGURE 2. Tamoxifen remodeling of PR subpopulations in the T47Dv cell line. Total (TB) and nonspecific (NSB) PR signals were measured by flow cytometry in T47Dv after eight weeks of growth in control med (-TAM) or in 1 μM tamoxifen (+TAM). PR FIU levels in 10,000 cells of each set were analyzed and are shown. In panels A and B, specific binding was calculated by curve subtraction and is shown by the shaded area. Using the 1-Par program, subpopulations with low ("a") and high ("b") PR levels were gated and analyzed as described in Graham et al (1992). 7A. PR in control cells, showing TB and NSB. 7B. PR in 1 μM tamoxifen-treated cells, showing TB and NSB. 7C. TB curves from control and tamoxifen-treated cells were superimposed. PR subpopulations present in +TAM, that are excluded in -TAM, were calculated by curve subtraction and are shaded. Note change in the vertical scale. These data were originally published in Graham et al (1992). Reproduced with permission.

the parent compound (Lerner and Jordan 1990), other metabolites may be estrogenic (Gottardis et al, 1989). Recent data indicate that tamoxifen and its anti-estrogenic metabolite, trans-4-hydroxytamoxifen, may be selectively excluded from tamoxifen-resistant breast cancers, or be further metabolized to relatively inactive forms (Osborne et al, 1992).

While, in different tumors, and different cells, one or both general mechanisms of resistance may become operative, we propose that tumor progression to the resistant state includes the selection and expansion of cell subpopulations, some of which remain strongly influenced by tamoxifen. That hormone treatment may itself provide the selective remodeling pressure is suggested by the studies described here, and by studies showing that human breast cancer cells change significantly in response to hormone deprivation (Katzenellenbogen et al, 1987; Daly et al, 1990) or stimulation (Gottardis and Jordan, 1988; Gottardis et al, 1989).

Our data suggest that subsets of cells may actually be stimulated by tamoxifen. Little is known about the mechanisms underlying these "agonist" actions of some antiestrogens. It is possible that binding of tamoxifen to specific types of ER mutants, establishes a transcriptionally productive receptor complex. The agonist activities of tamoxifen are usually expressed at low doses (Horwitz and McGuire, 1978), but they may also be tissue-specific (Turner et al, 1987). While tamoxifen at high doses suppresses PR, it induces PR at low doses (Horwitz and McGuire, 1978). The tumor "flare" that occurs during initiation of tamoxifen therapy in patients (Henderson et al, 1989), and the withdrawal response that occurs when the drug is stopped after tumors become resistant (Legault-Poisson et al, 1979), may also be explained by this property. Additionally, we have previously shown that pretreatment of cells with an antiestrogen can sensitize them to a subsequent challenge with estrogens. In this state, cells respond more rapidly and more extensively to estrogens; for example, superinduction of PRs is observed (Horwitz et al, 1981). It is possible that antiestrogen pre-treatment can sensitize tumor cells to low levels of estrogens, or to weak estrogens, to which, in other settings, they would be unresponsive. The molecular mechanisms underlying the phenomenon of superinduction remain unknown.

Progestin Resistance

The emergence of hormone-resistant cells eventually reduces the effectiveness of all therapies in advanced breast cancer, and progestin agonists or antagonists are unlikely to be exceptions. This is essentially an unexplored field. Unlike the case for other members of the steroid receptor family, no examples of natural PR mutants have yet been reported. It is possible that systemic mutations in PR are incompatible with life. It is likely, however, that acquired mutations can develop in tumors as one mechanism for the development of resistance, and that a systematic search would demonstrate them. To address possible mechanisms of progestin resistance Murphy et al (1991) generated a subline of T47D cells that are resistant to the growth-inhibitory effects of progestins. This was done by sequential selection in medium containing 1 µM MPA. The cells remained PR-positive, but receptor levels were halved. Transforming growth factor-α and EGF receptor mRNA levels were both increased. The investigators suggest that increased growth factor expression and action and decreased PR levels may be involved in the development of progestin resistance. Additionally, as shown above, it is likely that extensive heterogeneity exists in PR content within cell subpopulations of tumors that are PR-positive. Factors or treatments that lead to the selection and expansion of PR-poor or PR-negative populations would, in the long run, produce progestin resistance.

Progestin Resistance and the Natural PR Isoforms

Complementary DNAs for chicken PR were cloned by Jeltsch et al and Conneely et al (1986) and for human PR by Misrahi et al (1987).

The single-copy human PR gene encodes at least nine messenger RNA species ranging in size from 2.5-11.4 kilobases. The nine messages direct the synthesis of at least two and possibly three, structurally related receptor proteins. The two major protein species, the B- and A-receptors, were originally described by O'Malley and co-workers in the chick oviduct. Subsequent studies using the breast cancer cell line T47D showed that human PR also exist as two isoforms; the 116 kilodalton (kDa) B-receptors and N-terminally truncated 94 kDa A-receptors. While A-receptors were originally thought to be produced by a proteolytic artifact, it is now clear that the amino-truncated receptors, at least in chickens and humans, are a naturally synthesized form. In human endometrial carcinoma and breast cancer cell lines, the two receptor isoforms are expressed in approximately equimolar amounts. It is not known whether this quantitative relationship between the two isoforms is maintained in all human target tissues and tumors, and the mechanisms for their differential regulation are not known, but at least two of the nine mRNA species lack the translation initiation site for B-receptors and can therefore encode only A-receptors. These messages arise by transcription from an internal promoter in the human PR gene. Five other message species can potentially encode both receptor isoforms, by alternate translation initiation from two in-frame AUG codons. In theory, use of the upstream codon would generate the B-receptors and use of the downstream codon generates the A-receptors, but it is not known whether initiation at the downstream site actually occurs in intact cells (Horwitz, 1992 and references therein).

PR are unique among steroid receptors in having two naturally occurring hormone binding forms, and this structural feature may have important functional implications with respect to receptor function. Since both homo- and heterodimers can form between the A- and B-isoforms, three possible classes of receptor dimers (A:A, A:B, B:B) can bind to a PRE, each having a potentially different transcription regulatory capacity.

That this molecular heterogeneity is indeed translated into functional heterogeneity was first demonstrated by the study of Tora et al (1988) that assessed the cell-specific transcriptional activation of two different target genes by the chicken A- and B-receptors. Depending on the gene being modulated and the cell being analyzed, A-receptors can be stimulatory in a setting where B-receptors are inactive or are inhibitory. Additional functional differences are observed when receptors are occupied by progesterone antagonists. For example, RU486-occupied B-receptors, but not A-receptors, can act like transcriptional agonists. In fact, A-receptors can reverse the effects of B-receptors (KBH, unpublished). Similar functional switches have been observed with antagonist-occupied receptors in the face of elevated intracellular cAMP levels. Under these conditions agonist-like effects may be observed (KBH, unpublished).

SUMMARY

In summary, we propose that the molecular heterogeneity of ER in breast tumor cells characterized by the presence of mutant receptor forms, generates the cellular heterogeneity evident when PR or DNA ploidy are analyzed in cell subpopulations. Furthermore, it is likely that cellular heterogeneity leads to the lack of uniformity in response to tamoxifen that we have described. We find that heterogeneity of PR and DNA ploidy reflects existence of mixed subpopulations of breast cancer cells that are substantially remodeled under the influence of tamoxifen. It appears likely that rather than being "resistant", different subsets of cells can be inhibited or stimulated by tamoxifen and their suppression or outgrowth alters the phenotype of the tumor. PR heterogeneity in solid tumors of patients may predict for such a mixed, and potentially dangerous, response to antiestrogen treatment. Similarly, the molecular heterogeneity resulting from two PR isotypes can lead to inappropriate responses to hormones in certain genes or cell types. These responses may be additionally modulated by other transcription factors with which the receptors interact. As we learn more about the heterogeneity of PR, ER and other proteins in tumors, we may be able to recognize such lethal subpopulations or combinations of regulatory factors, some of which the FCM immunoassay can simply and rapidly measure. Specifically, with respect to tamoxifen, our data suggest that its use as a chemopreventant in women at high risk of developing breast cancer (Kiang, 1991) should be viewed with caution.

ACKNOWLEDGMENTS

We are grateful to the National Cancer Institute (NIH) for financial support.

REFERENCES

Adler S, Waterman ML, He X and Rosenfeld MG (1988): Steroid-receptor mediated inhibition of rat prolactin gene expression does not require the receptor DNA-binding domain. *Cell* 52: 685-695.

Baulieu E-E (1987): Steroid hormone antagonists at the receptor level: A role for the heat-shock protein MW 90,000 (hsp 90). *J Cell Biochem* 35: 161-174.

Castagnoli A, Maestri I, Bernardi F and Del Senno L (1987): PvuII RFLP inside the human estrogen receptor gene. *Nucleic Acids Res* 15: 866.

Conneely OM, Sullivan WP, Toft DO, Birnbaumer M, Cook RG, Maxwell BL, Zarucki-Schulz T, Green GL, Schrader WT and O'Malley BW (1986): Molecular cloning of the chicken progesterone receptor. *Science* 233: 767-770.

Daly RJ, King RJB and Darbre PD (1990): Interaction of growth factors during progression towards steroid independence in T47D human breast cancer cells. *J Cell Biochem* 43: 199-211.

Fawell SE, Lees JA, White R and Parker MG (1990): Characterization and colocalization of steroid binding and dimerization activities in the mouse estrogen receptor. *Cell* 60: 953-962.

Forman BM, Yang C, Au M, Casanova J, Ghysdael J and Samuels HH (1989): A domain containing leucine zipper-like motifs mediate novel in vivo interactions between the thyroid hormone and retinoic acid receptors. *Molec Endocrinol* 3: 1610-1626.

Foster BD, Cavener DR and Parl FF (1991): Binding analysis of the estrogen receptor to its specific DNA target site in human breast cancer. *Cancer Research* 51: 3405-3410.

Fuqua SAW, Fitzgerald SD, Chamness GC et al (1991): Variant human breast tumor estrogen receptor with constitutive transcriptional activity. *Cancer Res* 51: 105-109.

Garcia T, Sanchez M, Cox JL et al (1989): Identification of a variant form of the human estrogen receptor with an amino acid replacement. *Nucl Acids Res* 17: 8364.

Garcia T, Lehrer S, Bloomer W and Schachter B (1988): A variant estrogen receptor messenger ribonucleic acid is associated with reduced levels of estrogen binding in human mammary tumors. *Molec Endocrinol* 2: 785-791.

Gottardis MM and Jordan VC (1988): Development of tamoxifen-stimulated growth of MCF-7 tumors in athymic mouse after long-term tamoxifen administration. *Cancer Res* 48: 5183-5188.

Gottardis MM, Jiang S-Y, Jeng M-H and Jordan VC (1989): Inhibition of tamoxifen stimulated growth of an MCF-7 tumor variant in athymic mice by novel steroidal antiestrogens. *Cancer Res* 49: 4090-4093.

Graham II ML, Bunn PA Jr, Jewett PB, Gonzalez-Aller C and Horwitz KB (1989a) Simultaneous measurement of progesterone receptors and DNA indices by flow cytometry: Characterization of an assay in breast cancer cell lines. *Cancer Res* 49: 3934-3942.

Graham II ML, Dalquist KE and Horwitz KB (1989b): Simultaneous measurement of progesterone receptors and DNA indices by flow cytometry: Analyses of breast cancer cell mixtures and genetic instability of the T47D line. *Cancer Res* 49: 3943-3949.

Graham II ML, Krett NL, Miller LA et al (1990): T47Dco cells, genetically unstable and containing estrogen receptor mutations, are a model for the progression of breast cancers to hormone resistance. *Cancer Res* 50: 6208-6217.

Graham II ML, Smith JA, Jewett PB and Horwitz KB (1992): Heterogeneity of progesterone receptor content and remodeling by tamoxifen characterize subpopulations of cultured human breast cancer cells: Analysis by quantitative dual parameter flow cytometry. *Cancer Res* 52: 593-602.

Green S, Walter P, Kumar V et al (1986): Human oestrogen receptor cDNA: Sequence, expression and homology to v-erb-A. *Nature* 320: 134-139.

Greene GL, Gilna P, Waterfield M, Baker A, Hort Y and Shine J (1986): Sequence and expression of human estrogen receptor complementary DNA. *Science* 231: 1150-1154.

Guiochon-Mantel A, Loosfelt H, Ragot T et al (1988): Receptors bound to antiprogestins form abortive complexes with hormone responsive elements. *Nature* 336: 695-698.

Henderson IC, Harris JR, Kinne DW and Hellman S (1989): Cancer of the Breast. In: Cancer: Principles and Practice of Oncology. DeVita VT Jr, Hellman S and Rosenberg SA, eds. p 1252, Philadelphia: JB Lippincott Co.

Hill SM, Fuqua SAW, Chamness GC, Greene GL and McGuire WL (1989): Estrogen receptor expression in human breast cancer associated with an estrogen receptor gene restriction fragment length polymorphism. *Cancer Res* 49: 145-148.

Hirose T, Koga M, Matsumoto K and Sato B (1991): A single nucleotide substitution in the D domain of estrogen receptor cDNA causes amino acid alteration from Glu-279 to Lys-279 in a murine transformed Leydig cell line (B-1 F). *J Steroid Biochem Molec Biol* 39: 1-4.

Horwitz KB (1981): Is a functional estrogen receptor always required for progesterone receptor induction in breast cancer? *J Steroid Biochem* 15: 209-217.

Horwitz KB and McGuire WL (1978): Estrogen control of progesterone receptor in human breast cancer: Correlation with nuclear processing of estrogen receptor. *J Biol Chem* 253: 2223-2228.

Horwitz KB, Zava DT, Thiligar AK, Jensen EM and McGuire WL (1978): Steroid receptor analyses of nine human breast cancer cell lines. *Cancer Res* 38: 2434-2437.

Horwitz KB, Mockus MB and Lessey BA (1982): Variant T47D human breast cancer cells with high progesterone receptor levels despite estrogen and antiestrogen resistance. *Cell* 28: 633-642.

Horwitz KB, Aiginger P, Kuttenn F and McGuire WL (1981): Nuclear estrogen receptor release from antiestrogen suppression: Amplified induction of progesterone receptor in MCF-7 human breast cancer cells. *Endocrinology* 108: 1703-1709.

Horwitz KB, McGuire WL, Pearson OH and Segaloff A (1975): Predicting response to endocrine therapy in human breast cancer: A hypothesis. *Science* 189: 726-727.

Horwitz KB (1992): The molecular biology of RU486. Is there a role for antiprogestins in the treatment of breast cancer? *Endocrine Reviews* 13: 146-163.

Katzenellenbogen BS, Kendra KL, Normal MJ and Berthois Y (1987): Proliferation, hormonal responsiveness, and estrogen receptor content of MCF-7 human breast cancer cells grown in the short-term and long-term absence of estrogens. *Cancer Res* 47: 4355-4360.

Keaveney M, Klug J, Dawson MT et al (1991): Evidence for a previously unidentified upstream exon in the human oestrogen receptor gene. *J Molec Endocrinol* 6: 111-115.

Keydar I, Chen L, Karbey S et al (1979): Establishment and characterization of a cell line of human breast carcinoma origin. *Eur J Cancer* 15: 659-670.

Kiang, DT (1991): Chemoprevention for breast cancer: Are we ready? *J Natl Cancer Inst* 83: 462-463.

Kumar V and Chambon P (1988): The estrogen receptor binds tightly to its responsive element as a ligand-induced homodimer. *Cell* 55: 145-156.

Kumar V, Green S, Staub A and Chambon P (1986): Localisation of the oestradiol-binding and putative DNA-binding domains of the human oestrogen receptor. *EMBO J* 5: 2231-2236.

Legault-Poisson S, Jolivet J, Poisson R, Beretta-Piccoli M and Band PR (1979): Tamoxifen-induced tumor stimulation and withdrawal response. *Cancer Treat Rep* 63: 1839-1841.

Lehrer S, Sanchez M, Song HK et al (1990): Oestrogen receptor B-region polymorphism and spontaneous abortion in women with breast cancer. *Lancet* 335: 622-624.

Lerner LJ and Jordan VC (1990): Development of antiestrogens and their use in breast cancer. *Cancer Res* 50: 4177-4189.

Lippman, ME (1984): An assessment of current achievements in the systemic management of breast cancer. *Br Can Res Treat* 4: 69-77.

McGuire WL, Lippman ME, Osborne CK and Thompson EB (1987): Resistance to endocrine therapy. *Br Can Res Treat* 9: 165-173.

Misrahi M, Atger M, d'Auriol L, Loosfelt H, Meriel C, Fridlansky F, Guiochon-Mantel A, Galibert F and Milgrom E (1987): Complete amino acid sequence of the human progesterone receptor deduced from cloned cDNA. *Biochem Biophys Res Commun* 143: 740-748.

Murphy LC and Dotzlaw H (1989): Variant estrogen receptor mRNA species detected in human breast cancer biopsy samples. *Molec Endocrinol* 3: 687-693.

Murphy LC (1990): Estrogen receptor variants in human breast cancer. *Molec Cell Endocrinol* 74: C83-C86.

Murphy LC, Dotzlaw H, Johnson Wong MS, Miller T and Murphy LJ (1991): Mechanisms involved in the evolution of progestin resistance in human breast cancer cells. *Cancer Res* 51: 2051-2057.

Osborne, CK, Coronado EB and Robinson JR (1987): Human breast cancer in the athymic nude mouse. Cytostatic effects of long-term antiestrogen therapy. *Euro J Cancer Clin Onc* 23: 1189-1196.

Osborne CK, Coronado EB, Wiebe V and DeGregorio M (1992): Acquired tamoxifen resistance correlates with reduced breast tumor levels of tamoxifen and isomerization of trans-4-hydroxytamoxifen. *J Natl Cancer Instit* in press.

Parl FF, Cavener DR and Dupont WD (1989): Genomic DNA analysis of the estrogen receptor gene in breast cancer. *Br Cancer Res Treat* 14: 57-64.

Ponglikitmongkol M, Green S and Chambon P (1988): Genomic organization of the human oestrogen receptor gene. *EMBO J* 7: 3385-3388.

Scott GK, Kushner P, Vigne J-L and Benz CC (1991): Truncated forms of DNA-binding estrogen receptors in breast cancer. *J Clin Invest* 88: 700-706.

Shupnik MA, Gordon MS and Chin WW (1989): Tissue-specific regulation of rat estrogen receptor mRNAs. *Molec Endocrinol* 3: 660-665.

Tora L, Gronemeyer H, Turcotte B, Gaub M-P and Chambon P (1988): The N-terminal region of the chicken progesterone receptor specifies target gene activation. *Nature* 333: 185-188.

Turner RT, Wakely GK, Hannum KS and Bell NH (1987): Tamoxifen prevents the skeletal effects of ovarian deficiency in rats. *J Bone Min Res* 2: 449-456.

Washburn T, Hocutt A, Brautigan DL and Korach KA (1991): Uterine estrogen receptor in vivo: Phosphorylation of nuclear-specific forms on serine residues. *Molec Endocrinol* 5: 235-242.

LIGAND REQUIREMENTS FOR ESTROGEN RECEPTOR FUNCTION AND THE ACTIONS OF ANTIESTROGENS

Joseph C. Reese
Department of Biochemistry and Molecular Biology
University of Massachusetts Medical Center

Benita S. Katzenellenbogen
Department of Physiology and Biophysics
University of Illinois Champaign-Urbana

SUMMARY

We have undertaken studies to examine the ligand requirements for the DNA binding and functional activity of human estrogen receptor (ER) within intact cells. By employing an assay using competitive binding of ER with basal transcription factors on a constitutive promoter (CMV-(HRE)n-CAT, containing hormone response element(s) between the TATA-box and the start site of transcription of the Cytomegalovirus promoter upstream of the chloramphenicol acetyltransferase gene) we were able to examine the DNA binding ability of the human ER in whole cells. We used this promoter interference assay to examine the DNA binding of ER in cell lines containing high and low levels of endogenous ER, as well as in Chinese Hamster Ovary (CHO) cells expressing wild-type and mutant ERs from cotransfected expression vectors. We have found that the ER is capable of binding to reporter templates within

STEROID HORMONE RECEPTORS
V. K. Moudgil, Editor
© 1993 Birkhäuser Boston

intact cells in the absence of ligand; however, ligand enhances this interaction. We provide some evidence that this DNA binding may result in ligand-independent transactivation of transgenes by the receptor, and that DNA binding alone is not sufficient for full receptor activity. Likewise, receptors occupied with estradiol or two classes of antiestrogens were also capable of DNA binding. Receptors occupied with the pure antiestrogen ICI 164,384 bound to DNA in whole cells and were in sufficient quantity to occupy receptor binding sites on reporter templates, indicating that neither a loss of DNA binding of ICI-receptor complexes or reduction of ER protein in target cells could fully explain the antagonistic nature of this compound. Through the analysis of a temperature-sensitive ER mutant (C447A), we provide evidence that the ligand-independent DNA binding observed for the ER is the basis of functional differences observed between the ER and other members of the steroid receptor family.

INTRODUCTION

The estrogen receptor (ER) mediates gene expression by the hormone estrogen, and is primarily a nuclear localized protein in the presence or absence of ligand (King and Greene, 1987; Picard et al, 1990b). It has been uncertain whether or not ligand is necessary for DNA binding of the ER protein. Most studies to demonstrate DNA binding have utilized *in vitro* assays with oligonucleotides containing a consensus estrogen response element (ERE; Brown and Sharp, 1990; Fawell et al, 1990b; Lees et al, 1989; Martinez and Wahli, 1989; Reese and Katzenellenbogen, 1991b; Tzukerman et al, 1991). The prevailing picture is that the receptor binds to DNA in the presence or absence of ligand; however, it is not clear what occurs within mammalian cells. *In vitro* systems such as these utilize conditions that do not mimic faithfully the nuclear milieu, and studies have shown that the ionic conditions affect the ability of ER to bind to DNA at elevated temperatures (Brown and Sharp, 1990). It is also an issue whether or not ligand independent DNA binding plays an important role in ER function. Two ER mutants whose DNA binding abilities are rendered ligand dependent by temperature-sensitive mutations suggest that it does (Reese and Katzenellenbogen, 1992a; Tzukerman et al, 1991).

The ER contains two transactivation functions. The first, TAF-1, is hormone-independent and resides in the amino-terminal A/B region of the receptor, and the second, TAF-2, is hormone-dependent and is located in the hormone binding domain (HBD, region E) of the receptor (Tora et al,

1989b). It has been proposed that the partial agonistic activities of certain antiestrogens originate from the TAF-1 function (Berry et al, 1990). Also, it has been suggested that the hormone-independent CAT activity observed for the ER in transfected cells is the result of ER binding to DNA in the absence of ligand (Lees et al, 1989; Reese and Katzenellenbogen, 1992b; Tora et al, 1989a; Tzukerman et al, 1991). Therefore, it is important to determine if unliganded ER is capable of binding to DNA in mammalian cells, and if this interaction results in hormone-independent transactivation.

Besides raising questions regarding receptor occupancy requirements for DNA binding, *in vitro* studies have also provided conflicting information about the actions of certain antiestrogens (Fawell et al, 1990b; Lees et al, 1989; Reese and Katzenellenbogen, 1991b). Only recently have studies tried to examine the DNA binding abilities of antagonist-receptor complexes in intact cells (Elliston et al, 1990; McDonnell et al, 1991; Pham et al, 1991; Reese and Katzenellenbogen, 1992b; Wrenn and Katzenellenbogen, 1990). In particular, the actions of the pure estrogen antagonist ICI 164,384 (ICI) have been controversial. One study proposes that ICI blocks the dimerization, and consequently, the *in vitro* DNA binding abilities of receptor-ICI complexes (Fawell et al, 1990b); while other studies demonstrate that ICI causes a rapid reduction in ER protein in target tissues (Gibson et al, 1991) and cells (Dauvois et al 1992). This leaves two questions unanswered. The first being: Does ICI prevent DNA binding of receptor in intact cells? And the second: Does a reduction in the levels of ER protein account for the pure antagonistic nature of ICI?

Examination of the DNA Binding Ability of Human Estrogen Receptor within Intact Cells

We have developed an assay for examining estrogen receptor DNA binding within mammalian cells (Reese and Katzenellenbogen, 1992b). It utilizes the principle of competitive binding of ER with basal transcription factors on a constitutive (Cytomegalovirus, CMV) promoter. Insertion of consensus estrogen response elements (ERE) between the TATA box and the start site of transcription of the vector CMV-CAT (Fig. 1A) creates a vector whose expression of CAT activity is suppressed by the binding of ER by blocking the assembly of the transcription initiation complex. The effect of inserting oligonucleotides into the CMV-CAT vector on CAT expression was examined in cells lacking any measurable ER. The constructs containing one and two wild-type ERE's, and a construct containing two mutated ERE's, expressed CAT activity to levels of 107-90% of CMV-CAT lacking any inserted DNA sequences when transfected into ER negative breast carcinoma MDA-MB-231 cells and CHO cells (Fig. 1B). However, a construct which contains four ERE's displayed a significant reduction in the expression of CAT activity in both cell types (Fig. 1B), indicating a disruption in promoter activity or mRNA translation.

A

Promoter Interference Reporter Constructs

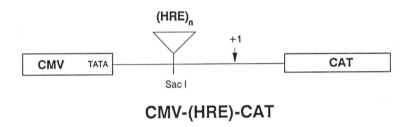

CMV-(HRE)-CAT

B

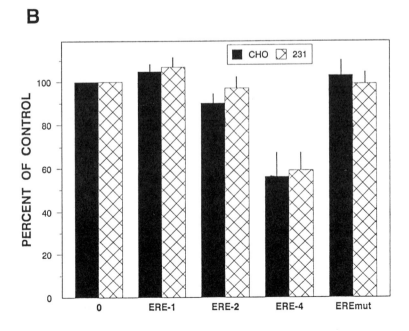

FIGURE 1. **A**. CMV-HRE-CAT reporter construct.

The functionality of the CMV-ERE-CAT constructs were verified by cotransfection of the CMV-(ERE)$_2$-CAT and CMV-(EREmut)$_2$-CAT constructs with the expression vectors for the wild-type receptor and previously identified ER DNA-binding mutants into CHO cells. Cotransfection of the CMV-(ERE)$_2$-CAT construct with increasing amounts of the ER expression vector pRER (Reese and Katzenellenbogen, 1991a), lead to a dose-dependent suppression of CAT activity (Fig. 2A). CAT activity was reduced in the presence or absence of added estradiol, although liganded receptor appears to be more effective.

To further verify the assay conditions, the ability of ER DNA binding mutants to suppress CAT activity from the CMV-(ERE)$_2$-CAT was examined in CHO cells. Two ER mutants, G400V (Tora et al, 1989a; Tzukerman et al, 1991) and C447A (Reese and Katzenellenbogen, 1992a, see below), that show hormone-dependent DNA binding *in vitro* at elevated temperatures were examined. The two ER mutants displayed a much reduced ability to suppress CAT expression from the CMV-(ERE)$_2$-CAT construct in the absence of ligand compared to the wild-type receptor (Fig. 2B); however, both were able to suppress the promoter to equal levels in the presence of ligand. An ER mutant which demonstrates higher ligand-independent transactivation ability than the wild-type receptor (E380Q, F. Pakdel, J. C. Reese, and B. S. Katzenellenbogen, in preparation) was also examined. As shown in Fig. 2B, the E380Q mutant ER demonstrates higher ligand-independent suppression of CAT activity, as well as a greater effectiveness than the wild-type receptor in the presence of ligand (Fig. 2B).

FIGURE 1. **A.** CMV-HRE-CAT reporter construct. A consensus estrogen response element and a mutant version were inserted into the Sac I site of CMV-CAT which expresses the chloramphenicol acetyltransferase gene constitutively from the Cytomegalovirus promoter. This site lies between the TATA box and the start site of transcription. **B.** Activity of the CMV-HRE-CAT constructs in cell lines containing exceedingly low or undetectable levels of ER. MDA-MB-231 (cross hatch bars) and CHO (black bars) cells were transfected by the calcium phosphate coprecipitation method as described in Reese and Katzenellenbogen 1992b with the indicated CMV-HRE-CAT constructs. The level of CAT activity in cells transfected with CMV-CAT was set at 100 percent and represents the control value. The values are the means and S.E.M.'s of 3-6 determinations. (From Reese and Katzenellenbogen, 1992b) Reprinted with permission.

The amounts of pRER needed to suppress CAT activity from the promoter interference plasmid were similar to those necessary to activate transcription from an estrogen-responsive plasmid containing two consensus ERE's (Fig. 3, under both assay conditions, half-maximal plasmid amounts were between 2.5-5ng pRER per plate), therefore indicating that promoter interference occurs at ER concentrations that are functional within the cells. This, and the lack of reduction of CAT activity expressed from the CMV-CAT constructs that do not contain ERE's (Fig. 2A), indicate that the reduction in CAT activity is not attributable to a "squelching" phenomenon which has previously been shown to occur with steroid receptors (Meyer et al, 1989), and can be seen to occur in Fig. 3A and 3B with pRER amounts greater than 25 ng per plate in the presence of hormone. A reporter plasmid containing a minimal promoter that fails to measure TAF-1 function in CHO cells (ERE-TATA-CAT, J. C. Reese and B. S. Katzenellenbogen unpublished), failed to show an increase in CAT activity in cells transfected with increasing amounts of pRER in the absence of added ligand (Fig. 3A), suggesting that the DNA binding of unliganded receptor to the promoter interference constructs is not attributable to contaminating steroids in the culture medium. In further support of this, the promoter interference constructs were suppressed in cells treated with control vehicle to a level of approximately 50% of that observed in cells treated with ligand. Therefore, if the DNA binding observed in vehicle treated cells is the

FIGURE 2. Examination of the DNA binding abilities of ER expressed in CHO cells. (A) CHO cells were cotransfected with 400 ng of the CMV-CAT construct containing no inserts (0, squares), two ERE's (ERE, circles), and two mutated ERE's (EREmut, triangles) with increasing amounts of the ER expression vector pRER. Following transfection, the cells were treated with control vehicle (open symbols) or with 10 nM estradiol (filled symbols) for 24 h. Control values were determined as the CAT activity in cells transfected with each CMV-HRE-CAT construct in the absence of pRER or ligand and were set at 100%. (B) The ability of previously identified ER DNA binding mutants to suppress CMV-$(ERE)_2$-CAT promoter activity was examined in CHO cells. The CMV-$(ERE)_2$-CAT reporter construct was cotransfected with 25 ng of the expression vector coding for the ER mutants indicated on the x axis. Cells were treated without (black bars) or with (cross hatch bars) 10 nM estradiol for 24 h and CAT activities measured. Results are the means and S.E.M.s of 4-6 determinations. *$p<0.05$ versus wild-type by Student's t-test. (From Reese and Katzenellenbogen, 1992b) Reprinted with permission.

result of contaminating steroids in the culture medium, one would predict that 50% of the receptors would be occupied by ligand, and thus, would result in the induction of the reporter construct to 50% of that observed in estradiol-treated cells. When a reporter plasmid with a more complex

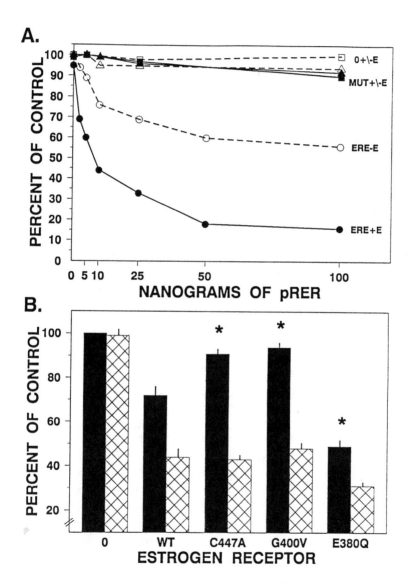

FIGURE 2. Examination of the DNA binding abilities of ER expressed in CHO cells.

384

promoter capable of measuring both TAF-1 and TAF-2 functions was used (ERE-vit-CAT, J. C. Reese and B. S. Katzenellenbogen, unpublished), an increase in CAT activity was observed in the absence of added ligand (Fig. 3B). Since the increase in CAT activity was not

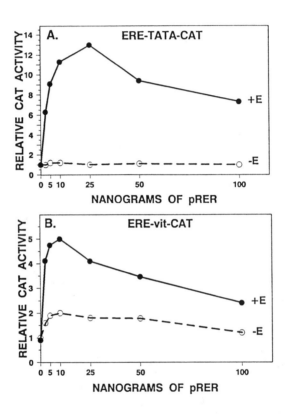

FIGURE 3. Examination of the transactivation abilities of receptor in cells transfected with two reporter plasmids. (A) CHO cells were cotransfected with increasing quantities of pRER (the ER expression vector) and 400 ng of ERE-TATA-CAT, a minimal promoter which measures TAF-2 function of the receptor. Cells were treated with (closed circles) or without (open circles) 10 nM estradiol for 24 h. CAT activity is presented as relative CAT activity compared to cells transfected with the reporter plasmid in the absence of pRER or ligand. (B) Cells were treated as described above, except that 400 ng of ERE-vit-CAT, a complex promoter capable of measuring TAF-1 and TAF-2 functions of ER, was used. (From Reese and Katzenellenbogen 1992b) Reprinted with permission.

observed in the absence of added ligand using an estrogen receptor whose TAF-1 function was eliminated by the deletion of the A/B region of the receptor (data not shown), it suggests that this ligand-independent transactivation is attributable to the TAF-1 function of the ER.

Examination of the DNA Binding State of ER in Cells Expressing High and Low Levels of Receptor

The activities of the CMV-HRE-CAT reporter constructs were examined when transfected into human breast cancer cells that contain high levels of ER (MCF-7-K1 and MCF-7-K3), as well as into a breast cancer cell line (ZR-75-1) and a rat pituitary tumor cell line (GH_4C1) which contain low levels of ER to determine if the DNA binding of unliganded ER observed in CHO cells is the result of superphysiological expression of receptor in the heterologous system, as suggested by others (McDonnell et al, 1991).

The activities of the various CMV-HRE-CAT constructs in these cells are presented in Figure 4 (panels A-D). In cells containing either high (panels A and B) or lower levels of ER (panels C and D), there was a decrease in promoter activity that was dependent on the number of inserted ERE's. Direct comparison of the ERE-4 constructs with the others is difficult, however, because this construct was less active in cell lines with undetectable levels of ER (Fig. 1B). The construct containing two mutated ERE's showed no promoter interference activity, similar to the construct lacking any inserted oligonucleotides (*ca.* 100% of control activity). Also, cotransfection of CMV-(ERE)$_2$-CAT with a five-fold excess of carrier DNA containing two ERE's reduced the magnitude of repression of the promoter in the cell types examined (data not shown). These results indicate that the repression is specific for ERE sequences, and most likely, specific for the binding of ER.

The promoters containing functional ERE sequences were suppressed in cells that were treated with control vehicle alone, indicating that unliganded ER is capable of repressing promoter function, presumably through binding to DNA in cells in the unliganded form. It is unlikely that DNA binding of receptor is the result of the activation of ER by serum components in the culture medium, because MCF-7 K1 cells transfected under serum concentrations ranging from 10% to 0.3%, displayed the same magnitude of suppression of the CMV-(ERE)$_2$-CAT construct (data not shown). Clearly, if serum components were responsible for the ligand-independent DNA binding observed in these cells, reduction of the serum content in the culture medium by 30-fold

386

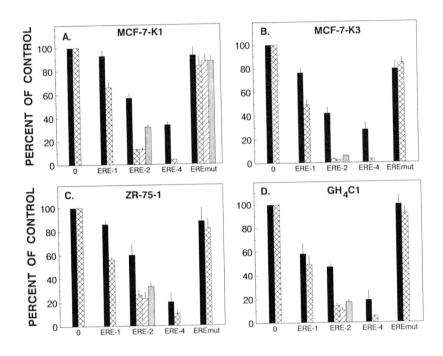

FIGURE 4. Examination of the DNA binding abilities of ER in cell lines expressing high and low levels of endogenous ER. MCF-7-K1 (panel A), MCF-7-K3 (panel B), ZR-75-1 (panel C) and GH$_4$C1 (panel D) cells were transfected with 2 µg of the CMV-HRE-CAT construct indicated on the x axis. Cells were treated with control vehicle (black bar), 10 nM estradiol (cross hatch), 10^{-6} M trans-OH tamoxifen (right hatch bars), or 10^{-6} M ICI 164,384 (gray bars) for 24 h. Data are presented as percent of control, with control being the level of CAT activity of the CMV-CAT construct (lacking any ERE's) in each particular cell line. Values are the means and S.E.M.'s of three determinations. The ER contents of these cells were measured by a whole cell hormone binding assay (Reese and Katzenellenbogen, 1991b) following transfection of the CMV-(ERE)$_2$-CAT construct and were found to be (in sites per cell, mean ± S.D.): 153,236 ± 10,061 (MCF-7-K1); 260,298 ± 21,113 (MCF-7-K3); 11,271 ± 2,305 (ZR-75-1); 8,413 ± 1,821 (GH$_4$C1). The MCF-7 cell subline MCF-7-K3 (Clarke et al, 1989) contains about twice as much ER as its parental line MCF-7-K1. (From Reese and Katzenellenbogen, 1992b) Reprinted with permission.

should have had a significant effect. The magnitude of suppression was, however, enhanced by treatment with ligand, suggesting that liganded receptor may form a more stable receptor-DNA complex. Since promoter function was reduced in cells containing either high or low levels of receptor, the DNA binding observed in transfected cells is not an artifact of the overexpression of receptor to superphysiological levels. Apparently, in this system, ER content does not affect the ability of unliganded receptor to bind to DNA, and ligand enhances receptor-DNA interaction within the cell.

Antiestrogen-Receptor Complexes Bind to DNA in Cells

Cells exposed to the antiestrogen trans-hydroxy-tamoxifen (OHT) displayed a similar reduction in $CMV-(ERE)_2$-CAT promoter activity as did cells treated with estradiol (Fig. 4), indicating that OHT-receptor complexes can bind to DNA. Similarly, cells treated with the pure antagonist ICI 164,384 (ICI) showed a decrease in promoter activity of the $CMV-(ERE)_2$-CAT construct, indicating that ER-ICI receptor complexes are capable of binding to DNA within the cell. While ICI treatment reduced the promoter activity in these cell lines well below the level of no added hormone, the extent of suppression was slightly, but consistently, less than that of cells treated with estradiol or OHT (Fig. 4).

The apparent reduced ability of the ICI-receptor complexes to suppress promoter activity, compared to estradiol (E_2) or OHT treated cells, most likely stems from the reduction of ER content in cells exposed to ICI. Western blots of extracts prepared from transfected cells following treatment with ligand show that ER levels in the cells treated with ICI were reduced by 2 h, and remained low for the 24 h period examined (Fig. 5) in several cell types. This phenomenon has been reported also in mouse uterus (Gibson et al, 1991) and breast cancer cells (Dauvois et al 1992). In contrast to the reduction in ER level seen upon cell exposure to ICI, ER content showed modest increases following OHT exposure, with slight decreases following E_2 exposure (Fig. 5). Of the four cell lines examined, GH_4C1 cells showed the least alteration in ER content in response to the ligands.

Transfection of quantities of an estrogen responsive reporter plasmid (ERE-vit-CAT, capable of measuring the TAF-1 activity of ER) equal to those of the CMV-HRE-CAT constructs used in the promoter interference assays, shows that ICI 164,384 behaves as a pure antagonist while OHT behaves as a partial agonist and E_2 is a strong stimulator of reporter gene transcription (Fig. 6). Hence, even though ICI-receptor complexes and

OHT- and E_2-receptor complexes bind to DNA, and the quantities of receptor in these cells are sufficient to bind to the CMV-$(ERE)_2$-CAT promoter construct, and are presumably abundant enough to bind to the transactivation reporter plasmid, ICI still behaves as a pure antiestrogen, evoking no increase in ERE-vit-CAT transactivation.

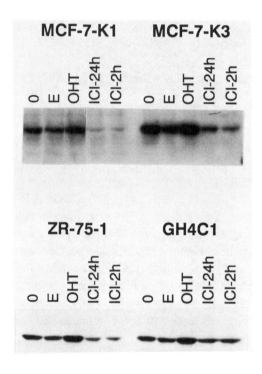

FIGURE 5. Western blot analysis of extracts from transfected cells treated as described in Figure 4. Whole cell extracts were prepared from transfected cells treated with or without the indicated ligands for 24 h, and from cells treated with ICI for 2 h. Proteins were fractionated and transferred to nitrocellulose, followed by Western blotting using the ER specific monoclonal antibody H222. Relative ER contents were estimated by densitometry of appropriately exposed autoradiograms and were found to be as follows. ZR-75-1 cells: 0, 100%; E, 60%; OHT, 198%; ICI-24h, 28%; ICI-2h, 20%. GH_4C1 cells: 0, 100%; E, 83%; OHT, 200%; ICI-24h, 77%; ICI-2h, 70%. MCF-7-K1 cells: 0, 100%; E, 58%; OHT, 189%; ICI-24h, 10%; ICI-2h, 10%. MCF-7-K3 cells: 0, 100%; E, 52%; OHT, 135%; ICI-24h, 13%, ICI-2h, 8%. (From Reese and Katzenellenbogen, 1992b) Reprinted with permission.

Analysis of a Temperature-Sensitive ER Mutant Reveals Functional Differences Between ER and Other Receptors

A previous study by us reported on an ER mutant (C447A) in which the cysteine at position 447 had been replaced by an alanine. This C447A ER mutant had similar affinity for estradiol *in vitro*, while displaying a dose response shift for estradiol in gene transfer/transactivation experiments (Reese and Katzenellenbogen, 1991a). To examine the alteration of this mutant, we conducted experiments in cellular extracts and in intact cells.

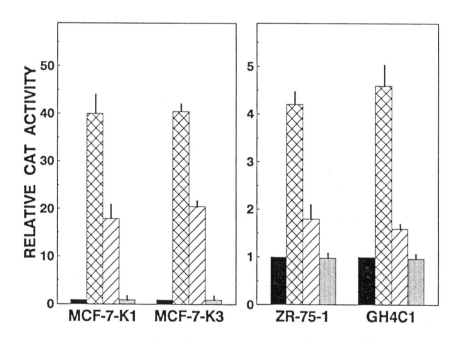

FIGURE 6. Examination of the antagonistic nature of two antiestrogens in transfected cells. Cells were transfected with 2 μg of ERE-vit-CAT expression vector and treated as described in Figure 5. Following 24 h of ligand exposure, CAT activity in extracts prepared from the cell lines indicated on the x axis was examined. Data are expressed as relative CAT activity compared to the control, which was set as the CAT activity in transfected cells treated with control vehicle. Values are the means and S.E.M.s of 3-4 determinations. (From Reese and Katzenellenbogen, 1992b) Reprinted with permission.

The binding affinity of the expressed wild-type (WT) and C447A receptor for estradiol was measured *in vitro* at 37°C and at 25°C. The affinity of both receptors for estradiol was comparable when examined under parallel conditions (Table 1), differing by less than two-fold. The stability of the C447A mutant was, therefore, compared to WT at different temperatures both in the occupied and unoccupied form. Figure 7A shows that while the WT receptor was stable throughout the 30 min period studied over the entire range of temperature conditions, the C447A mutant was unstable under some of these conditions. At 25°C both the mutant and wild-type receptors were equally stable in both the occupied and unoccupied state. However, the unliganded mutant receptor lost 50% of its ligand binding capability following a 15 min incubation in the unoccupied state at 30°C, and 80-90% of its specific hormone binding capacity following a 30 min incubation at 37°C (Fig. 7A). The thermal instability of the mutant receptor was fully prevented by ligand; thus, estradiol-occupied C447A receptor lost none of its binding capacity following incubation at 30°C or 37°C (Fig. 7A).

The reduction in the binding capacity of the mutant receptor was not due to proteolytic breakdown of the receptor, because Western blot analysis of both the wild-type and mutant receptor confirmed that the levels of intact (66,000 dalton) ER protein in the extracts containing unoccupied receptors remained constant throughout the elevated temperature treatments (Fig. 7B).

The Temperature-Sensitivity of the C447A Receptor Renders its DNA Binding Ability Ligand-Dependent

The DNA binding ability of the mutant receptor was examined under a variety of conditions by using estrogen response element DNA in a gel mobility shift assay. Whole cell extracts from transfected cells were incubated *in vitro* in the presence or absence of estradiol and were then incubated for 30 min at 25°C or 37°C prior to the DNA binding assay. Figure 8 shows that the wild-type receptor is able to bind to the DNA in the presence or absence of ligand under both conditions. While this is true for the C447A mutant at 25°C, at 37°C this mutant demonstrated strict hormone dependence for DNA binding. This is presumably a reflection of the instability of the receptor in the unoccupied state at 37°C (Fig. 7A), and it appears that the instability of the mutant receptor affects both the hormone binding ability and DNA binding ability of the receptor. Examination of the ability of the C447A receptor to bind to

TABLE I

HORMONE BINDING AND TRANSACTIVATION ABILITIES OF WILD-TYPE AND C447A ER AT VARIOUS TEMPERATURES

The affinity of the wild-type (WT) and C447A ER was measured in whole cell extracts (WCE) from transfected cells at 25°C and 37°C as described in Figure 1. Affinities are expressed as the equilibrium dissociation constants (K_d) and are the means ± S.E.M. of 3 determinations (ND= not determined). The estradiol concentration required for half-maximal CAT activity was determined for the WT and the C447A receptor at 37°C, 30°C, and 25°C as described in Figure 9. Values represent the means ± S.E.M. of 4 independent experiments. Fold indicates the ratio of the half-maximal E_2 concentrations for C447A vs. wild-type ER required for transactivation. (From Reese and Katzenellenbogen, 1992a). Reprinted with permission.

Receptor	K_d (nM) E_2 binding	Half-maximal [E_2], M (fold) Transactivation
	37°C	
Wild-type	0.63 ± 0.02	7.6 ± 0.9 x 10^{-12}
C447A	1.13 ± 0.04[a]	2.0 ± 0.5 x 10^{-10} (26X)
	30°C	
Wild-type	ND	1.1 ± 0.4 x 10^{-11}
C447A	ND	5.3 ± 1.8 x 10^{-11} (4.8X)
	25°C	
Wild-type	0.28 ± 0.06	7.4 ± 0.9 x 10^{-12}
C447A	0.41 ± 0.09[b]	1.8 ± 0.4 x 10^{-11} (2.4X)

[a]$p < 0.005$ and [b]$p < 0.10$ vs. WT ER by Student's t-test

DNA in intact cells using the promoter interference constructs also indicates an impaired DNA binding ability at elevated temperatures, in good agreement with the *in vitro* data presented here (Fig 2B).

Transactivation Ability of Receptor at Different Temperatures

The transactivating ability of both mutant and wild-type receptor was compared at 37°C, 30°C and 25°C. Following transfection, cells were given hormones in their media, and the plates of cells were then placed at 37°C, 30°C and 25°C for 24 h.

The results from one experiment comparing receptor activity as a function of culture temperature are shown in Fig. 9, and the results from several experiments are summarized in Table 1. When cells are exposed to hormone at 37°C, the C447A mutant displays a significant dose response shift compared to wild-type receptor, requiring 30-fold higher estradiol concentrations to achieve half-maximal stimulation of CAT activity (Fig. 9 and Table 1). Interestingly, as the incubation temperature is decreased, the difference in the half-maximal estradiol concentration between the mutant and wild-type receptor decreases (Fig. 9). This is caused by an observed shift of the dose response curve of the mutant receptor, as the E_2 concentration for half-maximal stimulation of transactivation activity of the wild-type receptor changed by less than

FIGURE 7. Stability of the Wild-type (WT) and C447A receptors *in vitro*. (A) WCE's containing WT (solid lines) or C447A (dashed lines) ER were pre-incubated on ice for 2 h in the absence of hormone (open symbols) or in the presence of 10 nM $[^3H]E_2$ ± 200-fold excess radioinert estradiol (closed symbols) to occupy the receptors with ligand followed by an incubation at 25°C (triangles), 30°C (circles) or 37°C (squares) for the indicated times. Hormone was then added to samples lacking ligand and all tubes were placed at 25°C for 2 h, to occupy the receptors. Data is expressed as the percent of specific binding with 100% equal to the specific binding measured in extracts kept at 4°C for 30 min, followed by the addition of hormone and incubation at 25°C. (B) WCE's were treated in the absence of hormone at 30°C and 37°C as described above, except that at the end of the 2 h incubation at 25°C, SDS-PAGE loading solution was added and samples were heated and loaded onto a 10% SDS-PAGE gel. ER was detected by Western blotting with the ER-specific antibody H222. (From Reese and Katzenellenbogen, 1992a) Reprinted with permission.

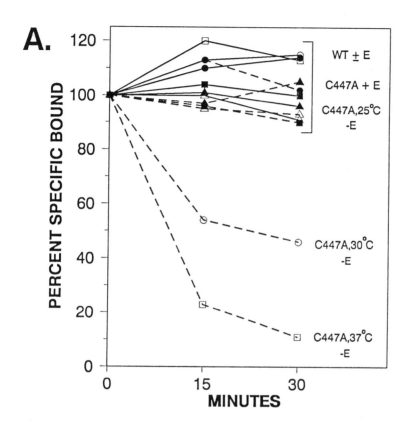

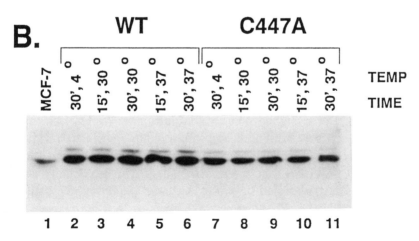

FIGURE 7. Stability of the Wild-type (WT) and C447A receptors *in vitro*.

two-fold among the three different temperatures examined (Fig. 8 and Table 1). At 25°C, a temperature where the stability of the mutant receptor is equal to that of the wild-type, the minimal difference in the dose-response curves for estradiol-stimulated CAT activity reflects the very minimal difference in the measured *in vitro* hormone binding affinity

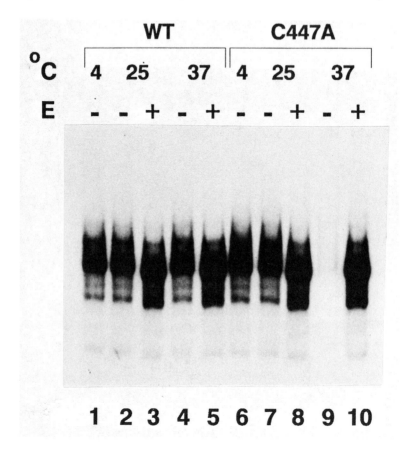

FIGURE 8. DNA binding ability of Mutant and Wild-type receptor. Whole cell extracts were prepared from cells transfected with the expression vector for the wild-type (lanes 1 to 5) or the C447A mutant (lanes 6 to 10). WCE's were pre-incubated at 4°C for 1 h in the presence or absence of estradiol and than placed at 4°C, 25°C and 37°C for 30 min. prior to the DNA binding assay. Protein-DNA complexes were separated on a 4% non-denaturing gel. The gel was run "long" (1.5X) to allow greater distinction between unliganded and liganded receptor-DNA complexes. Treatments are indicated above. Data is from, and methods are detailed in, Reese and Katzenellenbogen, 1992a.

of the receptors (Table 1). This data provides evidence that the temperature instability of the mutant receptor demonstrated *in vitro* in terms of hormone and DNA binding, is also manifested in transfected CHO cells in terms of impaired transactivation.

DISCUSSION

We utilized a promoter interference assay to demonstrate that the ER is capable of binding to DNA within whole cells in the absence of ligand, and that ligand enhances or stabilizes the interaction of the receptor with DNA response elements. We also provide evidence that ligand-independent DNA binding may provide the basis for the functional differences observed for the ER and the other steroid receptors.

We observed binding of ER to DNA within whole cells in the absence of ligand, in cells expressing both high and low levels of ER, as well as in CHO cells cotransfected with an ER expression vector (pRER). This contrasts with a study done in yeast, which suggested that the apparent ligand-independent DNA binding and transactivation functions of ER in transfected cells might be due to overexpression of the receptor to "superphysiological" levels (McDonnell et al, 1991). Examination of the DNA binding abilities of ER in cells expressing both higher (MCF-7-K3) and lower (ZR-75-1 and GH_4C1) levels of receptor than those expressed in the yeast used in that study, demonstrated that receptor concentration did not have an effect on the DNA binding state of the receptor. The differences in these two studies may reflect differences in the transformation of the receptor to the DNA binding state in the two organisms, or differences in the function of the two ER transactivation domains in yeast and mammalian cells (Berry et al, 1990).

It is unlikely that the DNA binding of ER observed in the absence of added estradiol is the result of residual estrogens in the culture media, because similar quantities of cotransfected ER expression vector failed to elicit any CAT activity from an estrogen-responsive reporter plasmid that measures the hormone-dependent transactivation function of ER (TAF-2) in CHO cells (i.e., ERE-TATA-CAT), and reduction of the serum content in the culture medium by 30-fold had no effect on the DNA binding observed in vehicle-treated cells. How ligand is capable of enhancing the suppression of these constructs by the estrogen receptor is not clear. It is unlikely that ligand acts to alter the fraction of the receptor in the nucleus, because unlike the glucocorticoid receptor, the estrogen receptor is nuclear in the absence of ligand (King and Greene, 1987; Picard et al, 1990b). It has been proposed for other steroid receptors that ligand may

396

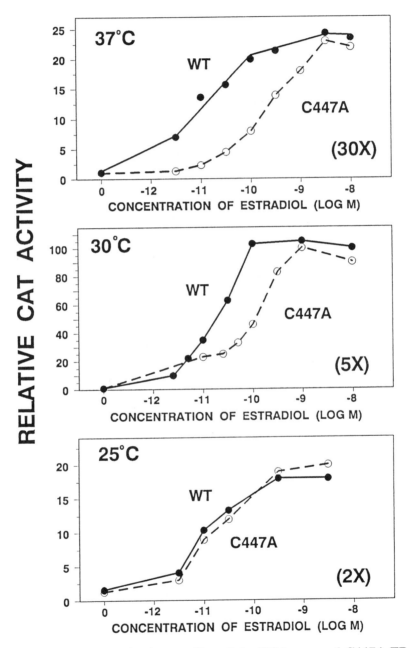

FIGURE 9. Transactivation profiles of the Wild-type and C447A ER at various temperatures.

dissociate inhibitory factors from the receptor within the cells, such as heat shock proteins (for review see Carson-Jurica et al, 1990). Our data provides evidence that at least 50% of the receptor is not bound by these inhibitory proteins, and since hsps are far more abundant in cells than is receptor, we see no reason why all of the receptor would not be bound by hsps and therefore prevented from binding DNA in the absence of ligand. Our data suggests that the receptor may not be bound by hsp90 in whole cells, and that the interaction of ER with hsps may represent an *in vitro* artifact. Our analysis of the C447A mutant would also support such a hypothesis (see below). One study, however, demonstrated diminished steroid receptor responsiveness of promoters in hsp deficient yeast (Picard et al, 1990a), but this effect has yet to attributed to the interaction of these proteins with the receptor *in vivo*. We also feel that the differences observed in yeast versus mammalian cells regarding ligand specificity and transactivation domains of the receptor (Berry et al, 1990; Schena and Yamamoto, 1988) may warrant some caution in the extrapolation of those results to mammalian cells. We favor a model where ligand stabilizes the receptor protein within the cells (with the noted exception of ICI 136,384, see below), or stabilizes receptor dimerization through the dimerization domain which lies within the hormone binding domain of the receptor (Fawell et al, 1990a), resulting in an increased stability of DNA binding.

When a reporter plasmid (ERE-vit-CAT) that is capable of measuring the hormone-independent transactivation function (TAF-1) was used, a slight increase in CAT activity was observed in the absence of added

FIGURE 9. Transactivation profiles of the Wild-type and C447A ER at various temperatures. Cells were co-transfected with the reporter plasmid $(ERE)_2$-TATA-CAT, the β-galactosidase expression plasmid, and the expression vector for the WT (closed circles) or the C447A mutant (open circles) ER at 37°C for 16 h. Following glycerol shock, fresh medium and hormone were added and the cells were incubated at 37°C, 30°C and 25°C for 24 h in the presence of increasing concentrations of estradiol. Details of the transfection procedure can be found in Reese and Katzenellenbogen, 1992a. The dose-response shift (fold change) of the mutant, compared to the wild-type ER, is indicated in parentheses in the bottom right hand corner of each panel. The relative CAT activity is the activity in cells transfected with the reporter plasmid and the ER expression vector compared to that in cells transfected with the reporter plasmid in the absence of the ER expression vector, which was set at 1. (From Reese and Katzenellenbogen, 1992a) Reprinted with permission.

estradiol. These results suggest that the ER is capable of binding to DNA in the unoccupied state, and that the CAT activity observed on certain promoters in the absence of ligand most likely stems from the activity of the TAF-1 function of receptor bound to DNA in the unliganded form. The consequences of this are far reaching. The data suggest that unliganded receptor may be capable of modulating gene transcription by binding to promoter regions, and the response observed would depend on the promoter and cell type. Promoters such as in the ERE-vit-CAT construct would be weakly stimulated by unliganded ER, while the ovalbumin promoter, which has been shown to be strongly stimulated by the TAF-1 function of ER in chicken fibroblasts (Berry et al, 1990), might display greater induction by the unliganded receptor. Alternatively, unliganded receptor bound to promoters that are unresponsive to the TAF-1 function of the receptor could possibly lead to the repression of the promoter by interfering with other trans-factors, leading to the control of promoters by a mechanism not previously studied for ER.

Taken together, these studies present evidence for an intermediate in the activation of the ER, and indicate that DNA binding alone is not sufficient for the full activation of gene transcription. Receptor bound to DNA is capable of weakly activating promoters through its TAF-1 function, and can be activated further, through its TAF-2 function by ligand, or by ligand independent pathways such as protein kinase mediated pathways. Ligand-independent DNA binding may be a prelude to the regulation of transcription by estrogen, and other steroid receptors, through protein kinase-mediated pathways (Aronica and Katzenellenbogen, 1991; Power et al, 1991).

Analysis of the C447A ER mutant also suggests that the observed ligand-independent DNA binding of ER has implications for hormone-dependent ER functions as well. The studies with receptor preparations *in vitro* as well as the studies in transfected cells indicate that the instability of the C447A receptor causes the dose response shift of this mutant for estradiol in transactivation studies. The dose response shift reflects the requirement for the C447A receptor to be occupied by ligand to function. The DNA binding assays *in vitro* and in intact cells indicate that the C447A receptor displays hormone dependence of DNA binding at elevated temperatures (Figs. 2B and Reese and Katzenellenbogen, 1992a). Hence, the transactivation profile of this mutant reflects the occupancy of the C447A mutant receptor by hormone, leading to a half-maximal stimulatory estradiol concentration (2×10^{-10} M) that is similar to the equilibrium dissociation constant for E_2 binding (K_d, 4×10^{-10} M). This addresses the issue of the disparity between the half-maximal E_2 concentrations in transactivation studies for wild-type ER (6-8 x 10^{-12} M) and the K_d (2-6 x 10^{-10} M) value for estradiol binding, which in theory should be similar.

Two ways in which the action of the estrogen receptor is unique from the other steroid hormone receptors is that (1) the receptor is capable of binding to DNA in the absence of ligand (Brown and Sharp, 1990; Fawell et al, 1990b; Lees et al, 1989; Martinez and Wahli, 1989; Reese and Katzenellenbogen, 1991b; Tzukerman et al, 1991) and (2) the estradiol concentration required for a half-maximal response in transactivation studies and the equilibrium dissociation constant (K_d) for hormone binding are roughly two orders of magnitude apart (Reese and Katzenellenbogen, 1991a; Reese and Katzenellenbogen, 1992a; Tora et al, 1989b). In contrast to the ER, both the progesterone receptor and the glucocorticoid receptor require hormone to bind to DNA (Bagchi et al, 1988; Denis et al, 1988; Edwards et al, 1989) and their half-maximal ligand concentrations reflect their K_d values (El-Ashry et al, 1989; Giguere et al, 1986). The C447A mutant, therefore, acts more like these receptors than does wild-type ER. Because the C447A receptor behaves in this fashion, the differences observed in the half-maximal ligand concentration and the K_d values between the ER and the other steroid receptors most likely do not stem from different protein sequences or dissimilar transactivation domains, but may reflect differences in their transformation to the DNA binding state; the implications being, that a rate-limiting step in the activation pathway for steroid hormone receptors may exist and require the conversion of the receptor to a DNA binding state. In the case of the glucocorticoid receptor and progesterone receptor, it may involve a hormone-dependent dissociation of non-receptor proteins from the receptor (Bagchi et al, 1988; DeMarzo et al, 1991; Denis and Gustafsson, 1989; Howard et al, 1990) or the recruitment of factors that assist in their DNA binding activities (Edwards et al, 1989). Recent studies have provided evidence of intermediate glucocorticoid and progesterone receptor complexes which assemble prior to the activation of the receptor complex (Bagchi et al, 1990; Cairns et al, 1991; DeMarzo et al, 1991). However, in the case of the ER, the receptor may already be "pre-bound" to DNA in the absence of ligand. By overcoming the rate-limiting step of transformation to the DNA binding state, maximal transactivation activity at lower ligand concentrations may be possible. The binding of ER to DNA in the absence of ligand may be one explanation for the hormone-independent activity of the receptor (Reese and Katzenellenbogen, 1992b; Tora et al, 1989a; Tzukerman et al, 1991) as discussed above.

Estrogen receptor occupied with antiestrogen or estrogen is capable of binding to ERE DNA in whole cells. ER-ICI receptor complexes also bound to DNA, as did receptor occupied by E_2 and OHT.

Treatment of cells with ICI caused a rapid reduction in the levels of cellular ER. The magnitude of the reduction was dependent on cell type. GH_4C1 cells treated with estradiol or ICI showed little difference in ER levels, while MCF-7 cells showed large differences. However, the ER levels remaining in all cells examined after ICI exposure were sufficient to markedly suppress promoter activity of the $CMV-(ERE)_2-CAT$ construct, and transfection of similar quantities of a transactivation reporter plasmid (ERE-vit-CAT) showed that ICI failed to activate transcription, in spite of the fact that there should be enough ER in these cells to bind to the ERE-vit-CAT construct. This point is underscored in transfected GH_4C1 cells, where ICI treatment led to a similar decline in ER levels as cells treated with estradiol, yet ICI still acted as a pure antagonist. These findings suggest that (1) ICI-ER complexes are capable of binding to DNA in whole cells, and (2) the reduction in ER levels caused by ICI may not fully explain the pure antagonistic nature of this ligand. In addition, the antiestrogen OHT-occupied receptors suppressed promoter activity as effectively as did E_2-occupied receptors, yet these OHT-occupied receptors gave only partial activation of the ERE-vit-CAT reporter template indicating that receptor binding to ERE DNA alone is not sufficient to ensure full receptor activity.

In conclusion, we demonstrate that ER is capable of binding to DNA templates in whole cells in the presence or absence of ligand, and that estrogen and antiestrogen are capable of enhancing or stabilizing this interaction. The binding of unoccupied receptor to DNA was observed in cells expressing high or low levels of receptor, and the binding of unliganded receptor causes hormone-independent transactivation of a reporter plasmid, presumably through the TAF-1 function in the amino terminus of the receptor. The "relaxed" ligand requirement for DNA binding of ER also alters the receptor's hormone-dependent activity as well. Also, ER-ICI receptor complexes are capable of binding to DNA, and the reduction in ER levels caused by ICI treatment may not fully explain the purely antagonistic nature of this ligand-receptor complex. Differences in the interaction of antiestrogen-receptor and unoccupied receptor complexes with components of the preinitiation complex may explain the ineffectiveness of these receptors in activating transcription despite binding to DNA response elements.

ACKNOWLEDGMENTS

We are grateful to Dr. G. Greene for providing the ER cDNA OR8, to Dr. D. W. Russell for providing the expression vector pCMV-5, and to Dr. D. Shapiro for providing us with the reporter plasmids. We also thank Abbott Laboratories for the ER monoclonal antibody H222. This work was supported by NIH grant CA18119 to B.S.K. and by a predoctoral fellowship to J.C.R. (USPHS GM07283).

REFERENCES

Aronica SM and Katzenellenbogen BS (1991): Progesterone receptor regulation in uterine cells: stimulation by estrogen, cyclic adenosine 3',5'-monophosphate, and insulin-like growth factor I and suppression by antiestrogens and protein kinase inhibitors. *Endocrinol* 128: 2045-2052.

Bagchi MK, Elliston JF, Tsai SY, Edwards DP, Tsai M-J and O'Malley BW (1988): Steroid hormone-dependent interaction of human progesterone receptor with its target enhancer element. *Mol Endo* 2: 1221-1229.

Bagchi MK, Tsai, SY, Tsai M-J and O'Malley BW (1990): Identification of a functional intermediate in receptor activation in progesterone-dependent cell-free extracts. *Nature* 345: 547-550.

Berry M, Metzger, M and Chambon P (1990): Role of the two activating domains of the oestrogen receptor in the cell-type and promoter-context dependent agonistic activity of the anti-oestrogen 4-hydroxytamoxifen. *EMBO J* 9: 2811-2818.

Brown M and Sharp, P (1990): Human estrogen receptor forms multiple-DNA complexes. *J Biol Chem* 265: 11238-11243.

Cairns W, Cairns C, Pongratz I, Poellinger L and Okret S (1991): Assembly of a glucocorticoid receptor complex prior to DNA binding enhances its specific interaction with a glucocorticoid response element. *J Biol Chem* 266: 11221-11226.

Carson-Jurica MA, Schrader WT and O'Malley BW (1990): Steroid Receptor Family: Structure and Function. *Endocrine Reviews* 11: 201-220.

Clarke R, Brunner N, Katzenellenbogen BS, Thompson BW, Norman MJ, Koppi C., Paik S, Lippman ME and Dickson RB (1989): Progression of human breast cancer cells from hormone-dependent to hormone-independent growth both in vitro and in vivo. *Proc Natl Acad Sci USA* 86: 3649-3653.

Dauvois S, Danielian PS, White R and Parker MG (1992): Antiestrogen ICI 164,384 reduces cellular estrogen receptor content by increasing its turnover. *Proc Natl Acad Sci USA* 89: 4037-4041.

Denis M, Poellinger L, Wikström AC and Gustafsson, JA (1988): Requirement of hormone for thermal conversion of the glucocorticoid receptor to a DNA binding state. *Nature* 333: 686-688.

Denis M and Gustafsson JA (1989): Translation of glucocorticoid receptor mRNA in vitro yields a nonactivated protein. *J Biol Chem* 264: 6005-6008.

DeMarzo AM, Beck CA, Oanate, SA and Edwards DP (1991): Dimerization of mammalian progesterone receptors occurs in the absence of DNA and is related to the release of the 90-kDa heat shock protein. *Proc Natl Acad Sci USA* 88: 72-76.

Edwards DP, Kuhnel B, Estes PA and Nordeen SK (1989): Human progesterone receptor binding to mouse mammary tumor virus deoxyribonucleic acid: dependence on hormone and nonreceptor nuclear factor(s). *Mol Endo* 3: 381-391.

El-Ashry D, Onate SA, Nordeen SK and Edwards DP (1989): Human progesterone receptor complexed with the antagonist RU 486 binds to hormone response elements in a structurally altered form. *Mol Endo* 3: 1545-1558.

Elliston JF, Tsai SY, O'Malley BW and Tsai MJ (1990): Superactive estrogen receptors. *J Biol Chem* 265: 11517-11521.

Fawell SE, White R, Hoare S, Sydenham M, Page M and Parker MG (1990a): Inhibition of estrogen receptor-DNA binding by the "pure" antiestrogen ICI 164,384 appears to be mediated by impaired receptor dimerization. *Proc Natl Acad Sci USA* 87: 6883-6887.

Fawell SE, Lees JA, White JA and Parker MG (1990b): Characterization and colocalization of steroid binding and dimerization activities in the mouse estrogen receptor. *Cell* 60: 953-962.

Gibson MK, Nemmers LA, Beckman Jr WC, Davis VL, Curtis SW and Korach KS (1991): The mechanism of ICI 164,384 antiestrogenicity involves rapid loss of estrogen receptor in uterine tissue. *Endocrinol* 129: 2000-2010.

Giguere V, Hollenberg SM, Rosenfeld MG and Evans RM (1986): Functional domains of the human glucocorticoid receptor. *Cell* 46: 645-652.

Howard KJ, Holley SJ, Yamamoto KR and Distelhorst CW (1990): Mapping the hsp90 binding region of the glucocorticoid receptor. *J Biol Chem* 265: 11928-11935.

King WJ and Greene GL (1984): Monoclonal antibodies localize estrogen receptor in the nuclei of target cells. *Nature (London)* 307: 745-749.

Lees JA, Fawell SE and Parker MG (1989): Identification of two transactivation domains in the mouse oestrogen receptor. *Nucleic Acids Res* 17: 5477-5487.

Martinez E and Wahli W (1989): Cooperative binding of estrogen receptor to imperfect estrogen-responsive DNA elements correlates with their synergistic hormone-dependent enhancer activity. *EMBO J* 8: 3781-3791.

McDonnell DP, Nawaz Z and O'Malley BW (1991): In situ distinction between steroid receptor binding and transactivation at a target gene. *Mol Cell Biol* 11: 4350-4355.

Meyer ME, Gronemeyer H, Turcotte B, Bocquel MT, Tasset D and Chambon P (1989): Steroid hormone receptors compete for factors that mediate their enhancer function. *Cell* 57: 433-442.

Pham TA, Elliston J, Nawaz Z, McDonnell DP, Tsai MJ and O'Malley BW (1991): Antiestrogen can establish nonproductive receptor complexes and alter chromatin structure at target elements. *Proc Natl Acad Sci USA* 88: 3125-3129.

Picard D, Kumar V, Chambon P and Yamamoto KR (1990a): Signal transduction by steroid hormones: nuclear localization is differentially regulated in estrogen and glucocorticoid receptors. *Cell Reg* 1: 291-299.

Picard D, Khursheed B, Garabedian MJ, Fortin MG, Lindquist S and Yamamoto KR (1990b): Reduced levels of hsp90 compromise steroid receptor action *in vivo*. *Nature* 348: 166-168.

Power RF, Mani SK, Codina J, Conneeley OM and O'Malley BW (1991): Dopaminergic and ligand-independent activation of steroid hormone receptors. *Science* 254: 1636-1639.

Reese JC and Katzenellenbogen BS (1991a): Mutagenesis of cysteines in the hormone binding domain of the human estrogen receptor: alterations in binding and transcriptional activation by covalently and reversibly attaching ligands. *J Biol Chem* 266: 10880-10887.

Reese JC and Katzenellenbogen BS (1991b): Differential DNA-binding abilities of receptor occupied with two classes of antiestrogens: studies using human estrogen receptor overexpressed in mammalian cells. *Nucleic Acids Res* 19: 6595-6602.

Reese JC and Katzenellenbogen BS (1992a): Characterization of a temperature-sensitive mutation in the hormone binding domain of the human estrogen receptor: studies utilizing cell-free extracts and intact cells and implications for hormone-dependent transcriptional activation. *J Biol Chem* 267: 9868-9873.

Reese JC and Katzenellenbogen BS (1992b): Examination of the DNA-Binding ability of estrogen receptor in whole cells: implications for hormone-independent transactivation and the actions of antiestrogens. *Mol Cell Biol* 12: 4531-4538.

Schena M and Yamamoto KR (1988): Mammalian glucocorticoid receptor derivatives enhance transcription in yeast. *Science* 241: 965-967.

Tora L, Mullick A, Metzger D, Ponglikitmongkol M, Park I and Chambon P (1989a): The cloned human oestrogen receptor contains a mutation which alters its hormone binding properties. *EMBO J* 8: 1981-1986.

Tora L, White J, Brou C, Tasset D, Webster N, Scheer E and Chambon P (1989b): The human estrogen receptor has two nonacidic transcriptional activation functions. *Cell* 59: 477-487.

Tzukerman M, Zhang X-K, Hermann T, Wills, KN, Graupner, G and Pfahl M (1991): The human estrogen receptor has transcriptional activator and repressor functions in the absence of ligand. *The New Biologist* 2: 613-620.

Wrenn CK and Katzenellenbogen BS (1990): Cross-linking of the estrogen receptor to chromatin in intact MCF-7 human breast cancer cells: optimization and effect of ligand. *Mol Endo* 4: 1647-1654.

Note added in proof: The manuscript by Pakdel, Reese, and Katzenellenbogen is now in press. *MOLEC ENDO.*

STEROID RECEPTORS AND CLINICAL CORRELATIONS

MOLECULAR BIOLOGICAL ASPECTS OF THE HUMAN ANDROGEN RECEPTOR RELATING TO DISEASE

Zhong-xun Zhou
Department of Pediatrics
University of North Carolina at Chapel Hill

Madhabananda Sar
Department of Cell Biology and Anatomy
University of North Carolina at Chapel Hill

Frank S. French
Department of Pediatrics
University of North Carolina at Chapel Hill

Elizabeth M. Wilson
Departments of Pediatrics and Biochemistry and Biophysics
University of North Carolina at Chapel Hill

INTRODUCTION

The androgen receptor (AR) is required for normal male sexual development. It acts as a transcriptional regulatory protein subsequent to binding either of the two biologically active forms of androgen, testosterone or dihydrotestosterone. Although knowledge of the molecular mechanisms by which this ligand activated transcription factor

STEROID HORMONE RECEPTORS
V. K. Moudgil, Editor
© 1993 Birkhäuser Boston

induces target gene expression remains incomplete, recent advances in understanding the molecular events in androgen regulation have stemmed from the use of recombinant DNA techniques in studies of this complex protein.

Certain key events were critical in advancing our understanding of the AR. These include the cloning of the human AR complementary DNA (cDNA) (Lubahn *et al*, 1988a,b; Chang *et al*, 1988) and the development of antibodies that recognize the native and denatured protein (Tan *et al*, 1988; Quarmby *et al*, 1990). Since purification of the low abundance AR protein has proved difficult, antibodies were developed against synthetic peptides with sequences derived from the cDNA. As shown in Fig. 1, two antibodies were raised against peptides with amino acid sequence corresponding to regions of the AR NH$_2$-terminal domain. The antibodies, designated AR52 and AR32, react specifically with denatured or native receptor proteins. Shown in Fig. 2 is the interaction of AR52 antibody with the [^{3}H]dihydrotestosterone-labeled AR extracted from the

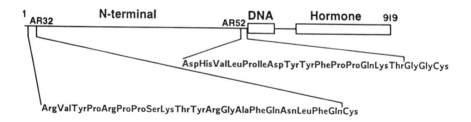

FIGURE. 1. Schematic diagram showing regions of human AR used in the design of synthetic peptide sequences for antibody production. AR32 and AR52 polyclonal antibodies were raised in rabbits in response to immunization with these synthetic peptides injected in the free form and coupled to Keyhole limpet hemocyanin. The carboxyl-terminal portion of the sequence outside the brackets represents spacer sequence to the coupling protein. The peptide for antibody AR32 includes amino acids 9-28 and the peptide for antibody AR52 includes amino acids 544-558 of human AR. The position of the peptides are shown relative to the three major domains of the AR: the NH$_2$-terminal, DNA binding, and steroid binding domains.

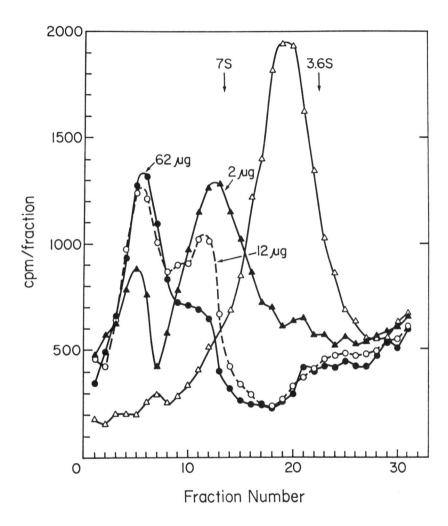

FIGURE. 2. Reactivity of AR52 IgG with endogenous rat AR. Increasing amounts of the IgG fraction of immune serum were added to [³H]dihydrotestosterone-labeled extracts of rat Dunning tumor R3327H cytosol. The AR52 IgG fraction was added in aliquots containing 2, 12 and 62 µg protein causing the sedimentation of the 4.5S receptor to increase to approximately 7 and 10S. Samples were analyzed on 2-20% sucrose gradients as previously described (Wilson *et al*, 1986).

Dunning tumor. With increasing antibody concentration, an increase in sedimentation of AR occurs as two discrete forms, approximately 8S and 10S. These antibody/receptor complexes may reflect receptor dimer formation or stabilization of nonsteroid binding components within the complex.

The peptide sequences used in antibody production were selected on the basis of their uniqueness to AR and for their identity with amino acid sequences present in AR from the rat, mouse and human. The antibodies were found useful not only for these species, but also for a variety of other species. For example, it was recently demonstrated that AR32 IgG localizes AR in the male songbird and quail brain using immunocytochemistry (Balthazart *et al*, 1992). The antipeptide antibodies also recognized AR in cross sections of guinea pig testis, showing AR localization in Sertoli, peritubular and interstitial cells (Sar *et al*, 1993). A limitation of the antibodies in immunostaining has been, however, that frozen sections are required as well as paraformaldehyde fixation (Sar *et al*, 1990).

An interesting aspect of work on AR is that natural mutations in the AR gene are associated with both congenital and acquired human diseases, including the androgen insensitivity syndrome, adult onset spinal/bulbar muscular atrophy, and prostate cancer. In addition to natural mutations, *in vitro* site directed mutagenesis has allowed a dissection of the functional domains of the AR protein. Two of these functional domains are discussed in further detail, one in the NH_2-terminal region required for transactivation, and the other, in the DNA binding and carboxyl-terminal regions that determine the steroid specificity of nuclear targeting. We have begun to investigate the mechanism of action of antihormones with the goal of establishing the molecular consequences of antiandrogen binding. It is conceivable that new therapeutic approaches will be identified, particularly in light of the possible association of AR gene mutations with prostate cancer.

Androgen Receptor Gene Defects Cause Abnormalities in Male Sex Development

Development of the male phenotype *in utero* and maturation of male reproductive function at puberty require the action of the male sex hormones, testosterone and dihydrotestosterone, and their high affinity binding protein, the androgen receptor (AR). Mutations in the AR gene can cause loss of receptor expression or function such that cells are unable to respond to androgen. Depending on the position and severity of the mutation, the resulting androgen insensitivity syndrome (AIS) prevents 46XY genetic male fetuses from masculinizing, causing them to develop a female phenotype or ambiguous genitalia. Affected individuals with a male phenotype and infertility have also been reported.

AR gene mutations occur throughout the coding sequence but appear most prevalent in the steroid binding domain. The incidence of AR mutations exceeds that for other steroid receptors, in part because loss of AR function is not life threatening, whereas some steroid receptors are involved in essential life sustaining processes. Moreover, the AR gene is X-chromosomal so that 46XY hemizygotes have only a single allele. Mutations in other steroid receptor genes, which are autosomal, may be less penetrant due to the presence of more than one allele.

The spectrum of single base mutations in the AR gene provides an opportunity to correlate a variety of single amino acid changes with AR function and phenotypic expression associated with AIS. Striking sequence similarities among the family of steroid receptors makes it possible to apply what is learned from natural AR mutations towards understanding the functional domains of other steroid receptors. Recent reports (French *et al*, 1990; Brown *et al*, 1990; DeBellis *et al*, 1992; Marchelli et al, 1991; McPhaul et al, 1991) relate the spectrum of AR gene defects in androgen insensitivity.

The first direct evidence linking complete AIS with an AR gene defect was a partial gene deletion of the AR steroid binding domain (Brown TR et al, 1988). It was subsequently reported that deletion of exon 3 in the DNA binding domain encoding the second finger region causes complete AIS (Quigley et al, 1992a). This mutant AR displayed normal androgen binding affinity and localized to the nucleus of cultured genital skin fibroblasts. The deletion resulted in reduced DNA binding affinity with failure of transactivation, indicating the importance of the second zinc finger region in receptor function. A complete deletion of the AR gene established the null phenotype of androgen insensitivity (Quigley et al, 1992b), characterized in the adult by a female sex phenotype with complete absence of sexual hair.

A number of single base mutations were identified in association with complete androgen resistance (Ris-Stalpers et al, 1991; Marcelli et al, 1991; Prior et al, 1992; DeBellis et al, 1992). Some mutations cause partial androgen insensitivity with a phenotype like that of the Reifenstein syndrome (Klocker et al, 1992). Azoospermia in an otherwise normal male was reported associated with a deletion mutation in the steroid binding domain (Akin et al, 1991). A single base somatic AR mutation was identified in stage B prostate cancer (Newmark et al, 1992).

While AR gene mutations affect the response to androgen in general, AR regulation of individual genes can be altered by mutations in androgen response elements (ARE). A form of X-linked hemophilia B that gradually improves after puberty was reported to be associated with a defect in the ARE of the human factor IX promoter (Crossley et al, 1992). Androgen stimulation of mouse β-glucuronidase (GUS) gene expression is inhibited by a mutation involving an ARE in intron IX (Lund et al, 1991).

Naturally occurring mutations are less common among other steroid receptors. A single base mutation in the glucocorticoid receptor steroid binding domain was reported to cause familial glucocorticoid resistance with autosomal inheritance (Hurley et al, 1991). Heterozygous members were only mildly affected while the homozygous were severely affected with the glucocorticoid resistant phenotype. Vitamin D resistant rickets was traced to mutations that change conserved amino acids in the DNA binding domain of the 1,25-dihydroxy-vitamin D3 receptor (Hughes et al, 1988).

Expression vectors containing coding sequences for full-length mutant receptors can be constructed to determine the transcriptional activation potential of AR mutations in mammalian cells by cotransfection with a reporter vector. This type of analysis allows a direct comparison of, for example, androgen binding properties of endogenous AR with recreated receptor mutants. In the testicular feminized (Tfm) rat, a single base mutation in the steroid binding domain was linked with androgen insensitivity in this rodent model of the disease (Yarbrough et al, 1990). When this mutation was recreated in an expression vector, it was found to display androgen binding characteristics different from wild type. As shown in Fig. 3, the recreated rat Tfm receptor retained high affinity binding but showed greatly reduced binding capacity. Immunoblot analysis revealed that these vectors were expressed to a similar extent (Yarbrough et al, 1990) confirming that the loss of binding capacity was due to the structurally altered steroid binding domain. An understanding of the reduction in binding capacity by this single base change in the steroid binding domain awaits further study.

Androgen Receptor Defects Cause Adult Onset Spinal/Bulbar Muscular Atrophy

The molecular basis of adult onset spinal/bulbar muscular atrophy, Kennedy's disease, was recently established (La Spada et al, 1991). The clinicopathology of this progressive neuronopathy (Sobue et al, 1989)

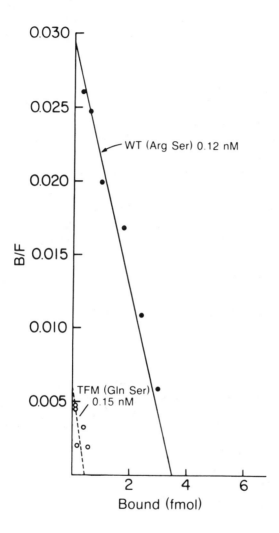

FIGURE. 3. Scatchard plot analysis of [³H]R1881 binding to wild type and Tfm rat mutant AR. The wild type and mutant ARs were constructed and expressed in monkey kidney COS cells and a whole cell binding assay performed as previously described (Yarbrough et al, 1990). The mutant receptor had an arginine 734 to glutamine change in the steroid binding domain to match the natural mutation that occurs in the testicular feminized rat AR.

includes gynecomastia and reduced fertility, clinical features consistent with androgen insensitivity (Arbizu et al, 1983). This clinical phenotype, together with the X chromosomal colocalization of the AR gene (Brown CJ et al, 1989) and the gene locus for spinal/bulbar muscular atrophy (Fischbeck et al, 1986), led to the identification of the AR gene defect in this disease.

Patients affected with adult onset muscular atrophy have an expanded tandem CAG repeat in the NH_2-terminal domain of the AR gene (La Spada et al, 1991). The CAG repeat in the normal AR gene encodes 21 glutamines with a normal range from 13 to 30 (Edwards et al, 1991) which amplifies to 39-60 glutamines in the AR of patients with X linked spinal/bulbar muscular atrophy (Fig. 4). The mutation segregated with the disease in 35 unrelated affected individuals but in none of 75 controls. The enlarged CAG repeat in the AR gene is the probable cause of the disorder, although the mechanism remains unknown.

The glutamine repeat is positioned at residues 58-78 in the human AR gene (Lubahn et al, 1988b) (Fig. 4) and immediately precedes the transactivation domain of AR (see below). Preliminary studies designed to test the functional consequences of the expanded repeat have not revealed clear differences in transient cotransfection assays using a luciferase reporter gene activated through the mouse mammary tumor

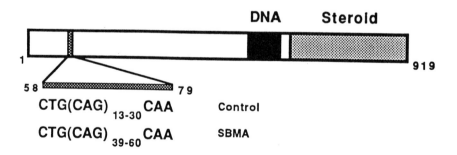

FIGURE. 4. Schematic diagram of the human androgen receptor highlighting the molecular defect associated with adult onset spinal/bulbar muscular atrophy (SBMA). The major domains of the AR are the NH_2-terminal and DNA and steroid binding regions. Between amino acids 58 and 79 occurs a glutamine repeat which is polymorphic in the normal population. In adult onset SBMA, the repeat expands to 39-60 glutamine residues and is thought to cause the disease (La Spada et al, 1991).

virus promoter. It is possible that the effects of the mutant AR are tissue specific and may require testing in motorneurons.

It was demonstrated years ago by autoradiography that AR is present in motorneurons of the spinal cord (Sar and Stumpf, 1977). Shown in Fig. 5 is immunocytochemical staining of AR in rat motorneurons of the spinal cord. Note the intense AR staining in nuclei of certain motorneurons of the rat thoracic spinal cord. More recently, abnormal androgen binding in genital skin fibroblasts (Warner et al, 1991) and lack of AR immunostaining in scrotal skin of patients with spinal/bulbar muscular atrophy from different kindreds (Matsuura et al, 1992) were reported. These mutant ARs in motor neurons may accelerate their normal rate of attrition with aging. However, motor neuron dysfunction with adult onset muscular atrophy is not a feature of other forms of androgen insensitivity.

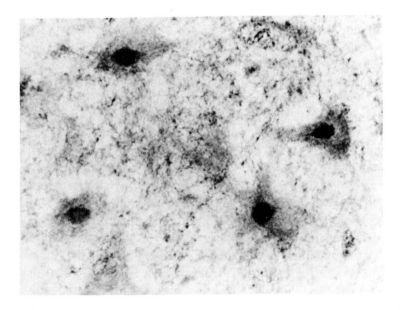

FIGURE 5. Immunohistochemical staining of AR in the thoracic region of the rat spinal cord. Thoracic spinal cord of an intact adult male rat was dissected and frozen in isopentane prechilled in liquid N_2. 8 µ thick frozen sections were fixed in 4% paraformaldehyde and immunostained using AR antipeptide antibody AR32 (5 µg/ml) and the avidin biotin peroxidase method as described (Sar et al, 1990). Note immunostaining in nuclei of motorneurons. Magnification, ×560.

Transcriptional Activation Domain of the Androgen Receptor

AR has an overall domain structure similar to other members of the family of steroid receptors. Its DNA binding domain shares sequence homology, particularly with the glucocorticoid, mineralocorticoid and progesterone receptors. Somewhat less sequence identity is found in the carboxyl-terminal steroid binding region. In earlier studies we reported that deletion of the AR steroid binding domain results in constitutive transcriptional activity (Simental et al, 1991). The full-length receptor, on the other hand, had no transcriptional activity in the absence of hormone. Thus, the steroid binding domain inhibits transactivation by AR in the absence of hormone. This inhibition is lost with deletion of the carboxyl-terminal region.

It is conceivable that inhibition by the steroid binding domain is accomplished by a direct interaction with another receptor domain mediated through its tertiary structure. Alternatively, the steroid binding domain may interact with inhibitory proteins. It was proposed that the steroid binding domain of the glucocorticoid receptor interacts with the 90 kDa heat shock protein to inhibit transcriptional activity (Pratt et al, 1992); however, this remains to be demonstrated for the AR.

We investigated whether the NH_2-terminal domain of AR is involved in transactivation functions using an androgen responsive reporter plasmid in transient cotransfection assays. Exonuclease treatment generated a series of mutants with increasing deletions of the NH_2-terminal region (Simental et al, 1991). It was established that a region between amino acids 141-338 was required for gene activation by AR. Subsequently we performed internal deletions within this area. It became clear that amino acids 142-239 were particularly important in AR transactivation. Deletion of this region resulted in inactivated AR transcriptional activity (Fig. 6). However, removal of only the 3' portion of this region had the opposite effect. Deletion of amino acids 199-239 caused a significant increase in transcriptional activity (Fig. 6). Thus complex interactions of the AR NH_2-terminal domain likely play a role in establishing its transactivation function. These interactions could be within the AR protein itself or with other transregulatory proteins.

Antiandrogens

The AR binds a variety of steroids with moderate affinity, including antihormones (Kemppainen et al, 1992). We asked whether AR undergoes functional activation when bound to nonandrogenic steroids for

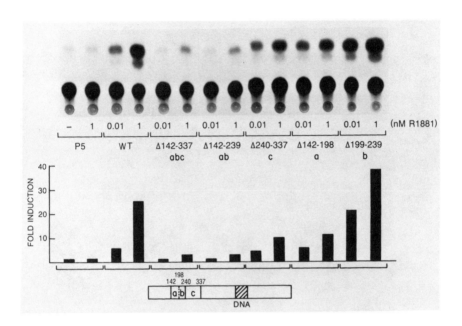

FIGURE 6. Chloramphenicol acetyl transferase activity of wild type and mutant receptors containing internal deletions within the NH$_2$-terminal transactivation domain. Monkey kidney CV1 cells were cotransfected using the calcium phosphate precipitation method using wild type or mutant AR expression vectors cloned in pCMV together with a reporter plasmid containing the chloramphenicol acetyltransferase (CAT) gene coupled to a strong androgen response element (Simental et al, 1991). Internal deletions were constructed using the polymerase chain reaction mutagenesis method. The deleted regions are indicated by amino acid numbers based on a full-length AR containing 919 amino acids. Transcriptional CAT activity was tested at two concentrations of R1881, 0.01 and 1 nM. Incubations were for 48 h and the cells were harvested and analyzed for CAT induction by determining the extent of conversion of ^{14}C-chloramphenicol to acetylated forms separated by thin layer chromatography. Below is a schematic diagram indicating the relative position of the deletions.

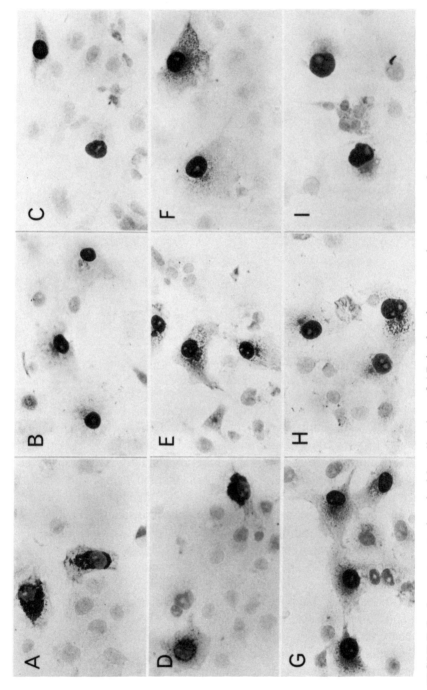

FIGURE 7. Immunocytochemical localization of AR in the absence and presence of steroids and antiandrogens.

which it displays only moderate binding affinity. The question was addressed by performing nuclear transport studies in monkey kidney COS cells using transiently transfected AR cDNA contained in a mammalian expression vector as previously described (Kemppainen et al, 1992). As shown in Fig. 7, we found that a variety of steroids cause nuclear transport of AR. In the absence of added hormone, the AR was distributed in a perinuclear region of the cytoplasm immediately surrounding the nucleus (Fig. 7A). Addition of androgen caused strong nuclear localization, shown in Fig. 7B for the synthetic androgen, R1881, and in Fig. 7C for dihydrotestosterone.

Nuclear localization (Fig. 7) was also observed following addition of hydroxyflutamide (Fig. 7E), cyproterone acetate (Fig. 7F), the antiprogestin, RU486 (Fig. 7G), estradiol (Fig. 7H) and progesterone (Fig. 7I). Flutamide was the only antiandrogen that caused persistence of perinuclear AR immunostaining like that observed in the absence of hormone (Fig. 7D). AR affinity for flutamide is low (K_I 1-4 µM) (Liao et al, 1974) and thus the perinuclear distribution in the cytoplasm was not unexpected. The antiandrogen effects of flutamide are thought to be mediated through its metabolite, hydroxyflutamide (Katchen and Buxbaum, 1975; Liao et al, 1974; Mainwaring et al, 1974; Varkarakis et al, 1975; Neri et al, 1979). Thus AR can bind a variety of steroids with moderate affinity and these hormones cause nuclear transport of AR.

The biological effects of antiandrogens were investigated further in a transient transfection assay using an androgen responsive reporter plasmid. Antiandrogens that bound AR with moderate affinity and

FIGURE 7. Immunocytochemical localization of AR in the absence and presence of steroids and antiandrogens. Full-length human AR (pCMVhAR) was transiently transfected into monkey kidney COS cells using the DEAE dextran method in 2 well chamber slides as previously described (Kemppainen et al, 1991). Twenty-four h after transfection and again 2 h prior to fixation, transfected cells were placed in serum-free, phenol red-free media containing the following: no hormone addition (A), 50 nM R1881 (a synthetic androgen agonist) (B), 50 nM dihydrotestosterone (C), 100 nM flutamide (D), 100 nM hydroxyflutamide (E), 100 nM cyproterone acetate (F), 100 nM RU486 (G), 100 nM estradiol-17B (H) or 100 nM progesterone (I). The slides were washed, fixed and reacted with the AR52 antipeptide IgG and visualized using the avidin biotin peroxidase method previously described (Sar 1985; Sar et al, 1990). Magnification, ×230

promoted nuclear transport also stimulated gene transcription when assayed at high concentrations (Kemppainen et al, 1992). An exception was hydroxyflutamide; as shown in Fig. 8, hydroxyflutamide displayed absence of agonist activity at a concentration of 100 nM. Hydroxyflutamide also inhibited the agonist activity of R1881, a strong synthetic androgen shown assayed at 0.05 nM (Kemppainen et al, 1992). Thus, hydroxyflutamide acts as a true antiandrogen; it causes nuclear transport of AR (shown above), lacks agonist activity, and has strong inhibitory activity.

Lack of agonist activity did not apply to the antiandrogen, cyproterone acetate. Cyproterone acetate inhibited R1881 activity (Kemppainen et al, 1992) but displayed agonist activity as well (Fig. 8). Like hydroxyflutamide, cyproterone acetate caused nuclear transport of AR but could not be considered a pure antagonist in this system.

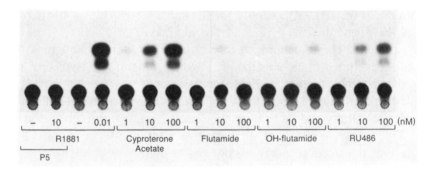

FIGURE 8. Transcriptional activation with androgen agonists and antagonists. CAT activity was determined by cotransfection in CV1 cells using the full-length human pCMVhAR expression and an androgen responsive CAT reporter plasmid as previously described (Kemppainen et al, 1992). After transfection cells were placed in media containing 0.2% fetal calf serum plus the hormones indicated. The first two lanes reflect lack of activity with the parent plasmid pCMV5 lacking AR coding sequence. Hormone concentrations are indicated for R1881 (a synthetic androgen agonist), cyproterone acetate, flutamide, hydroxyflutamide and RU486.

Prostate Cancer

It has long been recognized that androgens are essential for the growth and maintenance of prostate cell function. Most prostate cancers are androgen dependent and regress initially following androgen withdrawal.

Recent evidence suggests that AR gene mutations can occur in prostate cancer cells (Newmark et al, 1992), however, a causative relationship between somatic mutations in the AR gene and prostate cancer remains to be verified. AR mutations could be an early or late event during clonal selection in the development of prostate cancer. Mutations could occur at the gene level or result from mRNA processing errors.

Mutant forms of AR could have oncogenic potential through a variety of mechanisms. From mutagenesis studies we have learned that deletion of the steroid binding domain imparts constitutive activity on the AR (Simental et al, 1991). Such a mutant receptor could theoretically be a gene activator independent of androgen binding. Further studies are required to determine the frequency of AR mutations in prostate cancer and whether there is a correlation with the onset and/or development of prostate cancer.

The androgen dependent human prostate cancer cell line, LNCaP, expresses AR but lacks other steroid receptors. A single base mutation in the region of the gene encoding the AR steroid binding domain alters the steroid binding and transactivation properties of this naturally occurring mutant receptor (Harris et al, 1990; Harris et al, 1991; Veldscholte et al, 1990). LNCaP cells respond to hydroxyflutamide, estradiol and progesterone with an increase in cell proliferation (Sonnenschein et al, 1989; Wilding et al, 1989). Estradiol and progesterone binding was inhibited by androgen indicating that their effects are mediated through the AR (Sonnenschein et al, 1989). The single base mutation in the AR gene changes amino acid 877 from threonine to alanine. In cotransfection studies with the reconstructed mutant AR, a striking increase in transcriptional activity was observed with hydroxyflutamide indicating that alteration of a single amino acid within the steroid binding domain has the potential for changing the steroid binding specificity of AR, causing an antagonist to become an agonist. The LNCaP AR also increased transcriptional activation in response to progesterone and cyproterone acetate (Veldscholte, et al, 1990).

Conclusions

Analysis of mutations in the AR gene, occurring either naturally or by *in vitro* mutagenesis, can reveal important molecular properties of this regulatory protein. Recombinant DNA techniques have made possible not only the expression of high levels of receptor for analysis, but also the structural manipulation of its various functional domains. Future studies will undoubtedly reveal the complex mechanisms involved in gene activation by this ligand dependent transcription factor.

ACKNOWLEDGMENTS

This work was supported by the National Institute of Child Health and Human Development Center for Population Research Grants HD16910, HD04466, P30-HD18969, by National Institutes of Health Grant NS17479, and by the Andrew W. Mellon Foundation and by Basic Research Grant 1-FY92-0910 from the March of Dimes Birth Defects Foundation.

REFERENCES

Akin JW, Behzadian A, Tho SPT and McDonough PG (1991): Evidence for a partial deletion in the androgen receptor gene in a phenotypic male with azoospermia. *Am J Obstet Gynecol* 165: 1891-1894.

Arbizu T, Santamaria J, Gomez JM, Quilez A and Serra JP (1983): A family with adult spinal and bulbar muscular atrophy, X-linked inheritance and associated testicular failure. *J Neurol Sci* 59: 371-382.

Balthazart J, Foidart A, Wilson EM and Ball GF (1992): Immunocytochemical localization of androgen receptors in the male songbird and quail brain. *J Comp Neurol* 317: 407-420.

Brown CJ, Goss SJ, Lubahn DB, Joseph DR, Wilson EM, French FS and Willard HF (1989): Androgen receptor locus on the human X chromosome: regional localization to Xq11-12 and description of a DNA polymorphism. *Amer J Hum Genet* 44: 264-269.

Brown TR, Lubahn DB, Wilson EM, Joseph DR, French FS and Migeon CJ (1988): Deletion of the steroid binding domain of the human AR gene in one family with complete AIS; evidence of further genetic heterogeneity in this syndrome. *Proc Natl Acad Sci USA* 85: 8151-8155.

Brown TR, Lubahn DB, Wilson EM, French FS, Migeon CJ and Corden JL (1990): Functional characterization of naturally occurring mutant androgen receptors from subjects with complete androgen insensitivity. *Mol Endocrinol* 4: 1759-1772.

Chang C, Kokontis J and Liao S (1988): Molecular cloning of human and rat complementary DNA encoding androgen receptors. *Science* 240: 324-326.

Crossley M, Ludwig M, Stowell KM, De Vos P, Olek K and Brownlee GG (1992): Recovery from Hemophilia B Leyden: an androgen responsive element in the factor IX promoter. *Science* 257: 377-379.

DeBellis A, Quigley CA, Cariello NF, El-Awady MK, Sar M, Lane MV, Wilson EM and French FS (1992): Single base mutations in the human androgen receptor gene causing complete androgen insensitivity: rapid detection by a modified denaturing gradient gel electrophoresis technique. *Mol Endocrinal,* 6: 1909-1920.

Edwards A, Civitello A, Hammond HA and Caskey CT (1991): DNA typing and genetic mapping with trimeric and tetrameric tandem repeats. *Am J Hum Genet* 49: 746-756.

Fischbeck KH, Ionasescu V, Ritter AW, Ionasescu R, Davies K, Ball S, Bosch P, Burns T, Hausmanowa-Petrusewicz I, Borkowska J, Ringel SP and Stern LZ (1986): Localization of the gene for X-linked spinal muscular atrophy. *Neurology, Cleveland* 36: 1595-1598.

French FS, Lubahn DB, Brown TR, Simental JA, Quigley CA, Yarbrough WG, Tan JA, Sar M, Joseph DR, Evans BAJ, Hughes IA, Migeon CJ and Wilson EM (1990): Molecular basis of androgen insensitivity. *Rec Prog Horm Res* 46: 1-42.

Harris SE, Rong Z, Harris MA and Lubahn DB (1990): Androgen receptor in human prostate carcinoma LNCaP/Adep cells contains a mutation which alters the specificity of the steroid-dependent transcriptional activation region. *Endocrine Society Abstracts,* p. 93 (Atlanta, abstr).

Harris SE, Harris MA, Rong Z, Hall J, Judge S, French FS, Joseph DR, Lubahn DB, Simental JA and Wilson EM (1991): In: *Molecular and Cellular Biology of Prostate Cancer,* Karr JP, Coffey DS, Smith RG, Tindall DJ, eds. New York: Plenum Press, 315-330.

Hughes MR, Malloy PJ, Kieback DG, Kesterson RA, Pike JW, Feldman D and O'Malley BW (1988): Point mutations in the human vitamin D receptor gene associated with hypocalcemic rickets. *Science* 242: 1702-1705.

Hurley DM, Accili D, Stratakis CA, Karl M, Vamvakopoulos N, Rorer E, Constantine K, Taylor SI and Chrousos GP (1991): Point mutation causing a single amino acid substitution in the hormone binding domain of the glucocorticoid receptor in familial glucocorticoid resistance. *J Clin Invest* 87: 680-686.

Katchen B and Buxbaum S (1975): Disposition of a new nonsteroid antiandrogen, α,α,α-trifluoro-2-methyl-4'-nitro-m-propionotoluidide (flutamide) in men following a single oral 200 mg dose. *J Clin Endocrinol Metab* 41: 373-379.

Klocker H, Kaspar F, Eberle J, Uberreiter S, Radmayr C and Bartsch G (1992): Point mutation in the DNA binding domain of the androgen receptor in two families with Reifenstein syndrome. *Am J Hum Genet* 50: 1318-1327.

La Spada AR, Wilson EM, Lubahn DB, Harding AE and Fischbeck KH (1991): AR gene mutations in X-linked spinal and bulbar muscular atrophy. *Nature* 352: 77-79.

Liao S, Howell DK and Chang TM (1974): Action of a nonsteroidal antiandrogen, flutamide, on the receptor binding and nuclear retention of 5 α-dihydrotestosterone in rat ventral prostate. *Endocrinology* 94: 1205-1209

Lubahn DB, Joseph DR, Sullivan PM, Willard HF, French FS and Wilson EM (1988): Cloning of human androgen receptor cDNA and localization to the X chromosome. *Science* 240: 327-330.

Lubahn DB, Joseph DR, Sar M, Tan JA, Higgs HN, Larson RE, French FS and Wilson EM (1988b): The human androgen receptor: complementary DNA cloning, sequence analysis and gene expression in prostate. *Mol Endocrinol* 2: 1265-1275.

Lund SD, Gallagher PM, Wang B, Porter SC and Ganschow RE (1991): Androgen responsiveness of the murine B-glucuronidase gene is associated with nuclease hypersensitivity, protein binding, and haplotype-specific sequence diversity within intron 9. *Mol Cell Biol* 11: 5426-5434.

Matsuura T, Demura T, Aimoto Y, Mizuno T, Moriwaka F and Tashiro K (1992): Androgen receptor abnormality in X-linked spinal and bulbar muscular atrophy. *Neurology* 42: 1724-1726.

Mainwaring WIP, Mangan FR, Feherty PA and Freifeld M (1974): An investigation into the anti-androgenic properties of the nonsteroidal compound, SCH 13521 (4'-nitro-3'-trifluoromethylisobutyrylanilide). *Mol Cell Endocrinol* 1: 113-128.

Marcelli M, Zoppi S, Grino PB, Griffin JE, Wilson JD and McPhaul MJ (1991): A mutation in the DNA binding domain of the androgen receptor gene causes complete testicular feminization in a patient with receptor positive androgen positive androgen resistance. *J Clin Invest* 87: 1123-1126.

McPhaul MJ, Marcelli M, Tilley WD, Griffin JE and Wilson JD (1991): Androgen resistance caused by mutations in the androgen receptor gene. *FASEB J* 5: 2910-2915.

Neri R, Peets E and Watnick A (1979): Anti-androgenicity of flutamide and its metabolite Sch 16423. *Biochem Soc Trans* 7: 565-569.

Newmark JR, Hardy DO, Tonb DC, Carter BS, Epstein JI, Isaacs WB, Brown TR and Barrack ER (1992): Androgen receptor gene mutations in human prostate cancer. *Proc Natl Acad Sci USA* 89: 6319-6323.

Pratt WB, Scherrer LC, Hutchison KA and Dalman FC (1992): A model of glucocorticoid receptor unfolding and stabilization by a heat shock protein complex. *J Steroid Biochem Mol Biol* 41: 223-229.

Prior L, Bordet S, Trifiro MA, Mhatre A, Kaufman M, Pinsky L, Wrogeman K, Belsham DD, Pereira F, Greenberg C, Trapman J, Brinkman AO, Chang C and Liao S (1992): Replacement of arginine 773 by cysteine or histidine in the human androgen receptor causes complete androgen insensitivity with different receptor phenotypes. *Am J Hum Genet* 51: 143-155.

Quarmby VE, Kemppainen JA, Sar M, Lubahn DB, French FS and Wilson EM (1990): Expression of recombinant androgen receptor in cultured mammalian cells. *Mol Endocrinol* 4: 1399-1407.

Quigley CA, Evans BAJ, Simental JA, Marschke KB, Sar M, Lubahn DB, Davies P, Hughes IA, Wilson EM and French FS (1992a): Complete AIS due to deletion of exon C of the AR gene highlights the functional importance of the second zinc finger of the AR *in vivo*. *Mol Endocrinol* 6: 1103-1112.

Quigley CA, Friedman KJ, Johnson A, Lafreniere RG, Silverman LM, Lubahn DB, Brown TR, Wilson EM, Willard HF and French FS (1992b): Complete deletion of the AR gene: definition of the null phenotype of the AIS and determination of carrier status. *J Clin Endo Metab* 74: 927-933.

Ris-Stalpers C, Trifiro MA, Kuiper GGJM, Jenster G, Romalo G, Sai T, van Rooij HCJ, Kaufman M, Rosenfield RL, Liao S, Schweikert HU, Trapman J, Pinsky L and Brinkmann AO (1991): Substitution of aspartic acid 686 by histidine or asparagine in the human androgen receptor leads to a functionally inactive protein with altered hormone binding characteristics. *Mol Endocrinol* 5: 1562-1569.

Sar M and Stumpf WE (1977): Androgen concentration in motor neurons of cranial nerves and spinal cord. *Science* 197: 77-80.

Sar M (1985): Application of the avidin-biotin complex technique for the localization of estradiol receptor in target tissues using monoclonal antibodies. In: *Techniques in Immunocytochemistry*, Bullock GR and Petrusz P, eds. New York: Academic Press, Vol 3, 43-54.

Sar M, Lubahn DB, French FS and Wilson EM (1990): Immunohisto-chemical localization of the androgen receptor in rat and human tissues. *Endocrinology* 127: 3180-3186.

Sar M, Hall S, Wilson EM and French FS (1993): Androgen regulation of Sertoli cells. In: *The Sertoli Cell*, Russell LD, and Griswold MD, eds. Clearwater FL, Cache River Press, pp 509-516.

Simental JA, Sar M, Lane MV, French FS and Wilson EM (1991): Transcriptional activation and nuclear targeting signals of the human androgen receptor. *J Biol Chem* 266: 510-518.

Sonnenschein C, Olea N, Pasanen ME and Soto AM (1989): Negative controls of cell proliferation: human prostate cancer cells and androgens. *Cancer Res* 49: 3474-3481.

Tan JA, Joseph DR, Quarmby VE, Lubahn DB, Sar M, French FS and Wilson EM (1988): The rat androgen receptor: primary structure, autoregulation of its messenger RNA and immunocytochemical localization of the receptor protein. *Mol Endocrinol* 2: 1276-1285.

Varkarakis MJ, Kirdani RY, Yamanaka H, Murphy GP and Sandberg AA (1975): Prostatic effects of a nonsteroidal antiandrogen. *Invest Urol* 12: 275-284.

Veldscholte J, Ris-Stalpers C, Kuiper GGJM, Jenster G, Berrevoets C, Claassen E, van Rooij HCJ, Trapman J, Brinkman AO and Mulder E (1990): A mutation in the ligand binding domain of the androgen receptor of human LNCaP cells affects steroid binding characteristics and response to antiandrogens. *Biochem Biophys Res Commun* 173: 534-540.

Warner CL, Griffin JE, Wilson JD, Jacobs LD, Murray K, Fischbeck KH, Dickoff D and Griggs RC (1991): X-linked spinomuscular atrophy: a kindred with associated abnormal androgen receptor binding. *Neurology* 41(Suppl 1): 313.

Wilding G, Chen M and Gelmann EP (1989): Aberrant response *in vitro* of hormone-responsive prostate cancer cells to antiandrogens. *The Prostate* 14: 103-115.

Wilson EM, Wright BT and Yarbrough WG (1986): The possible role of disulfide bond reduction in transformation of the 10S androgen receptor. *J Biol Chem* 261: 6501-6508.

Yarbrough WG, Quarmby VE, Simental JA, Joseph DR, Sar M, Lubahn DB, Olsen KL, French FS and Wilson EM (1990): A single base mutation in the androgen receptor gene causes androgen insensitivity in the Tfm rat. *J Biol Chem* 265: 8893-8900.

ESTROGEN RECEPTOR VARIANTS IN BREAST CANCER

Carlos A. Encarnacion
Suzanne A.W. Fuqua
Division of Medical Oncology
University of Texas Health Science Center

INTRODUCTION

It has been known for nearly one hundred years that there is an important relationship between estrogen (E_2) and carcinoma of the breast. In the late 1800's, Beatson reported that patients with inoperable breast tumors frequently responded to surgical castration (Beatson, 1896). Other manipulations designed to lower the concentrations of circulating E_2 or to block its effects are also effective, inducing remissions in about 30-40% of patients (Osborne, 1991). Alternative treatments are also available for patients with breast carcinoma, thus the importance of having certain biological factors that can predict which women are likely to respond to hormonal treatments. The estrogen receptor (ER) has successfully fulfilled this role, becoming the most important single factor predicting response to hormonal treatments. Many clinical studies show that those patients whose tumors express ER have a 70-80% chance of response, compared to only 10% in those that do not (Edwards et al, 1979; McGuire, 1980; McGuire et al, 1991).

STEROID HORMONE RECEPTORS
V. K. Moudgil, Editor
© 1993 Birkhäuser Boston

Multiple studies published over the last few years have enhanced our knowledge of the structure and function of the ER. This is of utmost importance in understanding certain clinical characteristics shown by patients treated with hormonal therapies. For example, the development of resistance to antiestrogenic drugs could be caused by several possible ER alterations. Thus it is important to understand the normal function of the receptor in order to hypothesize on its possible relevance in the development of antiestrogen resistance.

Structure and Function of the Estrogen Receptor

The presence of an intracellular estrogen-binding protein in target tissues was first identified in the late 1960's, after radiolabeled estrogens became available for research. It became evident that this protein, initially called estrophillin, was extremely important in the intracellular accumulation of E_2 and the modulation of its cellular effects (Gorski, et al, 1968; Jensen and DeSombre, 1972). It is now known that once ER binds to estrogen, it dimerizes and, through a complex tridimensional alteration, acquires a transcriptional activating function with specificity toward certain estrogen-responsive genes. In recent years, the genomic sequence of the human estrogen receptor (hER) was determined by cloning experiments using MCF-7 human breast cancer cells. The gene consists of 6322 nucleotides and encodes a 595 amino acid molecule with a molecular weight of 66 kDa (Walter et al, 1985; Green et al, 1986). ER belongs to the steroid hormone receptor superfamily, which includes progesterone, mineralocorticoid, glucocorticoid, thyroid hormone, vitamin D, and retinoic acid receptors (Evans, 1988). It has 6 distinct functional domains, labeled A through F, many of which show marked interspecies homology (Figure 1). For example, domains A, C, and E of the hER have a respective homology of 87%, 100%, and 94% with their counterparts in the chicken ER (Krust et al, 1986). Furthermore, comparison of these domains with those from the mouse, rat, xenopus, and trout also discloses considerable homology (White et al, 1987). These different domains confer upon ER the ability to recognize and bind its ligand, identify specific DNA segments of estrogen-regulated genes (the estrogen response elements or EREs), leading to their transcription stimulation, probably through the interaction with other transcription factors.

A/B Region

This domain has intrinsic transcriptional activation function. This activity, known as TAF-1 does not require estrogen, and is considerably more active in certain cell types. Studies in which a mutated ER lacking the hormone binding domain (where the principal E_2-dependent transcription function resides) were expressed in HeLa and CV1 cells, showed that this mutant had 1-5% of the transcriptional activity of the wild-type receptor (Kumar et al, 1987). Similar studies showed 2-10% of wild-type activity when NIH-3T3 D4 cells were used (Lees et al, 1989). In contrast, 60-70% of wild-type activity was reported when chicken embryo fibroblasts were transfected with a vitellogenin ERE construct (Tora et al, 1989). Furthermore, 100% wild-type activity was seen with a similar construct in a Saccharomyces yeast model system (White et al, 1988). On the other hand, experiments using an ER construct without the A/B domain also showed variable results. In HeLa cells, this A/B-deleted construct had wild-type transcription activating function with a vitellogenin-CAT ERE, whereas its activity was only 17% of wild-type when a pS2-CAT ERE was used (Kumar et al, 1987). Similar results were obtained in experiments using a mutated murine ER, also missing the A/B domain, and vitellogenin ERE (Lees et al, 1989).

Thus, the relative importance of TAF-1 appears promoter and cell type-specific. As we shall discuss further, this independence from estrogen activation could be related to the phenomena of estrogen independence and resistance to antiestrogens.

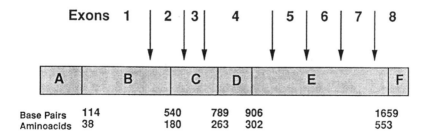

FIGURE 1. Schematic diagram of the estrogen receptor, displaying the functional domains and intron/exon boundaries. The amino acid and base pair positions are identified at the lower part of the diagram.

C Region

This 66 amino acid region is the most highly conserved between different species (Evans, 1988; Green et al, 1988). It is thought to fold into two zinc-binding structures (CI and CII) that interact with the DNA in a similar fashion to the zinc "fingers" of the ribosomal protein TFIIIA (Evans and Hollenberg, 1988). Each "finger" binds a zinc atom with 4 cysteine residues and is encoded by a different exon (Ponglikitmongkol et al, 1988). There are potentially functional differences between the two, since CI has several hydrophilic amino acids, whereas CII is richer in basic amino acids. These zinc "fingers" are important in the recognition of specific ERE sequences, shown by experiments in which a hER chimera, containing the C region of the glucocorticoid receptor, activated glucocorticoid-responsive genes in the presence of estrogen (Green an Chambon, 1987). Furthermore, experiments in which either CI or CII were substituted by their glucocorticoid counterparts demonstrated that CI largely determines the receptor's target specificity (Green et al, 1988). Only the amino acids in the base of this zinc "finger" appear to play a key role in this recognition (Mader et al, 1989; Unesono and Evans, 1989; Danielsen et al, 1989). It has thus been theorized that CI could be responsible for the recognition of specific nucleotide bases in the major groove of the DNA, while CII stabilizes this bond by associating to the phosphate backbone of the helix (Green and Chambon, 1991).

Albeit important, the two zinc "fingers" are not the only regions necessary for DNA binding. Experiments in which ERs with deletion of amino acids 250-264, or ER's truncated after amino acid 261 (both abnormalities distal to the zinc fingers) were used, showed that these receptors could not bind an ERE (Kumar and Chambon, 1988), suggesting that these segments have a role in further stabilizing the ER-DNA bond, either directly or perhaps by enhancing receptor dimerization.

Two additional functions are thought to be mediated by the C domain: collaboration with the E region for binding to heat shock protein 90 (hsp90) and signaling for the proper localization of the ER at the nucleus. Chambraud et al (1990) showed that ERs with deletions of amino acids 250-274 were unable to interact with heat shock protein 90 (hsp-90). Furthermore, these experiments also demonstrated that deletions of residues 250-270 resulted in receptors that were localized exclusively in the cytoplasm, whereas smaller deletions in the same area generated ERs that distributed evenly between the cytoplasmic and nuclear compartments.

D Region

This ER region is considered a molecular "hinge" which permits the hormone-mediated conformational change in ER's tridimensional structure. Mutations of this region induce no effects in the ER's estrogen binding or transcriptional activation capability in experiments using HeLa cells (Kumar et al, 1987). On the other hand, mutants of v-erbA (an oncoprotein with analogy to thyroid and estrogen receptors) with alterations in this region are inactive (Damm et al, 1987). Thus, at this time, is it not possible to disregard this region as one with only a purely allosteric function.

E Region

Domain E is a very large and complex region, composed of 250 amino acids and encoded by 5 different exons (Ponglikitmongkol et al, 1988). It encompasses subdomains implicated in hormone binding, transcriptional activation, dimerization, and estrogen-relieved transcriptional repression. It is commonly known as the hormone-binding domain, since it forms a hydrophobic pocket into which estrogen binds (Katzellenenbogen et al, 1987). It is believed that hormone binding induces a conformational change that rearranges important amino acid residues to the proper tridimensional position, activating ER's main transcriptional activating function (known as TAF-2). Once this happens, the active ER can interact with other transcription factors to initiate RNA synthesis. Most experts agree that this region is estrogen-dependent and thus requires the presence of ligand for its activity (Berry) et al, 1990; Kumar et al, 1987). Others have proposed that this segment also has some transactivating function in the unbound state, although this point is currently in debate (Tzukerman et al, 1990). New data suggest that not only E_2 binding, but also the association to DNA itself, induces conformational changes in the ER. These changes, detected by a decrease in the E_2 binding affinity, could account for part of the conformational alterations required to activate ER (Fritsch et al, 1992).

The fact that the unoccupied ER is incapable of binding DNA suggests that its C domain is protected, either by part of the ER itself or by other molecules. The E domain appears to play a major role in this phenomenon, since ER mutants lacking the E region will bind the ERE (Kumar et al, 1987). It has been suggested that the E region folds itself over the DNA-binding region, thus preventing the association between the unbound ER and the DNA (Martin et al, 1991). Alternatively, ER's

432

inability to bind DNA while unoccupied may be secondary to its association with other proteins, like hsp90. This protein, known to be associated to the unbound ER (Catelli et al, 1985), could physically obstruct the C region and prevent the association between ER and DNA (Binart et al, 1989). Both the E region and certain specific amino acids of the C region are apparently necessary for the interaction with hsp90 (Chambraud et al, 1990). Once the ligand binds to the ER, hsp90 dissociates from the receptor, exposing the zinc "fingers" at the DNA-binding domain. This is represented in Figure 2.

Finally, results from deletion and point mutagenesis analyses indicate that this domain is important in the dimerization process which occurs after hormone binding. This dimerization, in turn, appears necessary for the proper binding to DNA. Deletion mutants that lack the C-terminal part of the dimerization subdomain can not dimerize or bind DNA (Fawell et al, 1990). On the other hand, the E domain it is not absolutely required for DNA binding, since ERs with deletion of the entire E region can bind to the ERE, albeit with less affinity (Kumar et al, 1987).

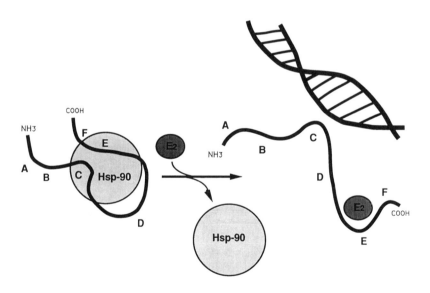

Figure 2. Representation of the ER alterations occurring after estrogen binding (see text). The functional domains are identified with the letters A-F.

Thus, the estrogen receptor is a very complex protein with multiple functions, all of which interact to initiate transcription of certain genes, ultimately resulting in cellular proliferation.

Antiestrogens and Breast Cancer

Since the binding of E_2 to its receptor activates certain processes that lead to tumor growth, blocking one or more of these steps should decrease or prevent this process. One of the most frequently used methods to achieve this purpose is the administration of antiestrogenic drugs like tamoxifen. Tamoxifen has become the mainstay hormonal treatment for post-menopausal women with breast carcinoma. A recent review by the Early Breast Cancer Trialist Group showed that postoperative tamoxifen, given for a period of more than two years decreased the mortality in women with early stage disease by 25% (Early Breast Cancer Trialist Group, 1992). In addition, tamoxifen can induce durable remissions in a good number of women with advanced ER positive breast cancer (Rose et al, 1985; Glauber and Kiang, 1992).

Tamoxifen is the transisomer of a triphenylethylene derivative with antifertility properties in laboratory animals (Harper and Walpole, 1967; Labhsetwar, 1970). It has a species-specific pharmacology, being purely antiestrogenic in chickens (Sutherland et al, 1977), partially estrogenic in rats (Jordan, 1987), and in short term studies, fully estrogenic in mice (Harper and Walpole, 1967; Terenius, 1971). With prolonged treatments, however, tamoxifen inhibits the stimulatory effects of E_2 in the target tissues of these animals. Besides being species-specific, tamoxifen is also tissue-specific, behaving as an antiestrogen in some tissues and as an estrogen in others (Jordan et al, 1987; Turner et al, 1987). This differential action permits the long term use of this drug in breast cancer patients without inducing adverse effects on bone or cholesterol profiles (Love et al, 1988; Torken et al, 1989, Bertelli et al, 1988).

Tamoxifen acts mainly by binding to the ER and inhibiting its action (Buckley and Goa, 1989) without preventing receptor dimerization or interaction to DNA. The recognition of the ERE by a tamoxifen-ER complex is similar to that by E_2-ER complex (Kumar and Chambon, 1988; Lees et al, 1989), but these tamoxifen-ER complexes have a different migration pattern than E_2-ER complexes in a gel-shift assay, suggesting that tamoxifen induces an alternate conformation that may not activate TAF-2.

A major problem related to the use of tamoxifen is the development of drug resistance. This phenomenon, which occurs in nearly all of the patients after a variable period of time (Katzellenenbogen, 1992), demands the use of alternative treatments, some of which have considerably more toxicity. Since a possible cause for this phenomenon is the selection of E_2-independent cells that no longer express functional ER, many researchers have evaluated the pattern of expression of both estrogen and progesterone receptors (PgR), using both biopsies from patients or cell cultures. Since most of these studies show inconsistent results, partly because of the different approaches and methods, we analyzed tumors from tamoxifen-treated patients who had recurred while taking the medication. ER and PgR were assayed both by ligand-binding assays (LBA) and by immunohistochemical (IHC) methods, the latter to eliminate potential false negative LBA's caused by receptor occupancy by endogenous or exogenous ligands. Our data show (Table 1) that a large percentage of patients remained ER positive, either by both LBA and IHC or by IHC alone. Furthermore, another group of patients were PgR positive, albeit ER negative, suggesting that these tumors had a functional, even if unmeasurable, ER. This ER could have been down regulated by the continued exposure to the ligand, in this case, tamoxifen (unpublished data). Thus, the development of tamoxifen resistance in most of our cases appears related to mechanisms other than a total loss of ER and estrogen dependence. This is supported by the clinical fact that patients who fail tamoxifen after an initial response, frequently do well with second or third line hormonal manipulations, thus their tumors are still under hormonal control.

TABLE 1. Receptor Phenotypes in Tamoxifen-Treated Patients

ER (LBA)	ER (IHC)	PgR	No. Patients
+	+	+	3
-	+	+	6
-	-	+	5
-	-	-	8

ER, Estrogen receptor; PgR, Progesterone receptor
LBA, Ligand binding assay; IHC, Immunohistochemistry

Our group has studied some of the possible mechanisms causing tamoxifen resistance. We have found that tamoxifen-resistant tumors have a lower intracellular concentration of the drug when compared to tamoxifen-sensitive tumors (Osborne et al, 1991). Furthermore, resistant tumors have higher concentrations of the cis isomer of tamoxifen as well as other estrogenic metabolites (Osborne, 1992; Wiebe, 1992). This suggests that these cells are able to eliminate the drug, either by expelling it out of the cell or metabolizing it to non-inhibitory or even stimulatory substances. In this review, however, we will concentrate on some of the ER RNA variants identified over the last years by several investigators, and their potential relationship to the development of tamoxifen resistance.

ER Variants

Three important things must be mentioned before detailing the specific ER variants. First, even though the data are suggestive, none of these variants has been definitely linked to the development of clinical antiestrogen resistance. Second, these variants frequently involve changes at exon/intron boundaries, suggesting RNA splicing variants rather than mutation of the DNA. In fact, some researchers have not identified ER gene mutations in many tumors showing these variants. Finally, in a large number of cases, wild-type ER is present together with the variants (Murphy, 1989; Graham, 1990; Fuqua, 1992; McGuire et al, 1991), which makes it even more difficult to predict their clinical relevance.

Changes within the A/B domain could be of clinical importance, since this domain has ligand-independent transcriptional activity. Tamoxifen-ER complexes can bind to DNA and thus an ER with an enhanced TAF-1 could start mRNA synthesis even if bound to an antiestrogen. Garcia et al (1988), using RNase protection assays, detected a variant ER mRNA with a substitution of a C for T at nucleotide 257. This created an amino acid substitution at position 86, resulting in a reduced E_2-binding affinity (Garcia et al, 1988; 1989). The clinical importance of this variant in the development of E_2 independence or tamoxifen resistance is unknown. At present, there are no additional data on ERs with mutations at the A/B domains that correlate with clinical antiestrogen resistance.

The C domain is another important area in which mutations could be clinically important. Point mutations in the zinc "fingers" could affect the ability of ER to properly recognize DNA sequences. Also, nonsense mutations or splicing abnormalities in this area could result in ERs that lack the E domain, and thus are able to bind freely to DNA and initiate transcription through TAF-1. A study in which gel shift assays were used to indirectly evaluate the C domain showed that more than 60% of the tumors expressing >100 fmol/mg of ER did bind to DNA, whereas the majority of those with < 100 fmol/mg did not bind (Scott, 1991). This binding capacity correlated with the expression of PgR by those tumors, suggesting that this differential binding had functional relevance. Whether this fact is clinically important, however, is not known. We have analyzed ER positive tumors and have detected a 6 base pair insertion at the exon 2/intron boundary (McGuire, 1991). This insertion generates an ER with decreased transcriptional activity. More recently, other ER alterations at exon 2 were detected by single strand conformational polymorphisms in two tumor samples from patients on tamoxifen treatment. The sequence and functional importance of these mutant ERs remains to be explored (Karnik et al, 1992).

Other variants have been found in which deletions and other abnormalities are seen at the exon/intron boundaries. A study with T47D human breast cancer cells found two variant ERs in which either exon 2 or 3 were deleted. The exon 3 variant, when transfected into normal cells, acted as a dominant-negative factor, inhibiting transcription (Wang and Misksech, 1991). Again, the clinical relevance of these variants is unclear, since similar mutations isolated from human breast cancer biopsies did not act in a similar fashion in other *in vitro* models (McGuire, 1991). Other changes in the C region have also been detected in breast cancer biopsies. Using Northern analysis, Murphy et al (1989) detected three variant mRNAs of 4.5, 3.8 and 2.5 kb, respectively. A cDNA library produced from these tumors also showed abnormalities at intron/exon boundaries in all three variants (Dotzlaw el al, 1992). One of them, clone 24, has a stop codon following an abnormal sequence of amino acids at the exon 3/intron boundary. This clone can bind to DNA and, since it lacks the E domain, theoretically could be able to activate transcription via TAF-1. *In vitro* transfection experiments with this clone, however, show that it is nonfunctional. The cDNA of the smaller species (2.5kb) also showed a change at the intron/exon boundary, this time at exon 2 (Dotzlaw, 1992). Again, *in vitro* experiments suggest that this clone is nonfunctional.

Thus, at present there is no clinical evidence that mutations in the C region are associated with tamoxifen resistance, but knowing that TAF-1 is tissue and promoter-specific, it is possible that these nonfunctional ER variants lacking the E domain may be clinically active and important *in vivo* as detailed below.

Table 2. Estrogen Receptor Variants at the Different Functional Domains

Domain	Reference	Source	RNA Alteration	Protein Alteration	ER Activity
A/B	Garcia, 1988,1989	Human tumors	C to T substitution at nucleotide 257	Alanine to valine substitution	Decreased E2 binding
C	Murphy, 1989	Human Tumors	Exon 3/intron variant(4.5kb)	Truncated protein	Non-functional
			Exon 2 deletion, exon3/intron variant(3.8kb)	Truncated protein	Unknown
			Exon 2/intron variant(2.5kb)	Truncated protein	Non-functional
	Wang, 1991	T47D cells	Exon 3 deletion	Truncated protein	Dominant negative
			Exon 2 deletion	Truncated protein	Unknown
	McGuire, 1991	Human tumors	6 bp insertion at exon 2/intron	Insertion of aspartic acid and arginine	Decreased activity
	Karnik, 1992	Tamoxifen-resistant tumors	Exon 2 variants	Unknown	Unknown
E	Fuqua, 1991	ER-/PgR+ human tumors	Exon 5 deletion	Truncated protein	Constitutive activating function
	Fuqua, 1992	ER+/PgR- human tumors	Exon 7 deletion	Truncated protein	Dominant negative

We have focused upon the E domain by analyzing tumors that show a discordance in their receptor expression (McGuire et al, 1991). Since PgR expression correlates with ER activity, tumors which are ER-positive/PgR-negative, or particularly those that are ER-negative/PgR-positive, could be more likely to have ER mutations of clinical importance.

Our group found an ER variant with a deletion of exon 5 in ER-negative/PgR-positive breast cancer biopsies (Fuqua, 1991). The variant mRNA codes for a 40 kd protein that lacks a functional hormone binding domain. Since it can not bind its ligand, this could explain why these tumors are ER-negative, since the LBA methods used routinely to measure ER would fail to detect these receptors. It was shown that this receptor variant has constitutional transcriptional activity which neither requires nor is enhanced by E_2. This suggests that this ER's activity is mediated by TAF-1. Interestingly, stable MCF-7 transfectants with the exon 5 deletion ER show *in vitro* antiestrogen resistance (Fuqua, unpublished data).

Finally, while studying ER-positive/PgR-negative tumors, a 3' truncated ER was discovered. This variant RNA, which lacks exon 7, was capable of binding DNA, as assessed by gel-shift assays, and acted as a dominant-negative factor when expressed by yeast cells (Fuqua, 1992). This suggests that a tumor with such a variant could develop independence to E_2 and perhaps show no antiestrogen inhibition. Similar variants assayed in T47D cells at much higher quantities, however, did not exhibit a dominant-negative behavior (Wang et al, 1991). A summary of ER variants at the different functional domains is presented in Table 2.

Future Directions

The acquisition of new insights into the structure and function of the ER has made a major impact in understanding the role of steroid hormones in the modulation of growth of carcinoma of the breast. This knowledge could also contribute to the theoretical explanation of the phenomena of E_2 independence and antiestrogen resistance. The presence of ER variants in both cell lines and human tumor samples suggests that these ERs may be important in the clinical behavior of these tumors. It is well known that many spontaneous mutations can confer a growth advantage, as well as drug resistance. Furthermore, some of the reported

ER variants represent alternative splicing rather than genetic mutations. Whether these splicing variations are true errors or highly-orchestrated alterations designed to gain selective advantage remains to be elucidated.

Unfortunately, the studies of receptor function done on *in vitro* models has many limitations and can frequently mislead. ERs with no activity in a cell line may be extremely active *in vivo*. By the same token, ERs with abnormal function *in vitro* may be completely unimportant *in vivo*. The additional confounding factor of the coexpression of both variant and wild-type ERs poses many questions on the real meaning of these variant receptors.

In summary, data is still insufficient to definitely link any of these ER alterations with clinical syndromes, such as tamoxifen resistance. However, there is a potential association between the two, sufficient to warrant further study of this matter. Additional knowledge in this field may prove important for the design of future treatment modalities directed to treat this disease.

REFERENCES

Beatson GT (1896): On the treatment of inoperable cases of carcinoma of the mamma: Suggestions for a new method of treatment with illustrative cases. *Lancet* 3: 104-107.

Berry M, Metzger D and Chambon P (1990): Role of the two activating domains of the oestrogen receptor in the cell-type and promoter-context dependent agonistic activity of the anti-oestrogen 4-hydroxytamoxifen. *EMBO J* 9: 2811-2818.

Bertelli G, Pronzoto P and Amoroso D (1988): Adjuvant tamoxifen in primary breast cancer: Influence on plasma lipids and antithrombin III. *Breast Cancer Res Treat* 12: 307-310.

Binart N, Chambraud B, Dumas B, Rowlands DA, Bigogne C, Levin JM, Garnier Baulieu EE and Castelli MG (1989): The cDNA-derived amino acid sequence of chick heat shock protein Mr 90,000 (Hsp 90) reveals a 'DNA like' structure: Potential site of interaction with steroid receptors. *Biochem Biophys Res Commun* 159: 140-147.

Buckley M and Goa KL (1989): Tamoxifen: A reappraisal of its pharmacodynamic and pharmacokinetic properties and therapeutic use. *Drugs* 37: 451-490.

Catelli MG, Binart N, Jung-Testas I, Renoir J-M, Balieu EE, Feramisco JR and Welch WJ (1985): The common 90-kd protein component of non-transformed '8S' steroid receptors is a heat shock protein. *EMBO J* 4: 3131-3135.

Chambraud B, Berry M, Redeuilh G, Chambon P and Baulieu EE (1990): Several regions of the human estrogen receptor are involved in the formation of receptor-heat shock protein 90 complexes. *J Biol Chem* 265: 20686-20691.

Damm K, Berg H, Graft T and Vennestrom B (1987): A single point mutation in erbA restores the transforming potential of a mutant avian erythroblastosis virus (AEV) defective in both erbA and erbB oncogenes. *EMBO J* 6: 375-382.

Danielsen M, Hinck L and Ringold GM (1989): Two amino acids within the knuckle of the first zinc finger specify DNA responsive element activation by the glucocorticoid receptor. *Cell* 57: 1131-1138.

Dotzlaw H, Alkhalaf M and Murphy LC (1992): Characterization of estrogen receptor variant mRNA from human breast cancer. *Mol Endocrinol* 6: 773-785.

Early Breast Cancer Trialists Collaborative Group (1992): Systemic treatment of early breast cancer by hormonal, cytotoxic or immune therapy. *Lancet* 339: 1-15.

Edwards DP, Chamness GC and McGuire WL (1979): Estrogen and progesterone receptor proteins in breast cancer. *Biochim Biophys Acta* 560: 457-486.

Evans RM (1988): The steroid and thyroid hormone receptor family. *Science* 240: 889-895.

Evans RM and Hollemberg SM (1988): Zinc fingers: Gilt by association. *Cell* 52: 1-3.

Fawell SE, Lees JA White, R and Parker MG (1990): Characterization and co-localization of steroid-binding and dimerization activities in the mouse estrogen receptor. *Cell* 60: 953-962.

Fritsch M, Welch RD, Murdoch FE, Anderson I and Gorski J (1992): DNA allosterically modulates the steroid-binding domain of the estrogen receptor. *J Biol Chem* 267: 1823-1828.

Fuqua SAW, Fitzgerald SD, Allred DC, Elledge RM, Nawaz Z, McDonell DP, O'Malley BW, Greene GL and McGuire WL (1992): Inhibition of estrogen receptor action by a naturally occurring variant in human breast tumors. *Cancer Res* 52: 483-486.

Fuqua SAW, Fitzgerald SD, Chamness GC, Tandon AK, McDonell DP, Nawaz Z, O'Malley BW and McGuire WL (1991): Variant human breast tumor estrogen receptor with constitutive transcriptional activity. *Cancer Res* 51: 105-109.

Garcia T, Lehrer S, Bloomer WD and Schachter B (1988): A variant estrogen receptor messenger ribonucleic acid is associated with reduced levels of estrogen binding in human mammary tumors. *Mol Endocrinol* 2: 785-791.

Garcia T, Sanchez M, Cox JL, Shaw PA, Ross JB, Lehrer S and Schachter B (1989): Identification of a variant form of the human estrogen receptor with an amino acid replacement. *Nucleic Acids Res* 17: 8364-8368.

Glauber JG and Kiang DT (1992): The changing role of hormonal treatment in advanced breast cancer. *Semin Oncol* 19: 308-316.

Graham ML II, Krett NL, Miller LA, Leslie KK, Gordon DF, Wood WM, Wei LL and Horwitz KB (1990): T47DCO cells, genetically unstable and containing estrogen receptor mutations, are a model for the progression of breast cancer to hormone resistance. *Cancer Res* 50: 6208-6217.

Green S and Chambon P (1987): Oestradiol induction of a glucocorticoid-responsive gene by a chymeric receptor. *Nature* 325: 75-78.

Green S and Chambon P (1991): The oestrogen receptor: From perception to mechanism. In: *Nuclear Hormone Receptors. Molecular Mechanisms, Cellular Functions, Clinical Abnormalities.* Parker, MG ed. San Diego: Academic Press.

Green S, Kumar V, Theulaz I, Wahli W and Chambon P (1988): The N-terminal DNA-binding 'zinc finger' of the oestrogen and glucocorticoid receptors determines target gene specificity. *EMBO J* 7: 3037-3044.

Green S, Walter P, Kumar V, Krust A, Bornet J-M, Argos P and Chambon P (1986): Human oestrogen receptor cDNA: Sequence, expression, and homology to v-erb-A. *Nature* 320: 134-139.

Gorski J, Toft DO and Shyamala G (1968): Hormone receptors: Studies on the interaction of estrogen with uterus. *Recent Prog Horm Res* 24: 45-80.

Harper MJK and Walpole AL (1967): A new derivative of triphenylethylene: Effect on implantation and mode of action in rats. *J Reprod Fertil* 13: 101-119.

Jensen EU and DeSombre ER (1972): Mechanisms of action of the female sex hormones. *Ann Rev Biochem* 41: 203-230.

Jordan VC, Phelps E and Lindgren JU (1987): Effects of antiestrogens on bone in castrated and intact female rats. *Breast Cancer Res Treat* 10: 31-35.

Jordan VC and Robinson SP (1987): Species-specific pharmacology of antiestrogens: Role of metabolism. *Fed Proc* 40: 1870-1874.

Karnik PS, Smith C, Tubbs RR and Bukowski RM (1992): Variant estrogen receptors in tamoxifen resistant human breast cancer. [abstract] *Breast Cancer Res Treat* 23: 184.

Katzellenenbogen BS (1992): Antiestrogen resistance: Mechanisms by which breast cancer cells undermine the effectiveness of endocrine therapy. *J Natl Cancer Inst* 83:1434-1435.

Katzellenenbogen BS, Elliston JF, Monsma FJ, Springer PA, Ziegler YS and Greene GL (1987): Structural analysis of covalently labeled estrogen receptors by limited proteolysis and monoclonal antibody reactivity. *Biochemistry* 26,2364-2373.

Krust A, Green S, Argos P, Kumar V, Walter P, Bornet J-M and Chambon P (1986): The chicken oestrogen receptor sequence: Homology with v-erb-A and the human oestrogen and glucocorticoid receptors. *EMBO J* 5: 891-897.

Kumar V and Chambon P (1988): The estrogen receptor binds tightly to its responsive element as a ligand-induced homodimer. *Cell* 55:145-156.

Kumar V, Green S, Stack G, Berry M, Jin J-R and Chambon P (1987): Functional domains of the human estrogen receptor. *Cell* 51: 941-951.

Labhsetwar AP (1970): Role of estrogen in ovulation. A study using the estrogen antagonist ICI 46,474. *Endocrinology* 87: 542-551.

Lees JA, Fawell SE and Parker MG (1989): Identification of two transactivation domains in the mouse oestrogen receptor. *Nucl Acids Res* 17: 5477-5488.

Love RR, Mazess RB, Tormey DC, Barden HS, Newcomb PA and Jordan VC (1988): Bone mineral density in women with breast cancer treated for at least two years with tamoxifen. *Breast Cancer Res Treat* 12: 297-301.

Mader S, Kumar V, DeVerneuil H and Chambon P (1989): Three amino acids of the oestrogen receptor are essential to its ability to distinguish an estrogen from a glucocorticoid-responsive element. *Nature* 338: 271-274.

Martin MB, Saceda M and Lindsey RK (1991): Estrogen and Progesterone Receptors. In: *Regulatory Mechanisms in Breast Cancer*. Lippman, M and Dickson, R eds. Boston: Kluwer Academic Publishers.

McGuire WL (1980): The usefulness of steroid hormone receptors in the management of primary and advanced breast cancer. In: *Breast Cancer: Experimental and Clinical Aspects.* Mouridson HT and Palshof, T eds. New York: Pergamon Press.

McGuire WL, Chamness GC and Fuqua SAW (1991): Estrogen receptor variants in clinical breast cancer. *Mol Endocrinol* 5: 1571-1577.

Murphy L and Dotzlaw H (1989): Variant estrogen receptor mRNA species detected in human breast cancer biopsy samples. *Mol Endocrinol* 3: 687-693.

Osborne CK (1991): *Receptors.* In: *Breast Diseases.* Harris, JR, Hellman, S, Henderson, IC, and Kinne, DW eds. Philadelphia. J.B. Lippincott Co.

Osborne CK, Coronado E, Allred DC, Wiebe V and DeGregorio M (1991): Acquired tamoxifen resistance: Correlation with reduced breast tumor levels of tamoxifen and isomerization of trans-4 hydroxytamoxifen. *J Natl Cancer Inst* 83:1477-1482.

Osborne CK, Wiebe VJ, McGuire WL, Ciocca DR and DeGregorio MW (1992): Tamoxifen and the isomers of 4-hydroxytamoxifen in tamoxifen-resistant tumors from breast cancer patients. *J Clin Oncol* 10: 304-310.

Ponglikitmonkgol M, Green S and Chambon P (1988): Genomic organization of the human oestrogen receptor gene. *EMBO J* 7: 3385-3388.

Rose C, Thorpe SM, Anderson KW, Pedersen BV, Mouridsen HT, Blichert-Toft M and Rasmussen BB (1985): Beneficial effect of tamoxifen in primary breast cancer patients with high estrogen receptor values. *Lancet* 1: 16-18.

Scott GK, Kushner P, Vigne JL and Benz CC (1991): Truncated forms of DNA-binding estrogen receptors in human breast cancer. *J Clin Invest* 88: 700-706.

Sutherland RL, Merter J and Baulieu EE (1977): Tamoxifen is a potent "pure" antiestrogen in the chick oviduct. *Nature* 267: 434-435.

Terenius L (1971): Structure-activity relationship of antiestrogens with regard to interaction with 17b-oestradiol in the mouse uterus and vagina. *Acta Endocrinol* 66: 431-447.

Tora L, White J, Brou C, Tasset D, Webster N, Scheer E and Chambon P (1989): The human estrogen receptor has two independent nonacidic transcriptional activation functions. *Cell* 59: 477-487.

Torken S, Siris E, Seldin D, Flaster E, Hyman G and Lindsay R (1989): Effects of tamoxifen on spinal bone density in women with breast cancer. *J Natl Cancer Inst* 87: 1086-1088.

Turner RT, Wakely GK, Hannon KS and Bell NH (1987): Tamoxifen prevents the skeletal effects of ovarian deficiency in rats. *J Bone Miner Res* 2: 449-456.

Tzukerman M, Zhang X-K, Hermann T, Wills KN, Graupner G and Pfahl M (1990): The human estrogen receptor has transcriptional activator and repressor functions in the absence of ligand. *The New Biologist* 2: 613-620.

Umesono K and Evans R (1989): Determinants of target gene specificity for steroid/thyroid hormone receptors. *Cell* 57: 1139-1146.

Walter P, Green S, Greene G, Krust A, Bornet J-M, Jeltsch JM, Staub A, Jensen E, Scrace G, Waterfield M and Chambon P (1985): Cloning of the human estrogen receptor cDNA. *Proc Natl Acad Sci USA* 82: 7889-7893.

Wang Y and Micksicek RJ (1991): Identification of a dominant negative form of the human estrogen receptor. *Mol Endocrinol* 5: 1707-1715.

White R, Lees JA, Needham M, Ham J and Parker MG (1987): Structural organization and expression of the mouse estrogen receptor. *Mol Endocrinol* 1: 735-744.

White JH, Metzger D and Chambon P (1988): Expression and function of the human estrogen receptor in yeast. *Cold Spring Harbor Symp Quant Biol* 53: 819-828.

Wiebe VJ, Osborne CK, McGuire WL and DeGregorio MW (1992): Identification of estrogenic tamoxifen metabolite(s) in tamoxifen resistant human breast tumors. *J Clin Oncol* 10: 990-994.

EMERGING AND NOVEL SYSTEMS

HUMAN BONE CELLS: NEWLY DISCOVERED TARGET CELLS FOR SEX STEROIDS

Thomas C. Spelsberg
Merry Jo Oursler
James P. Landers
Malayannam Subramaniam
Steven A. Harris
Department of Biochemistry and Molecular Biology
Mayo Clinic and Graduate Schools of Medicine

B. Lawrence Riggs
Department of Medicine, Division of Endocrinology
Mayo Clinic and Graduate Schools of Medicine

ABSTRACT

This presentation was planned to introduce the novice to bone tissue as a target tissue for sex steroids. Recent advances in this and other laboratories have now implicated both osteoblasts and osteoclasts as target cells for steroid hormones including estrogen (E_2). It is, therefore, the hypothesis of this laboratory that E_2 plays a major role in the coupling of bone resorbing osteoclasts and bone forming osteoblasts. This coupling appears to involve chemical signals which maintain homeostasis between the activities of these two cell types and abnormalities in coupling have been implicated in osteoporosis. Our laboratories discovered E_2 receptors (ER) (as well as progesterone and androgen receptors) in normal human

STEROID HORMONE RECEPTORS
V. K. Moudgil, Editor
© 1993 Birkhäuser Boston

osteoblast-like (hOB) cells, avian osteoclasts (avOC), and recently in human osteoclast-like giant cell tumors (hGCTs). We have identified no E_2 effects on cell proliferation or matrix protein production in hOB cells, but have identified a significant E_2 induction of TGF-β in these cultured hOB cells. TGF-β is an important cytokine likely to play a key role in E_2 action on bone. Since TGF-β is reported to be regulated by Fos/Jun, we then investigated and have preliminary results which demonstrate that E_2 rapidly and markedly regulates the steady state mRNA levels of the c-fos gene in the hOB cells and, to a lesser degree, the other nuclear proto-oncogenes. We are currently examining the nature of the E_2 regulation of the c-fos gene in hOB cells and the activation of the TGF-β factor by steroids.

We have previously demonstrated that avOC contain high levels of ER and respond to E_2 treatment with a dose-dependent decrease in *in vitro* resorption of [^{3}H]proline labelled bone particles. To more accurately assess E_2 influence on OC activity, the specificity of the E_2 modulation of resorption levels was recently determined using a quantitative pit resorption assay. Treatment with E_2 significantly decreased the number of avOC resorption pits formed when compared with either vehicle or 17-αE_2 treatment. Cotreatment with E_2 and hydroxytamoxifen (a complete E_2 antagonist in birds) abrogated the influence of E_2 on resorption activity. We have recently found that hGCTs respond to E_2 similarly to the avOC. These *in vitro* actions of E_2 in reducing bone resorption mimics the biological responses of E_2 on human bone resorption *in vivo*. In order to elucidate the mechanism by which E_2 inhibits OC resorbing activity, which involves OC lysosomal activity, the effects of E_2 on the steady state mRNA levels of expression of certain lysosomal proteins has been demonstrated. It is the ultimate goal to identify the growth factors involved in the coupling process between OB and OC and determine which are regulated by E_2.

INTRODUCTION

Although it has been recognized for many years that E_2 is a key component in the maintenance of normal bone balance, the mechanisms by which E_2 exerts its influence on bone have remained unresolved. Recent identification of ER in both bone-forming osteoblasts and bone-resorbing osteoclasts has opened up exciting new areas of research on the direct effects of E_2 on both osteoblasts and osteoclasts. This chapter outlines the research performed in the authors' laboratories. Additional sections outlining the work performed by other investigators' laboratories is included to give a more comprehensive review of the field to the readers.

Outline of Bone Turnover (Remodeling)

Throughout life, bone is continuously remodeled at discrete locations. These locations have been termed bone remodeling units (Parfitt, 1979). The rate of turnover of bone mass is determined by the number of bone remodeling units in the skeleton. The maintenance of the bone mass requires a balance of remodeling at each of these units. At the beginning of each remodeling cycle, the bone lining cells retract and expose the bone surface at the site to be remodeled, osteoclast precursors migrate to the bone, attach, and fuse to form bone resorbing osteoclasts. After bone resorption is completed, the osteoclasts are replaced by osteoblasts which then synthesize new bone matrix to refill the resorption space. Bone imbalance is due to a remodeling imbalance whereby one of two or both processes may occur: the osteoclasts could produce an unusually deep resorption space, or resorb a greater area of bone, or the osteoblasts could fail to completely refill the normal resorption space. Thus, either an enhanced activity of the bone resorbing osteoclasts or a decreased activity of bone forming osteoblasts is involved in the imbalance, or both. In general, if a remodeling imbalance exists, the degree of bone loss will be exacerbated if there are more bone remodeling units, that is, if bone turnover activity is high.

General Action of Estrogen on Bone Remodelling

Since E_2 suppresses bone remodeling and encourages a balance between osteoblastic and osteoclastic activity, it is believed to play an important role in maintaining bone mass in the adult women (Parfitt, 1979). When E_2 levels are deficient such as in post-menopausal females, there is an increase in the frequency of new bone remodeling units as well as an increase in remodeling imbalance. This remodeling imbalance is believed to be the result of an increase in osteoclastic activity wherein the osteoclasts construct deeper resorption spaces (Parfitt, 1979). However, there is also some evidence that the ability of the osteoblasts to refill these spaces may be impaired by E_2 deficiency (Parfitt, 1979). In any case, the deeper resorption spaces result in perforation of trabecular plates and loss of architectural elements especially in cancellous bone, further weakening the skeleton in regions such as the vertebrae and distal forearm which contain large amounts of cancellous bone (Parfitt et al, 1983). Interestingly, these abnormalities in postmenopausal women appear to be more severe in women with postmenopausal osteoporosis resulting in the increase in the frequency of bone fractures (Eriksen et al, 1990).

Over a lifetime, women loose about 50% of their cancellous bone and about 35% of their cortical bone (Riggs et al, 1981). It is unclear how much of this bone loss is related to E_2 deficiency and how much is related to age- and environmentally-related processes. Certainly, the genetics of an individual plays a role in the extent of bone loss and the degree of osteoporosis. In any case, based on currently available data, an estimate that 25% of cancellous bone loss and 15% of cortical bone is lost due to E_2 deficiency. It is now believed that excessive bone loss due to E_2 deficiency is a major cause of osteoporosis (Lindsay, 1981). Long-term E_2 treatment reduces the incidence of fractures of the vertebrae, distal forearm, and hip by about 50% from the incidence in patients receiving no E_2 (Ettinger et al, 1985; Hutchinson et al, 1979; Weiss et al, 1980).

Many studies in humans have demonstrated that the early and predominant effect of E_2 on bone remodeling is a decrease in the amount of bone resorption (Heaney, 1965; Parfitt, 1979; Riggs et al, 1972). These late-occurring decreases in bone formation following E_2 treatment probably represents secondary consequences of the initial decrease in bone resorption. The late events are likely to be related to the coupling of bone formation to bone resorption (Ivey & Baylink, 1981). Figure 1 represents the authors' model of the general process of direct coupling between the bone forming OB and bone resorbing OC via paracrine and autocrine factors. An indirect model had originally been proposed whereby paracrine factors trapped in bone matrix would be released during resorption to regulate osteoclast activity (McSheedy and Chambers, 1986a).

Studies in the Authors' Laboratories on Estrogen Receptors and Estrogen Action on Osteoblasts

Despite the overwhelming evidence that E_2 modulates bone remodeling, until recently it had not been possible to demonstrate ER in bone cells (Chen & Feldman, 1979; van Paassen et al, 1978). These negative results led to the belief that the effect of E_2 on bone was indirect. In 1988, two groups (Eriksen et al, 1988; Komm et al, 1988) reported the presence of high affinity ER in cultured bone cells of the osteoblast lineage. Thus, sensitive and specific techniques have demon-

Sex Steroids in Bone Biology

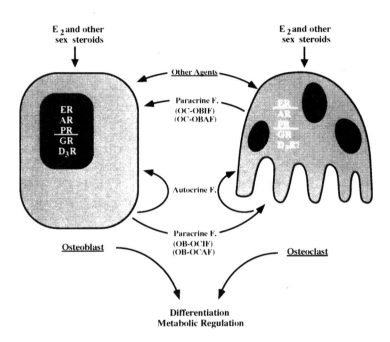

FIGURE 1. *Model of role of paracrine and autocrine factors in the estrogen action on bone cells.* E_2 = estrogen; F = factors; OB-OCIF = osteoblast derived osteoclast inhibitory factors; OB-OCAF = osteoblast derived osteoclast activating factors; OC-OBIF = osteoclast derived osteoblast inhibitory factors; OC-OBAF = osteoclast derived osteoblast activating factors. ER = estrogen receptor complex; AR = androgen receptor complex; PR = progesterone receptor complex; GR = glucocorticoid receptor complex; D_3R = 1,25 dihydroxyvitamin D_3 receptor complex.

452

strated that bone cells of the osteoblast lineage have definite, albeit relative low, concentrations of ER.

In subsequent studies we have demonstrated that the presence of progesterone receptors (Eriksen et al, 1988), androgen receptors (Colvard et al, 1989), and glucocorticoid receptors (Subramaniam et al, 1992) in the normal human osteoblasts (hOB). In the latter case, the glucocorticoids were shown to regulate the expression of the c-fos, c-myc,

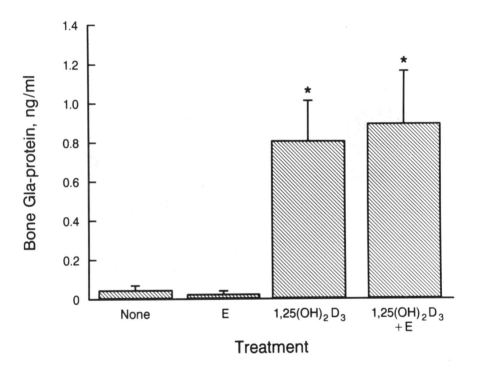

FIGURE 2. *Effect of various treatments of BGP release from hOB cells as measured by radioimmunoassay.* Treatments included 17β-estradiol (E) at 1 pM to 10 nM (data shown for 10 nM), 1,25-dihydroxyvitamin D_3, and both together (data shown for 1,25-dihydroxyvitamin D_3 in combination with 10 nM 17β-estradiol). * = Significantly ($P<0.05$) different from control. Control release was 0.04 ± 0.02 ng/ml per 10^4 cells per 24 h; after 1,25-dihydroxyvitamin D_3 treatment it increased to 0.83 ± 0.21 ng/ml ($P<0.06$; n=11). (Taken with permission from Keeting et al, 1991; *J Bone and Min Res* 6:297-304).

and c-jun nuclear proto-oncogenes and the gene for the bone matrix protein, alkaline phosphatase. We have performed an extensive analysis of E_2 action on normal human osteoblasts (Keeting et al, 1991). Although osteoblasts contain ER, it is unclear whether estrogen has direct effects on osteoblast proliferation and differentiation. We evaluated the effects of E_2 treatment (1 pM to 10 nM) on the proliferation and differentiation of cultured normal adult human cells that expressed many of the phenotypic characteristics and hormonal sensitivities of mature osteoblasts (hOB cells). Treatment of hOB cells with E_2 for as long as 144 h did not affect the rate of DNA synthesis and had minimal, if any, effects on differentiated functions. Whereas alkaline phosphatase activity was increased by nearly twofold ($P<0.01$) when the hOB cells were treated with 1 nM 1,25-dihydroxyvitamin D_3 ([1,25-$(OH)_2D_3$], treatment with E_2 had no effect when treated alone and did not affect the cells' response to 1,25-$(OH)_2D_3$. Similarly, as shown in Figure 2 the release of bone gla protein (BGP) (osteocalcin), was increased by treatment with 1,25-$(OH)_2D_3$ ($P<0.05$), but E_2 treatment did not affect this response. Cellular levels of mRNA for alkaline phosphatase and BGP were also not altered by E_2 treatment. We conclude that E_2 treatment does not have major effects on the growth or bone forming activities of cultured hOB cells. These results are consistent with previous observations *in vivo* that indicate that E_2 acts principally to decrease bone resorption, not to modulate its formation. More recent studies, however, have shown the E_2 does regulate the expression at the level of transcription of the nuclear proto-oncogene c-fos (Harris, et al, 1992). This nuclear protein probably regulates, in turn , the TGF-β gene expression in the hOB cells. The latter is described later in this chapter.

Studies in Other Laboratories on the ER and E_2 Action on Osteoblasts.

Subsequent to the demonstration of ER in bone cells, various investigators have reported a variety of direct effects of E_2 on proliferation as well as the synthesis of enzymes and bone matrix proteins by osteoblast-like cells *in vitro*. However, the reported changes resulting from E_2 treatment have been relatively small and, often, contradictory. In cells of the osteoblastic lineage from a variety of sources, E_2 has been reported to enhance (Ernst et al, 1989; Ernst et al, 1988), to inhibit (Gray et al, 1987; Watts et al, 1989), or to have no effect on proliferation. E_2 has been reported to increase (Ernst et al, 1989; Ernst et al, 1988; Gray et al, 1987; Komm et al, 1988), to decrease (Watts et al, 1989), or to have no effect on differentiated function (Keeting et al, 1991). Several factors may contribute to these divergent and conflicting results. The main factor causing these differences is most likely the wide variety of cell model systems used in these studies and their different states of differentiation. Further, transformed cell lines often give responses that are not representative of the mature osteoblast phenotype. Similarly, fetal or neonatal cells may give responses that differ from those of mature adult osteoblastic cells. Also, the low and variable numbers of ER in some of the cell lines may have confounded results. Finally, because the hOB cells were derived from unrelated donors, there is a large variability in responses. These variations may obscure the relatively small responses that were found in the clonal cell lines. The lack of large and consistently reproducible responses, however, suggest that a direct modulation of osteoblast function may not be the major effect of E_2 action on bone cells.

Estrogen Receptors and Estrogen Action on Osteoclasts

Recently, using an immunomagnetic separation method to isolate highly purified viable avian osteoclasts (Oursler et al, 1991a; Collin-Osdoby et al, 1991), our laboratories, in collaboration with those of Dr. Phil Osdoby of Washington University, St. Louis, Missouri, demonstrated that not only do avOC have ER but they respond to E_2 by reducing their bone resorbing activity using the bone particle assay (Oursler et al, 1991b). The bone resorption activity by the isolated avOC was inhibited within 24 hours by E_2 at concentrations as low as 10^{-10}M E_2. This inhibition was reversible upon E_2 withdrawal.

As shown in Figure 3 (Panel A), using the more sensitive quantitation pit resorption assay, Oursler et al, (Oursler et al, 1993a) recently demonstrated that E_2 inhibits the bone resorptive activities of avOC by reducing the number of pits on the bone slices. The pit areas were unaffected. The inactive steroisomers of E_2, $17\alpha E_2$, showed no activity and the anti-E_2 tamoxifen (TAM), or cycloheximide individually blocked the E_2 inhibition of OC resorption. These data indicate that E_2 inhibits OC initiation of bone resorption (Pit No.), but does not alter the rate of bone resorption (pit area) once the OC have initiated resorption.

In attempts to understand the mechanism of E_2 inhibition of avOC bone resorption, the action of E_2 on lysosome activity was examined. Osteoclasts tightly attach to bone and form a sealed compartment into which both lysosomal enzymes and proteins are secreted (see review by Vaes, 1988). Thus, lysosomal enzyme synthesis and release may have a prominent role in the process of bone resorption. As shown in Figure 3 (Panel B), Oursler et al (1993a) reported that E_2 treatment resulted in the corresponding decrease in the mRNA levels of LEP 100 (a lysosomal membrane protein) and lysozyme, a lysosomal protease. Again, the 17α E_2 was ineffective and the TAM and cycloheximide blocked the E_2 effects on the lysosomal protein gene expression. Further, the E_2-induced reduction of these mRNA levels was reversible once E_2 was removed (Figure 3).

As shown in Figure 4, the effects of E_2 on the expression of the avOC lysosomal protein genes was typical of a late responding structural gene with a significant reduction noted between 8 and 18 hours with a nadir of ~35% of control values after 50 hr of continuous E_2 treatment (Oursler et al, 1993a).

The AP-1 transcription factor complex composed of the FOS/JUN protein homo-hetero-duplexes, is known to regulate many genes as well as the steroid regulation of genes (for a review, see Ransone and Verma, 1990; Landers and Spelsberg, 1992). It was of interest to determine whether or not E_2 regulates the expression of these genes in the avOC as well as the rapidity of this action. As shown in Figure 5, E_2 increased the steady state mRNA levels of the *c-fos* and *c-jun* nuclear proto-oncogenes within 30 minutes in a steroid dose-dependent manner. Thus, in a chronological view, E_2 acts on avian OC by immediately regulating the *c-fos/c-jun* gene expression, followed by the inhibition of lysosome protein gene expression and lysosomal activities which then results in the inhibition of bone resorbing activities of the OC cells.

Studies are now underway to discern whether human OC are also target cells for E_2 and would respond to E_2 as do the avian OC. While this is probable, it is also possible that significant differences in bone physiology exists between animals which undergo cyclical egg laying with a corresponding increase in skeletal turnover and mammalian species which do not exhibit these cycles. We have begun to analyze human OC. Recent studies by Oursler et al (1993b) indicate the multinucleated OC-like cells isolated from human giant cell tumors, not only contained functional ER but also responded to E_2 via inhibiting lysosomal protein gene expression and bone resorption. It was shown that other non-OC cell types from these same tumors did not contain ER and did not display any E_2 effects on the lysosomal protein gene expression. These cells also showed no bone resorptive activities with or without E_2 treatment (Oursler et al, 1993b). Thus, the OC-like cells in these tumors are surely human osteoclasts causing the bone resorption around these tumors. If these results are substantiated, it would support the hypothesis that E_2 acts directly on mature human and animal OC to inhibit bone resorption.

Role of Growth Factors in Estrogen Action on Bone Cells
Studies in other laboratories

As depicted in Figure 1, studies in the field of bone metabolism and bone diseases suggest that growth factors and cytokines in general are mediators of a complex intercellular chemical communication network between osteoblasts and osteoclasts in a process termed coupling. Coupling of resorption and formation occurs during bone turnover in normal young adults wherein the amount of bone removed by osteoclasts

FIGURE 3. *Summary action of estrogen on avian osteoclasts. Panel A* represents summary data from the quantitative pit formation assay. Isolated osteoclasts were cultured on bone slices with either ethanol (vehicle), 10^{-8}M 17α estradiol (17α E_2), 10^{-8}M 17β estradiol (17β E_2) or combinations of 10^{-8}M of 17β E_2 and hydroxytamoxifen (17β E_2/TAM) or 10^{-8}M 17β E_2 + 3μ moles cycloheximide. After 18 hours, bone slices were harvested and processed as described in the Methods section. The number of pits per OC and pit area were determined on each individual slice as described elsewhere (Oursler et al, 1993a). Data presented are the mean (n=16) resulting from four separate experiments. *Panel B* represents the effects of estrogen on Lep-100 and lysozyme mRNA levels in the OC. Isolated osteoclasts were incubated for 18 hours with one of the following: vehicle, 10^{-8}M 17β E_2, 10^{-8}M 17α E_2 (inactive isomer), or a combination of 10^{-8}M 17β E_2 + TAM, or 10^{-8}M E_2 + 3μ moles cycloheximide. The total RNA isolated and blotted as described in the Methods section. Blots were probed to determine the levels of lysozyme and Lep-100 mRNA, followed by probing for the amount of 18S ribosomal RNA in each lane as described elsewhere (Oursler et al, 1993a). Studies were carried out in four separate experiments. Densitometric analysis was performed on Northern blots as described elsewhere (Oursler et al, 1993a). Quantitative analyses of the four Northern blots were standardized to 18S ribosomal RNA (rRNA) levels in each blot and the resultant ratios of mRNA/rRNA plotted as the mean (n=4). Statistical analyses of the replicate values in the original paper showed statistical significance. (Taken with permission from Oursler et al, 1993, *Endocrinology,* 132: 1373-1380).

E_2 Action on Avian OC

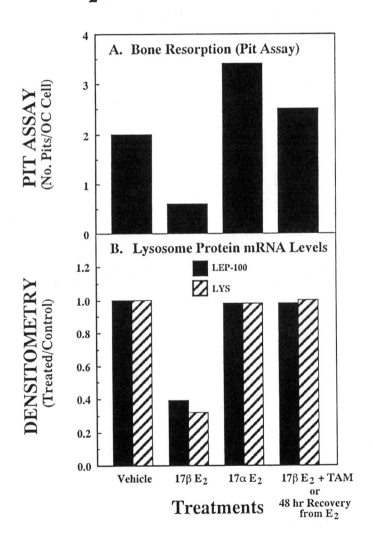

FIGURE 3. Summary action of estrogen on avian osteoclasts.

458

is precisely replaced by osteoblasts. The two cell types work together in the phenomenon which is fundamental to maintenance of normal healthy bone. Progress towards understanding the coupling between osteoblast-mediated bone formation and osteoclast-mediated bone resorption has begun. A future important goal in this area is the identification of the osteoclast- or osteoblast-derived regulatory factors (cytokines) which act on osteoclast or osteoblast as autocrine or paracrine factors (see Figure 1). It is possible that estrogen regulates the production of one or more of these regulatory factors in osteoblasts and osteoclasts. Thus, cytokines involved in osteoblast-osteoclast intercellular

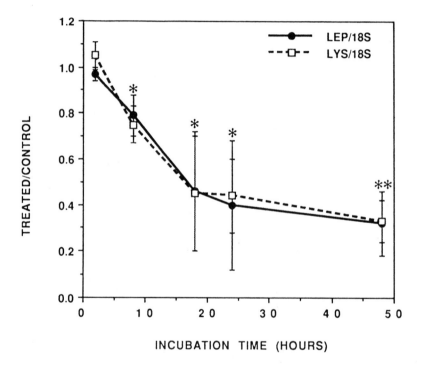

FIGURE 4. *Chronology of the effects of estrogen on the lysosomal protein mRNA levels.* The experiment was carried out essentially as described in the legend of Figure 3 (Panel B) except samples were taken at various times of incubations of E_2 with the OC cells. The (*) represents a significant difference from control (0 time) values of $P<.01$ and (**) represents a significance of $P<.001$. (Taken with permission from Oursler et al, 1993, *Endocrinology,* 132: 1373-1380).

communication are potential coupling factors (Felix et al, 1989; Fuller et al, 1991; Garrett & Mundy, 1989; Ishimi et al, 1990b; Lerner & Gustafson, 1980; Mundy & Bonewald, 1989; Norrdin et al, 1990). Coupling factors were originally defined as those factors which were synthesized by osteoblasts and other cells, and then incorporated into the bone matrix during bone formation for later release and activation during osteoclast bone resorption. However, a more direct communication between the osteoblast and osteoclast very likely exists, involving effects

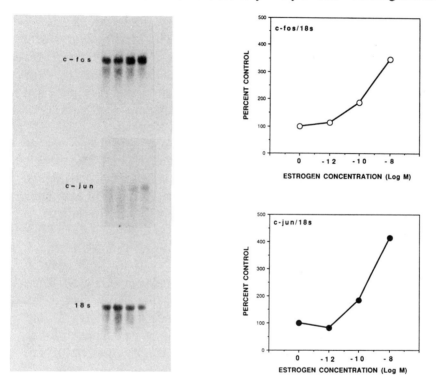

FIGURE 5. *Northern blot analysis for estrogen responses.* Isolated osteoclasts were cultured for 30 min in the presence of rat bone particles with the indicated dose of estrogen. Total RNA was isolated and analyzed by Northern blotting and probing with cDNAs to determine RNA levels for *c-fos, c-jun* and 18S rRNA as described previously (Oursler et al, 1991). Densitometric quantitation was performed on each blot and the ratio of *c-fos* or *c-jun* mRNA levels to 18S rRNA was performed and plotted against the dose of E_2 which was incubated with the OC cells. (Taken with permission from Oursler et al, 1991, *Proc Natl Acad Sci USA*, 88: 6613-6617).

on cell activities including differentiation. Thus, some factors may act directly on the bone cells immediately after secretion whereas factors deposited in the bone matrix would influence bone metabolism at a later time after release and/or activation during bone resorption of that matrix. One can envision multiple tight controls with both a positive (activating) and negative (inhibitory) regulation in this process as depicted in Figure 1.

Past studies suggest that osteoblast are the target for a variety of hormones and growth factors, even those whose major action is on bone resorptive activity in osteoclasts (Chambers, 1991; McSheedy & Chambers, 1986a; McSheedy & Chambers, 1986b; Murrills et al, 1990). However, more recent studies have shown that osteoblasts are target cells for both E_2 and other steroids (Eriksen et al, 1988; Colvard et al, 1989; Subramaniam et al, 1992). As depicted in Figure 1, it is speculated that agents, such as E_2, which regulate bone resorption, also act via the osteoblast-derived osteoclast regulatory growth factors.

Because of the major effect of E_2 on bone is to decrease bone resorption rather than to modulate bone formation, it is possible that the action of E_2 on osteoblasts could well regulate osteoclast function through indirect mechanism such as cytokines. With the exception of calcitonin, most factors that regulate bone resorption do so indirectly through an osteoblast-mediated pathway (Akatsu et al, 1991; Fujita et al, 1990; Gowen & Mundy, 1986; Ishimi et al, 1990a; Jilka, 1986; McSheedy & Chambers, 1986a; McSheedy & Chambers, 1986b; McSheedy & Chambers, 1987; Pfeilschifter et al, 1989; Thomson et al, 1987). Thus, E_2 could inhibit the release of osteoclast stimulatory factors or, alternatively, could enhance the release of osteoclast inhibitory factors. Based on their action in experimental animals, in organ culture, or in cell culture, a number of cytokines do modulate the coupling process in bone resorption. Cytokines that have been reported to increase osteoclastic activity include granulocyte-macrophage stimulatory factor (GM-CSF) (Lorenzo et al, 1987), macrophage-colony stimulating factor (M-CSF) (Corboz et al, 1992), tumor necrosis factor-α (TNF-α) (Ishimi et al, 1990a), interleukin-1 (IL-1) (Kawanto et al, 1989; Sabatini et al, 1988), and IL-6 (Ishimi et al, 1990a). The cytokines that have been reported to decrease bone resorption include TGF-β (Dieudonne et al, 1991b; Pfeilschifter et al, 1988) and IL-4 (Watanabe et al, 1990). Also, IL-8 has been shown to be involved in chemotaxis and cell-cell interactions and, thus, could affect osteoblast-osteoclast interactions and osteoclast recruitment (Streiter et al, 1989).

A role for activator verses inhibitor factors is a common theme in cell biology and is readily exemplified by the transforming growth factor alpha (TGF-α) activation of - and transforming growth factor beta (TGF-β) inhibition of - cell proliferation of MCF-7 breast cancer cells (Dickson et al, 1986; Knabbe et al, 1987). The "stimulators" of osteoclast functions could be TGF-α, or the low molecular weight factor X of Chambers and coworkers (Chambers, 1991; Chenu et al, 1988), or other soluble factors such as bone morphogenic proteins. TGF-β could be an "inhibitor" of osteoclast function. Interestingly, TGF-α has been shown to stimulate bone resorption and related osteoclast activities while TGF-β has been shown to either stimulate or inhibit bone resorption, depending on the particular target bone involved and the concentration of this growth factor (Bonewald & Mundy, 1990; Marcelli et al, 1990a; Oreffo et al, 1988; Oreffo et al, 1990; Pfeilschifter et al, 1988; Tashjian Jr. et al, 1985).

Bone matrix-derived factors, such as TGF-β and insulin-like growth factor-II (IGF-II), which regulate osteoblast cell differentiation and activity, have been implicated as paracrine factors which might also regulate osteoclasts when released by them during resorption (Bonewald & Mundy, 1990; Dieudonne et al, 1991a; Ernst & Froesch, 1988; Marcelli et al, 1990a; McCarthy et al, 1989; Oreffo et al, 1988; Oreffo et al, 1990; Pfeilschifter et al, 1988; Tashjian Jr. et al, 1985). The bone environment is a rich source for a number of growth factors (Farley et al, 1982; Hauschka et al, 1986; Linkhart et al, 1986; Mohan et al, 1990; Mohan et al, 1988; Seyedin et al, 1986). Many growth factors are synthesized and secreted by bone cells (Chaudhary et al, 1992). Some are also synthesized elsewhere in the body, carried to the bone by the circulatory system, and deposited in the extracellular milieu of the bone. These cytokines are released at a later time by osteoclastic resorption activity to regulate bone cell activities as described above. As with the osteoblast-derived factors discussed previously, osteoclast-derived autocrine factors may also be important in modulating osteoclast activity.

Studies in the authors' laboratories.

A systematic survey of cytokine modulation by E_2 was conducted in our laboratories by Chaudhary et al, (1992). Our laboratories assessed the effect of E_2 on the production of various cytokines as assessed by specific ELISAs of hOB cells with a mature osteoblast phenotype. They found that IL-6, IL-8, and GM-CSF, but not IL-4 or TNF-α, were produced by these cells; however, treatment with 10 nM E_2 had no effect on either constitutive production or production after stimulation with IL-1β. In contrast, Rickard et al, (1992) used bioassays and immunometric assays to identify TNF-α production by hOB cells and to show that its release was inhibited by E_2. The reason for the discrepant results of the study of Rickard et al, (1992) with that of Chaudhary et al, (1992), who found no production of TNF-α, is unclear. However, it should be noted that although both groups used cultured osteoblast-like cells obtained from normal human bone, there were differences in the methods for obtaining and culturing the cells as well as different detection methods. Further, the differences in cell culture conditions followed by the multiple population doublings could well create cells of different stage of differentiation.

Among potential osteoclast-inhibiting factors, a major candidate for mediating estrogen action is TGF-β. This factor is a multifunctional protein that is produced by many cells including osteoblasts and has a wide range of biological activities (Massague, 1992; Robey et al, 1987). In most biological systems, its effects are inhibitory, but for osteoblasts it is generally a potent mitogen (Marcelli et al, 1990b; Noda & Camilliere, 1989; Robey et al, 1987). TGF-β has potent influences on osteoclastic function and has been reported to either increase or decrease osteoclastic resorptive activity and inhibit osteoclast recruitment (Dieudonne et al, 1991a; Marcelli et al, 1990b; Pfeilschifter et al, 1988).

Oursler et al, (1991c) showed that E_2 increased both the steady-state levels of mRNA as well as the release of TGF-β protein in hOB cells in a dose-dependent manner as assessed by the bioassay method. Figure 6 shows that both E_2 and parathyroid hormone (PTH) induce the production of TGF-β, but the induction was abolished when the cells are treated by both hormones. Further, the factor is totally inactive (latent) with both hormones as per the bioassay (Oursler et al, 1991c). Therefore, in order for the hormone to regulate TGF-β activity, some other factors (low pH or a protease) are needed to activate the TGF-β (i.e. remove the binding proteins).

This finding, that the TGF-β production by PTH is abolished by E_2 treatment, is of particular interest because of the well known antagonism of the bone resorbing effect of PTH by E_2 (Gallagher & Wilkinson, 1973; Marcus et al, 1984). Thus, one mechanism by which E_2 decreases bone resorption may be an E_2-mediated release of cytokines by osteoblasts which act as paracrine factors to modulate osteoclasts. Although there is no general agreement on the precise pathway and factors involved in this osteoblast and osteoclast intercellular communication, it appears to be complex and may involve multiple cytokines which function to either increase or decrease bone resorption.

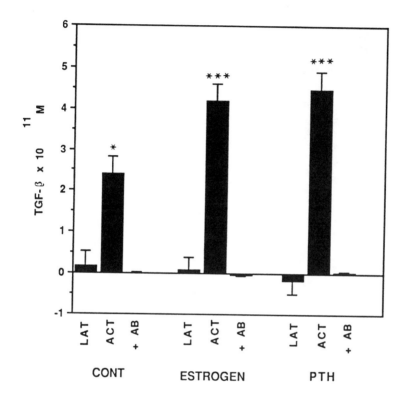

FIGURE 6. *TGF-β protein activity in untreated and acid-activated samples.* TGF-β protein assay was performed on nonactivated (LAT), acid-activated (ACT), and acid-activated then blocking antibody treated (+AB) samples as described in Materials and Methods. Samples were vehicle control, 10^{-8}M E$_2$, and 10^{-8}M PTH-treated hOBs. Data are means ± SD for four replicate TGF-β assay points. Acid activation studies were repeated six times, and these data are representative of the observed response. *, P<0.05; ***P<0.0005 when comparing LAT to ACT samples. (Taken with permission from Oursler et al, 1991, *Endocrinology* 129, 3313-3320).

CONCLUSIONS

Figure 7 outlines the results of the authors' studies on the E_2 action of OB and OC. The finding that E_2 modulates the nuclear proto-oncogene expression and TGF-β production by osteoblasts, taken together with data which indicates that E_2 modulates lysosomal protein mRNA levels and resorptive activity in osteoclasts, suggests that both osteoblasts and osteoclasts are capable of responding directly to estrogen. The presence of progesterone, androgen, and glucocorticoid receptors in osteoblasts and in osteoclasts (data not presented here) supports that the bone cells are target cells for all sex steroids and glucocorticoids. Since the authors have demonstrated that avOCs are target cells for estrogen *in vitro*, it is also possible that E_2 directly influences osteoclasts *in vivo*,

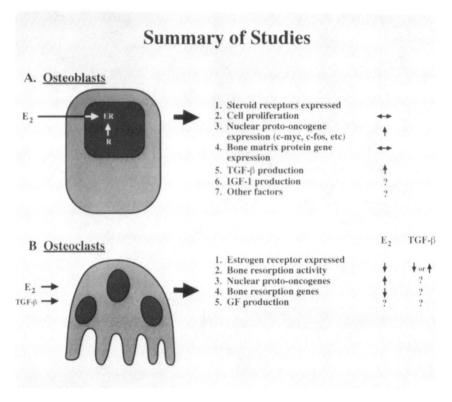

FIGURE 7. *Summary of studies from the authors' laboratories.* This figure summarizes the results of studies in the authors' laboratories on the estrogen action on isolated bone cells. ($\leftrightarrow$) = no effect; $\uparrow$ means an increase; $\downarrow$ means a decrease; (?) = unknown.

by decreasing the rate of bone resorption activity. The E_2 modulation of TGF-β production by osteoblasts could also indirectly influence osteoclast development and activity. The action of steroids on osteoblast and osteoclast differentiation is another unexplored area. Jilka et al, (1992) have already demonstrated an E_2 inhibition of osteoclast-like cell development *in vitro* which appears to involve a decrease in stromal cell IL-6 production. Viewed together, these observations demonstrate that E_2 and perhaps, all sex steroids, directly modulate bone cell functions.

ACKNOWLEDGMENTS

This research was sponsored by NIH grants AG04875, and RO1 AR41652 and the Mayo Foundation.

REFERENCES

Akatsu T, Takahashi N, Udagawa N, Imamura K, Yamaguchi A, Sato K, Nagata N and Suda T (1991): Role of prostaglandins in interleukin-1-induced bone resorption in mice in vitro. *J Bone Min Res* 6: 183-190.

Bonewald LF and Mundy GR (1990): Role of transforming growth factor-beta in bone remodeling. *Clin Ortho Rel Res* 250: 261-276.

Chambers TJ (1991): In *Bone Volume 2: The Osteoclast* (B. K. Hall, Eds.) pp. 141-173, CRC Press, Boca Raton, Florida.

Chaudhary LR, Spelsberg TC and Riggs BL (1992): Production of various cytokines by normal human osteoblast-like cells in response to interleukin-1β and tumor necrosis factor-α: Lack of regulation by 17β-estradiol. *Endocrinology* 130: 2528-2534.

Chen T and Feldman D (1979): Distinction between alpha-fetoprotein and intracellular estrogen receptors: evidence against the presence of estradiol receptors in rat bone. *Endocrinology* 102: 236-244.

Chenu C, Pfeilschifter J, Mundy G and Roodman G (1988): Transforming growth factor beta inhibits formation of osteoclast-like cells in long-term human marrow cultures. *Proc Natl Acad Sci USA* 85: 5683-5691.

Collin-Osdoby P, Oursler MJ, Webber D and Osdoby P (1991): Osteoclast-specific monoclonal antibodies coupled to magnetic beads provide a rapid and efficient method of purifying avian osteoclasts. *J Bone and Min Res* 6: 1353-1365.

Colvard DS, Eriksen EF, Keeting PE, Wilson EM, Lubahn DB, French FS, Riggs BL and Spelsberg TC (1989): Identification of androgen receptors in normal human osteoblast-like cells. *Proc Natl Acad Sci USA* 86: 854-857.

Corboz VA, Cecchini MG, Felix R, Fleisch H, van der Pluijm G and Lowik CWGM (1992): Effect of macrophage colony-stimulating factor on in vitro osteoclast generation and bone resorption. *Endocrinology* 130: 437-442.

Dickson. RB, Bates SE, McManaway ME and Lippman ME (1986): Characterization of estrogen responsive transforming activity in human breast cancer cell lines. *Cancer Res* 46: 1707-1712.

Dieudonne SC, Foo P, van Zoelen EJJ and Burger EH (1991a): Inhibiting and stimulating effects of TGF-β_1 on osteoclastic bone resorption in fetal mouse bone organ cultures. *J Bone Min Res* 6: 479-487.

Dieudonne SC, Foo P, van Zoelen EJJ and Burger EH (1991b): Inhibiting and stimulating effects of TGF-beta 1 on osteoclastic bone resorption in fetal mouse bone organ cultures. *J Bone Min Res* 6: 479-486.

Eriksen E, Colvard D, Berg N, Graham M, Mann K, Spelsberg TC and Riggs BL (1988): Evidence of estrogen receptors in normal human osteoblast-like cells. *Science* 241: 84-87.

Eriksen EF, Hodgson SF, Eastell R, Cedel SL, O'Fallon WM and Riggs BL (1990): Cancellous bone remodeling in type I (postmenopausal) osteoporosis: Quantitative assessment of rates of formation, resorption, and bone loss at tissue and cellular levels. *J Bone Min Res 5:* 311-319.

Ernst M and Froesch ER (1988): Growth hormone dependent stimulation of osteoblast-like cells in serum-free cultures via local synthesis of insulin-like growth factor I. *Biochem Biophys Res Comm* 151: 142-147.

Ernst M, Heath JK and Rodan GA (1989) Estradiol effects on proliferation, messenger ribonucleic acid for collagen and insulin-like growth factor-I, and parathyroid hormone-stimulated adenylate cyclase activity in osteoblastic cells from calvariae and long bones. *Endocrinology* 125: 825-833.

Ernst M, Schmid CH and Froesch ER (1988): Enhanced osteoblast proliferation and collagen gene expression by estradiol. *Proc Natl Acad Sci USA* 85: 2307-2310.

Ettinger B, Genant HK and Cann CE (1985): Long-term estrogen replacement therapy prevents bone loss and fractures. *Ann Intern Med* 102: 319-324.

Farley J, Masada T, Wergedal J and Bayling D (1982); Human skeletal growth factor: characterization of the mitogenic effect on bone cells in vitro. *Biochem* 21: 3508-3512.

Felix R, Fleisch H and Elford PR (1989): In *Calcified Tissue International* pp. 356-360, Springer-Verlag, New York, NY.

Fujita T, Matsui T, Nakao Y, Shiozawa S and Imai Y (1990): Cytokines and osteoporosis. *Ann New York Acad Sci* 587: 371-375.

Fuller K, Gallagher AC and Chambers TJ (1991): Osteoclast resorption-stimulating activity is associated with the osteoblast cell surface and/or the extracellular matrix. *Biochem Biophys Res Comm* 181: 67-73.

Gallagher JC and Wilkinson R (1973): The effect of ethinyl oestradiol on calcium and phosphorus metabolism of post-menopausal women with primary hyperparathyroidism. *Clin Sci 45:* 785-802.

Garrett IR and Mundy GR (1989): Relationship between interleukin-1 and prostaglandins in resorbing neonatal calvaria. *J Bone Min Res* 4: 789.

Gowen M and Mundy G (1986): Actions of recombinant interleukin 1, interleukin 2, and interferon-gamma on bone resorption in vitro. *J Immunol* 136: 2478-2483.

Gray TK, Flynn TC, Gray KM and Nabell LM (1987): 17β-estradiol acts directly on the clonal osteoblastic cell line UMR106. *Proc Natl Acad Sci USA* 84: 6267-6271.

Harris SA, Subramaniam M, Riggs BL and Spelsberg TC (1992): Estrogen induces c-fos promoter activity in normal osteoblast-like cells. *J Bone and Min Res* 7 (Suppl. 1): S-94.

Hauschka PV, Mavrakos AE, Iafrati MD, Doleman SE and Klagsbrun M (1986). Growth factors in bone matrix. Isolation of multiple types of affinity chromatography on heparin-Sepharose. *J Biol Chem* 261: 12665-12672.

Heaney RP (1965): A unified concept of osteoporosis. *Amer J Med* 39: 877-880.

Hutchinson TA, Polansky SM and Feinstein AR (1979): Postmenopausal oestrogens protect against fractures of hip and distal radius. *Lancet* 2: 705-709.

Ishimi Y et al (1990a): IL-6 is produced by osteoblasts and induces bone resorption. *J Immunol* 145: 3297-3306.

468

Ishimi Y et al (1990b): IL-6 is produced by osteoblasts and induces bone resorption. *J Immunol* 145: 3297-3303.

Ivey JL and Baylink DJ (1981): Postmenopausal osteoporosis: proposed roles of defective coupling and estrogen deficiency. *Met Bone Dis Relat Res* 3: 3-8.

Jilka RL (1986): Are osteoblastic cells required for the control of osteoclast activity by parathyroid hormone. *J Bone Min Res* 1: 261-266.

Jilka RL, Hangoc G, Girasole G, Passeri G, Williams DC, Abrams JS, Boyce B, Broxmeyer H and Manolagas SC (1992): Increased osteoclast development after estrogen loss: mediation by Interleukin-6. *Science* 257: 88-91.

Kawanto M, Yamamoto I, Iwato K, Tanaka H, Asaoku H, Tanabe O, Ishikawa H, Hobuyoshi M, Ohmoto Y, Hirai Y and Kuramoto A (1989): Interleukin-1β rather than lymphotoxin as the major bone resorbing activity in human multiple myeloma. *Blood* 73: 1646-1649.

Keeting PE, Scott RE, Colvard DS, Han IK, Spelsberg TC and Riggs BL (1991): Lack of a direct effect of estrogen on proliferation and differentiation of normal human osteoblast-like cells. *J Bone Min Res* 6: 297-304.

Knabbe C, Lippman ME, Wakefield LM, Flanders KC, Kasid A, Derynck R and Dickson RB (1987): Evidence that transforming growth factor-beta is a hormonally regulated negative growth factor in human breast cancer cells. *Cell* 48: 417-424.

Komm BS, Terpening C, Benz D, Graeme K, Gallegos A and Korc M (1988): Estrogen binding, receptor mRNA, and biologic response in osteoblast-like osteosarcoma cells. *Science* 241: 81-84.

Landers JP and Spelsberg TC (1992): New Concepts in Steroid Hormone Action: Transcription Factors, Proto-oncogenes and the Cascade Model for Steroid Regulation of Gene Expression. CRC Press, Inc., G. Stein, J. Stein, and L. Lian, eds., *Crit Rev Euk Gene Exp* 2(1): 19-63.

Lerner U and Gustafson GT (1980): A revised role for cyclic AMP in prostaglandin-induced bone resorption. *Med Biol* 58: 18-24.

Lindsay R (1981): In *Osteoporosis: Recent Advances in Pathogenesis and Treatment* pp. 481, University Park Press, Baltimore.

Linkhart TA, Jennings JC, Mohan S, Wakely GK and Baylink D (1986): Characterization of mitogenic activities extracted from bovine bone matrix. *Bone* 7: 479-482.

Lorenzo JA, Sousa SL, Fonseca JM, Hock JM and Medlock ES (1987): Colony stimulating factors regulate the development of multinucleated osteoclasts from recently replicated cells in vitro. *J Clin Invest* 80: 160-164.

Marcelli C, Yates AJ and Mundy GR (1990a): In vivo effects of human recombinant transforming growth factor β ov bone turnover in normal mice. *J Bone Min Res* 5: 1087-1096.

Marcelli C, Yates AJ and Mundy GR (1990b): In vivo effects of human recombinant transforming growth factor beta on bone turnover in normal mice. *J Bone Min Res* 5: 1087-1096.

Marcus R, Madvig P, Crim M, Pont A and Kosek J (1984): Conjugated estrogens in the treatment of postmenopausal women with hyperparathyroidism. *Ann Intl Med* 100: 633-640.

Massague J (1992): Receptors for the TGF-β family. *Cell* 69: 1 067-1070.

McCarthy TL, Centrella M and Canalis E (1989): Regulatory effects of insulin-like growth factors I and II on bone collagen synthesis in rat calvarial cultures. *Endocrinology* 124: 301-309.

McSheedy PMJ and Chambers TJ (1986a): Osteoblast-like cells in the presence of parathyroid hormone release soluble factor that stimulates osteoclastic bone resorption. *Endocrinology* 119: 1654-1662.

McSheedy PMJ and Chambers TJ (1986b): Osteoblastic cells mediate osteoclastic responsiveness to parathyroid hormone. *Endocrinology* 118: 824-830.

McSheedy PMJ and Chambers TJ (1987): 1,25-Dihydroxyvitamin D_3 stimulates rat osteoblastic cells to release a soluble factor that increases osteoclastic bone resorption. *J Clin Invest* 80: 425-429.

Mohan S, Bautista CM, Herring SJ, Linkhart TA and Bayling DJ (1990) Development of valid methods to measure insulin-like growth factors-I and -II in bone cell-conditioned medium. *Endocrinology* 126: 2534-2542.

Mohan S, Jennings JC, Linkhart TA, Wergedal JE and Baylink D (1988): Primary structure of human skeletal growth factor (SGF): homology with IGF-II. *J Bone Min Res* 3: S-218.

Mundy GR and Bonewald LF (1989): In *Macrophage-Derived Cell Regulatory Factors* (C. Sorg, Eds.) pp. 38-53.

Murrills RJ, Stein LS, Fey CP and Dempster DW (1990): The effects of parathyroid hormone (PTH) and PTH-related peptide on osteoclast resorption of bone slices in vitro: an analysis of pit size and the resorption focus. *Endocrinology* 127: 2648-2653.

Noda M and Camilliere JJ (1989): In vivo stimulation of bone formation by transforming growth factor-β. *Endocrinology* 124: 2991-2994.

Norrdin RW, Jee WSS and High WB (1990): The role of prostaglandins in bone in vivo. *Prostaglandins Leukotrienes and Essential Fatty Acids* 41: 139-149.

Oreffo R, Bonewald L, Garrett I, Seyedin S and Mundy G (1988): Transforming growth factors beta I and II inhibit osteoclast activity. *J. Bone Min. Res.* 3: S-178.

Oreffo ROC, Bonewald L, Kukita A, Garrett IR, Seyedin SM, Rosen D and Mundy GR (1990): Inhibitory effects of the bone-derived growth factors osteoinductive factor and transforming growth factor-β on isolated osteoclasts. *Endocrinology* 126: 3069-3075.

Oursler MJ, Collin-Osdoby P, Anderson F, Li L, Webber D and Osdoby P (1991a): Isolation of avian osteoclasts: Improved techniques to preferentially purify viable cells. *J Bone and Min Res* 6: 375-385.

Oursler MJ, Osdoby P, Pyfferoen J, Riggs BL and Spelsberg TC (1991b): Avian osteoclasts as estrogen target cells. *Proc Natl Acad Sci USA* 88: 6613-6617.

Oursler MJ, Cortese M, Keeting P, Anderson M, Bonde S, Riggs BL and Spelsberg TC (1991c): Modulation of transforming growth factor-beta production in normal human osteoblast-like cells by 17 beta-estradiol and parathyroid hormone. *Endocrinology* 129: 3313-3320.

Oursler MJ, Pederson L, Pyfferoen J, Osdoby P, Fitzpatrick L and Spelsberg TC (1993a): Estrogen modulation of avian osteoclast lysosomal gene expression. *Endocrinology,* 132: 1373-1380.

Oursler MJ, Pederson L, Fitzpatrick L, Riggs BL and Spelsberg TC (1993b): Human giant cell tumors of the bone (osteoclastomas) are estrogen target cells. *Proc Natl Acad Sci USA* (in press).

Parfitt AM (1979): Quantum concept of bone remodeling and turnover: implications for the pathogenesis of osteoporosis. *Calcific Tissue International* 28: 1-5.

Parfitt AM, Mathews CHE, Villaneuva AR, Kleerekoper M, Frame B and Rao DS (1983): Relationships between surface, volume, and thickness of iliac trabecular bone in aging and in osteoporosis. *J Clin Invest* 72: 1396-1409.

Pfeilschifter J, Chenu C, Bird A, Mundy GR and Roodman GD (1989): Interleukin 1 and tumor necrosis factor stimulate the formation of human osteoclast-like cells in vitro. *J Bone Min Res* 4: 113-118.

Pfeilschifter J, Seyedin S and Mundy G (1988): Transforming growth factor beta (TGF-beta) inhibits bone resorption in fetal rat long bones. *Calcified Tissue International* 42: A-34.

Ransone LJ and Verma IM (1990): Nuclear proto-oncogenes fos and jun. *Ann Rev Cell Biol* 6: 539-544.

Rickard D, Russell G and Gowen M (1992): Oestradiol inhibits the release of tumor necrosis but not interleukin 6 from adult human osteoblasts in vitro. *Osteoporosis International* 2: 94-102.

Riggs BL, Jowsey J, Goldsmith RS, Kelly PJ, Hoffman DL and Arnaud CD (1972): Short- and long-term effects of estrogen and synthetic anabolic hormone in postmenopausal osteoporosis. *J Clin Invest* 51: 1659-1663.

Riggs BL, Wahner HW, Dunn WL, Mazess RB, Offord KP and Melton LJ III (1981): Differential changes in bone mineral density of the appendicular skeleton with aging: relationship to spinal osteoporosis. *J Clin Invest* 67: 328-335.

Robey PG, Young MF, Flanders KC, Roche NS, Kondaiah P, Reddi AH, Termine JD, Sporn MB and Roberts AB (1987): Osteoblasts synthesize and respond to transforming growth factor-type beta (TGF-β) in vitro. *J Cell Biol* 105: 457-463.

Sabatini M, Boyce B, Aufdemorte T, Bonewald L and Mundy GR (1988): Infusions of recombinant human interleukins 1a and 1b cause hypercalcemia in normal mice. *Proc Natl Acad Sci USA* 85: 5235-5239.

Seyedin SM, Thompson AY, Bentz H, Rosen D, McPherson J, Conti A, Siegel N, Gallupi G and Piez K (1986): Cartilage-inducing factor-A. Apparent identity to transforming growth factor-beta. *J Biol Chem* 261: 5693-5702.

Streiter RM, Phan SH, Showel HJ, Remick DG, Lynch JP, Genord M, Raiford C, Eskandari M, Marks RM and Kunkel SL (1989): Monokine-induced neutrophil chemotactic factor gene expression in human fibroblasts. *J Biol Chem* 264: 10621-10626.

Subramaniam M, Colvard D, Keeting PE, Rasmussen K, Riggs BL and Spelsberg TC (1992): Glucocorticoid regulation of alkaline phosphatase, osteocalcin, and proto-oncogenes in normal human osteoblast-like cells. *J Cell Biochem* 50: 411-424.

Tashjian Jr. AH, Voelkel EF, Lazzaro M, Singer FR, Roberts AB, Derynck R, Winkler ME and Levine L (1985): α and β human transforming growth factors stimulate prostaglandin production and bone resorption in cultured mouse calvaria. *Proc Natl Acad Sci USA* 82: 4535-4538.

Thomson BM, Mundy GR and Chambers TJ (1987): Tumor necrosis factors alpha and beta induce osteoblastic cells to stimulate osteoclastic bone resorption. *J Immunol* 138: 775-779.

Vaes G (1988): Cellular biology and biochemical mechanism of bone resorption. A review of recent developments on the formation, activation, and mode of action of osteoclasts. *Clinical Orthopedics and Related Research* 231: 239-271.

van Paassen HC, Poortman J, Borgart-Creutzburg IHC, Thijssen JHH and Duursma SA (1978): Oestrogen binding proteins in bone cell cytosol. *Calcific Tissue Res* 25: 249-254.

Watanabe K, Tanaka Y, Morimoto I, Yahata K, Leki K, Fujihara T, Yamashita U and Eto S (1990): Interleukin-4 as a potent inhibitor of bone resorption. *Biochem Biophys Res Comm* 172: 1035-1041.

Watts CKW, Parker MG and King RJB (1989): Stable transfection of the oestrogen receptor gene into a human osteosarcoma cell line. *J Steroid Biochem* 34: 483-490.

Weiss NS, Ure CL, Ballard JH, Williams AR and Daling JR (1980): Decreased risk of fractures of the hip and lower forearm with postmenopausal use of estrogen. *New Eng J Med* 303: 1195-1198.

CHARACTERISTICS OF THE HUMAN ESTROGEN RECEPTOR PROTEIN PRODUCED IN MICROBIAL EXPRESSION SYSTEMS

James L. Wittliff
Jing Dong
Christine Schaupp
Hormone Receptor Laboratory
University of Louisville

Petr Folk
Department of Physiology & Developmental Biology
Charles University, Prague

Tauseef Butt
SmithKline Beecham Pharmaceuticals
King of Prussia

INTRODUCTION

Steroid and thyroid hormone actions are communicated via intracellular receptors which are members of a complex family of transcriptional regulatory proteins (Evans, 1988; Beato, 1989; O'Malley, 1990). When these regulatory proteins bind their ligands, they recognize DNA-response elements on target cell genes and coordinate their expression. In general, proteins in the steroid receptor superfamily are expressed as a single polypeptide containing three discrete functional

STEROID HORMONE RECEPTORS
V. K. Moudgil, Editor
© 1993 Birkhäuser Boston

domains (Green and Chambon, 1988; McDonnell et al, 1988; Conneely et al, 1988; Farrell et al, 1990; Guichon-Mantel et al, 1989 and Howard et al, 1990). The central domain consists of two zinc-binding finger motifs which interact specifically with hormone-response elements (HRE) on the target genes (Green et al, 1986; Klein-Hitpass et al, 1988; Tsai et al, 1988). Several regions adjacent to the amino-terminus of evolutionarily advanced receptors, i.e. those binding estrogens, progestins, and glucocorticoids, interact with other factors of the transcriptional process as well as with the DNA and hormone-binding domains of the receptor to facilitate transcription (Rusconi and Yamamoto, 1987; Tora et al, 1989a; Dobson et al, 1989; O'Malley, 1990). The hormone-binding domain occupies a large sequence near the carboxyl-terminus of these proteins (Miesfeld, 1989). Inactive (unliganded) receptor proteins are associated with heat shock proteins (hsp90) which dissociate upon binding of the steroid. This event results in active ligand-receptor complexes which then may associate tightly with the HRE stimulating gene transcription (Green and Chambon, 1988; Beato, 1989; Burger and Watson, 1989; Fuller, 1991; Wahli et al, 1991).

MICROBIAL EXPRESSION SYSTEMS FOR STEROID RECEPTORS

Normal mammalian cells rarely express large amounts of steroid receptors, although many neoplastic lesions express extensive quantities of both estrogen and progestin receptors (Wittliff, 1984, 1987). Recently, human estrogen, progesterone, and glucocorticoid receptors have been expressed in yeast in which hormone-responsive transcription has been retained (Metzger et al, 1988; Schena and Yamamoto, 1988; Mak et al, 1989; McDonnell et al, 1989). Attachment of the ubiquitin gene to the gene sequence for the amino-terminus of human vitamin D receptor, progesterone receptor (hPR), and estrogen receptor (hER) improved both the activity and quantity of these proteins when expressed in yeast (Mak et al, 1989; Shena and Yamamoto, 1989; McDonnell et al, 1991; Schaupp et al, 1992). However, expression of intact and biologically active steroid receptors in *Escherichia coli* has been difficult until our recent success (Wittliff et al, 1990). Furthermore, it is unknown if mammalian steroid receptors require post-translation modifications for either hormone- or DNA-binding properties to perform their regulatory activities. The advantage of microbial expression systems is the production of biologically active receptors retaining characteristics of the wild-type protein. Large scale production of these labile regulatory proteins would provide a means to study the details of their structure and its relationship to functional properties.

Previously we proposed that expression of foreign genes as a ubiquitin fusion in *E. coli* facilitated stabilization and increased the efficiency of translation (Butt et al, 1988, 1989). All eukaryotes contain ubiquitin which is a 76-amino acid protein found either as a free molecule or as a covalently-attached component of certain proteins in the nucleus, cytoplasm, or membranes (Rechsteiner, 1987). The protein sequence of ubiquitin is the most conserved discovered to date in that only three conservative amino acid substitutions separate yeast ubiquitin from that found in humans. Ubiquitin or the ubiquitin pathway has not been detected in prokaryotes. Ubiquitin, which contains no cysteine residues, is highly water soluble, extremely heat-stable, and resistant to protease action. The crystal and solution structures of ubiquitin elucidated by x-ray and NMR techniques revealed that the protein is a compact molecule with a hydrophobic core (Vijay-Kumar et al, 1987; Weber et al, 1987). The site of covalent attachment to other proteins is the carboxyl-terminus which contains four amino acids protruding from the core of the molecule. Conjugation of the carboxyl-terminal glycine residue of ubiquitin to the ε-amino group of lysine residues in acceptor proteins of eukaryotes has been proposed as a signal for protease action (Hershko and Ciechanover, 1986).

Certain eukaryotes contain genes encoding ubiquitin carboxyl-terminus extension proteins (UBCEP) whereby the translation frame of ubiquitin is extended from carboxyl-terminal glycine to 52 or 76-80 amino acids (Findley et al, 1989). The UBCEPS, which bind ribonucleic acid, are considered to be ribosomal proteins. We have proposed that ubiquitin acts as a chaperon to the acceptor proteins as a result of genetic and biochemical experiments of ubiquitin-fusion proteins in *E. coli* and considerations of natural ubiquitin fusion proteins in yeast (Butt et al, 1988, 1989).

Production of intact, active human steroid receptors in *E. coli* using recombinant DNA technology has been difficult to achieve. However, truncated versions of hPR fused to β-Gal and ubiquitin have been expressed in *E. coli* (Eul et al, 1989; Power et al, 1990). One of the problems occurring in the microbial expression of hER is that genes may be cloned which express mutations altering receptor properties. An example of this is the cloning on a hER which contained a mutation altering the properties of hormone association with the ligand-binding domain (Tora et al, 1989b).

Recently we demonstrated the expression of intact hER as ubiquitin-fusion using *E. coli* (UB-ER, Wittliff et al, 1990). The hER protein expressed as a ubiquitin-fusion had virtually identical properties to an isoform of the wild-type receptor regarding ligand and DNA-binding properties as well as its recognition by monoclonal antibodies directed against two different epitopes of the hER (Wittliff et al, 1989). The ubiquitin moiety can be cleaved from the fusion protein to release an authentic intact receptor protein. Expression in *E. coli* of large quantities of ER retaining ligand and DNA-binding domains and immunorecognition properties provides an opportunity to study structure-function relationships of the human steroid hormone receptors and other transcriptional factors previously difficult to ascertain. Our recent work with a yeast expression system (Butt et al, 1988; Schaupp et al, 1992) suggests this approach is even more efficient for production of steroid receptor proteins.

E. coli Strains and Construction of the Vectors

E. coli strain N99 cI$^+$ (Shatzman et al, 1986) was used for cloning purposes. Strains AR58 [a cryptic λ lysogen derived from N99 that is gal E::Tn10, Δ-8(chlD-pg1), Δ-Hl (Cro-ChlA), N$^+$ and cI857], AR68 (protease deficient strain, derived from AR58) and AR120 [a λ cryptic N99 derivative that is cI$^+$, Δ gal, and ::Tn10; the left arm of λ distal to the N gene is substituted by host DNA carrying lac Z; the right arm contains the T11 lesion that inhibits Cro expression and Δ-(CII-UVrB)] were employed for expression studies (Mott et al, 1985).

The hER cDNA clone PGem ER35 (Greene et al, 1986) was digested with BamHl and TthIIIl to release the complete translation frame. The TthIIIl site is 30 bases upstream of the ER initiator methionine codon. To facilitate the construction of ubiquitin fusion with ER, an AFIII-TthIIIl oligonucleotide linker was synthesized that contained six amino acids of the carboxy-terminus of ubiquitin followed by a sequence that reconstructed the TthIIIl site, thus, the amino terminus of the ER protein. The oligonucleotide linker sequence is:

'5 GATCCC TTAAGACTAAGAGGTGGT TCCATGGGGACA 3'
      AFl II   UB-ER Junction   TthIIIl

The resulting fusion has an authentic carboxy-terminus of ubiquitin but contains an extra 14 amino acid sequence SMGTRSAPCPRSRT before the authentic initiator methionine of the ER. The resulting plasmid was called BCPE1. A kpnl site was engineered at the 3' end of the cDNA by linearizing BCPE1 with EcoR1 and filling the site with a Klenow DNA polymerase fragment. The plasmid was recircularized in the presence of an oligonucleotide that contained a kpnl site. The resulting plasmid BCPE2 was digested with AflII-kpnl and the ER DNA was inserted into an AflII-kpnl digested *E. coli* human ubiquitin expression vector (Butt et al, 1989). In the final construct, the human ubiquitin-ER fusion (PUB-ER) gene was under the control of the λ phage P_L promoter. (See Figure 1.)

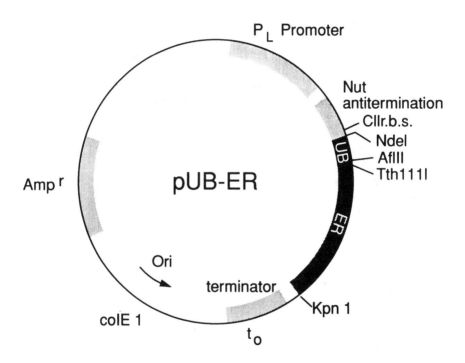

FIGURE 1. *Restriction Map of the Vector (pUB-ER) Used in E. coli to Express the Human Ubiquitin-Estrogen Receptor Fusion Protein.* Only important restriction enzyme sites are shown. Various DNA fragments are not drawn according to scale. The expression of the gene is driven by a λ P_L promoter. (Whittliff et al, 1990) The λ C11 gene sequence containing the ribosome-binding site is present before the ATG of the ubiquitin sequence. The direction of transcription is clockwise. The Nut antitermination sequence and the t_o sequence containing the transcription terminator are shown.

478

Yeast Strains and Construction of the Vector

The *S. cerevisiae* strain used was BJ3505 (Mat α, pep 4; his3 PR6-1'5Δ1.6R his 3lys 2-208 trp 1-Δ1 ura 3-52 gal2 CUP1^R). Growth and transformation of the yeast was performed according to standard procedures. The construction of the human estrogen receptor-ubiquitin fusion vector for expression in yeast (Lyttle et al, 1992) has been described previously (Metzger et al, 1988). In the final construct, called YEpE10, the human ubiquitin-ER is under the control of a yeast copper metallothionein promotor. (See Figure 2.)

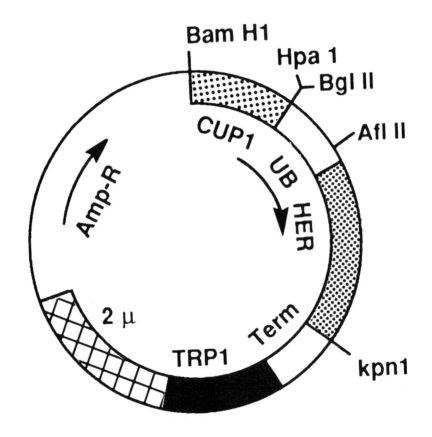

FIGURE 2. *Restriction Map of the Plasmid (YEpE10) Used in S. cerevisiae to Express the Human Ubiquitin-Estrogen Receptor Fusion Protein.* Only important restriction enzyme sites are designated and the various DNA fragments are not drawn according to scale. Expression of the hER gene is under the control of the CUP1 promoter as described earlier (Lyttle et al, 1992). Reprinted with permission.

Preparation of Microbial Extracts Containing Receptors

All procedures were performed on ice or at 4°C during the preparation and subsequent manipulations as reported earlier (Wittliff et al, 1990; Schaupp et al, 1992). Homogenization of human tissue was performed as described previously (Hyder and Wittliff, 1988; Wittliff et al, 1987, 1989).

Hormone Binding Determinations and Enzyme Immunoassay

To evaluate ligand binding characteristics by titration analyses, extracts were incubated 16-20 h at 4°C with increasing concentrations of radiolabeled ligands in the presence and absence of radioinert steroid at a 200-fold excess (Wittliff, 1987). [³H]Estradiol-17β or [¹²⁵I]iodo-estradiol-17β served as specific radiolabeled ligands for evaluation of both kinetic and molecular properties of the recombinant receptor. Diethylstilbestrol or estradiol-17β were used as the unlabeled inhibitor to insure specific (high affinity) binding. Dextran-coated charcoal suspension was used to remove unbound steroid. A representative ligand titration curve and Scatchard plot of binding data from hER expressed in *E. coli* are shown in Figure 3. In all bacterial preparations studied thus far, the Kd values ranged from $2\text{-}7 \times 10^{-10}$ M and only one type of binding site was observed (Wittliff et al, 1990). As high as 1 pmol of hER has been obtained using the *E. coli* system.

When hER was prepared from the yeast expression system and titrated with [³H]diethylstilbestrol (Figure 4), Kd values of 10^{-9} to 10^{-10}M were observed (Lyttle et al, 1992). Similar results were observed when either [³H]estradiol-17β or [¹²⁵I]iodoestradiol-17β were used as ligand with yeast-expressed hER (Table 1 and Schaupp et al, 1992).

To ascertain if the hER protein expressed in the *E. coli* and in the *S. cerevisiae* systems retained epitopes recognized by two different monoclonal antibodies prepared against human ER from breast cancer cells, an enzyme immunoassay was performed concurrently using the Abbott Laboratories ER-EIA kit according to their instructions. Intactness of the expression product may be determined since the integrity

480

of an epitope near the DNA-binding domain and of an epitope near the carboxy-terminus (steroid-binding domain) must be retained to exhibit a positive reaction (Sato et al, 1986). This method also allows a comparison of mass measurements with ligand binding properties as a further assessment of integrity as illustrated in Table 1. Note the excellent agreement of the ligand-binding data with the results obtained using the monoclonal antibody-based EIA with extracts from both microbial expression systems.

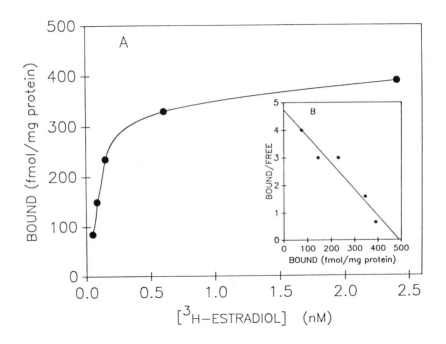

FIGURE 3. *Representative Ligand Titration and Scatchard Plot Analysis of hER Expressed in E. coli.* *E. coli* induced by nalidixic acid were harvested and extracted as described in Wittliff et al (1990). In A, hER was detected by titration with increasing concentrations of [³H]estradiol-17β in the presence and absence of unlabeled diethylstilbestrol. Only specific binding is plotted showing saturation of the receptor at approximately 1.0 nM ligand. Scatchard analysis of these data, shown in (B), gave a straight line indicating one type of binding site.

TABLE 1. Quantity of Estrogen Receptors in Extracts of *E. coli* and *S. cerevisiae* Assessed by Ligand Binding and Immunorecognition

Extract	Quantification by EIA (fmol/mg cytosol/protein)	Quantification by MTA (fmol/mg cytosol/protein)	Kd Value (M)
Uninduced *E. coli*	27	8	5×10^{-10}
Induced *E. coli*	487	484	3×10^{-10}
Uninduced Yeast	505	447	4×10^{-10}
Induced Yeast	2,643	2,567	5×10^{-10}

Extracts were prepared as described by Wittliff et al (1990) using uninduced and nalidixic acid-induced *E. coli*. Yeast strain BJ3505 containing the YEpE10 plasmid with a copper metallothionein promoter (CUP1) was evaluated in the uninduced state and after induction with copper sulfate (Butt et al, 1988). EIA = enzyme immunoassay; MTA = multipoint titration assay using radiolabeled ligands.

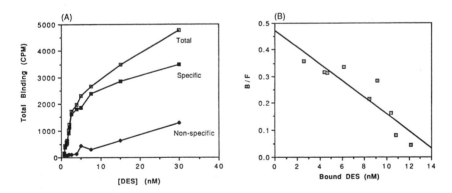

FIGURE 4. *Representative Ligand Titration and Scatchard Plot Analyses of hER Expressed in Yeast.* *S. cerevisiae* were induced by copper sulphate, harvested and extracted as described by Lyttle et al (1992). The graphs shown in A represent the binding curves using [^{3}H]diethylstilbestrol as ligand. Radioinert ligand was used to measure nonspecific binding. Scatchard analysis of the specific binding data, shown in B, gave a straight line indicating one type of binding site.

Ligand Binding Specificity

Aliquots of preparations of both *E. coli*-expressed hER and wild-type hER were incubated simultaneously with [^{125}I]iodoestradiol-17β and a 200-fold excess of the following unlabeled compounds: estradiol-17β, diethylstilbestrol, 4-hydroxytamoxifen, toremifene, R5020, dihydro-testosterone and dexamethasone. The results in Figure 5 clearly indicate the hER expressed in *E. coli* retains specific estrogen-binding properties virtually identical to those of the wild-type receptor protein. (Wittliff et al, 1990).

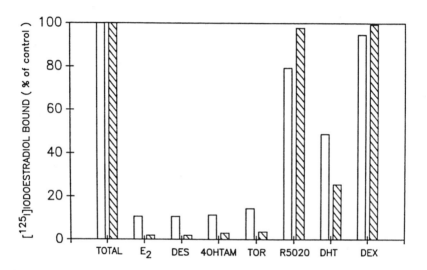

FIGURE 5. *Ligand Binding Specificity of hER Expressed in E. coli. Extracts of soluble proteins from E. coli induced by nalidixic acid were prepared as described in Wittliff et al (1990).* [^{125}I]Iodoestradiol-17β at 0.5 nM was incubated with aliquots of the *E. coli* extract containing a 200-fold excess of the following unlabeled compounds (hatched bars): E_2 = estradiol-17β; DES = diethylstilbestrol; 4OHTAM = 4-hydroxytamoxifen; TOR = toremifene; R5020 = promogestone; DHT = dihydrotestosterone; DEX = dexamethasone. Total binding was ascertained in the absence of inhibitor and expressed as 100% of control. Simultaneously, a preparation of wild-type hER was analyzed under identical conditions (open bars).

Preparation for High Performance Liquid Chromatography (HPLC)

Cell extracts were prepared in various buffers as indicated earlier (Wittliff et. al 1990; Schaupp et al, 1992) and labeled with either [^{3}H]estradiol-17β or [^{125}I]iodoestradiol-17β for 6-20 hr at 4°C. For nonspecific binding, a 200-fold excess of unlabeled diethylstilbestrol or estradiol was included in a separate reaction. Labeled cytosol was treated with a dextran-coated charcoal pellet, mixed and processed as described in our previous papers (Wiehle et al, 1984; Hyder and Wittliff, 1987, 1988; Wittliff, 1987; Wittliff et al, 1981, 1989).

High Performance Hydrophobic-Interaction Chromatography (HPHIC)

To obtain information on the surface hydrophobicity of the recombinant hER, HPHIC was performed on a Synchropak-Propyl 300 column (6 x 100 mm; Synchrom, Inc., IN) with a silica guard column. Immediately before injection, the extract was mixed quickly with an equal volume of P_{10}EGD-buffer containing 3M ammonium sulfate to establish the hydrophobic environment in the sample (Hyder and Wittliff 1987, 1988; Wittliff et al, 1990). Elution was initiated with phosphate-buffer containing 2 M ammonium sulphate at a flow rate of 1 ml/min. The gradient programs have been described in detail (Hyder and Wittliff, 1987, 1988). The recovery of both total radioactivity and protein was in the range of 72-106% but usually greater than 85%.

A representative elution profile of hER expressed in the bacterial system used in an HPHIC experiment is presented in Figure 6. Note that a single isoform was detected with characteristics identical to the MI isoform of hER from human breast carcinoma (Hyder and Wittliff, 1987, 1988). Thus the ligand specificity results given in Figure 5 are representative of this hER isoform indicating further the retention of properties of the receptor from human breast tumors (Wittliff et al, 1989).

Yeast-expressed hER was purified on an HPHIC column using protocols for elution which we established using human wild-type ER (Hyder and Wittliff, 1987; Wittliff, 1989). Typically, three hydrophobic hER isoforms are identified in human breast carcinoma (Figure 7). When yeast-expressed hER was compared under the same chromatography conditions, a major isoform eluted at 13 ml in a position similar to the MI isoform detected in breast cancer. Unlike the bacterial-expressed hER, yeast extracts contained additional hydrophobic components (Figure 7). These data indicate that both systems express hER retaining properties of the naturally-occurring receptor.

Enzyme Immunoassay for ER Isoforms Separated by Chromatography

After identification of ER-isoforms separated by HPLC, fractions from peaks containing ligand-binding protein and from adjacent areas were concentrated and an enzyme immunoassay of hER was performed with aliquots using the Abbott Laboratories ER-EIA kit according to their instructions (Wittliff et al, 1990). An example of this type of experimental approach comparing properties of immunorecognition with those ascertained by ligand binding is shown in Figure 6.

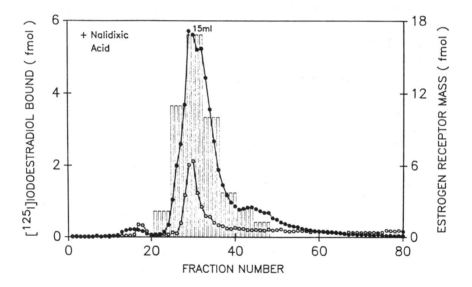

FIGURE 6. Hydrophobic-Interaction Chromatography of hER Isoforms in Extracts of *E. coli* Induced by Nalidixic Acid. Preparations of hER were incubated with 2 nM [^{125}I]iodoestradiol-17β in the absence (○) and presence (●) of a 200-fold excess of unlabeled DES and separated on a SynchroPak propyl-300 HIC column using an ammonium sulfate gradient (Hyder and Wittliff, 1988). Only a single ligand binding isoform of hER (●) was detected (15 ml) which retained immunorecognition properties (bar graph) identical with those of the full-length wild-type ER found in human breast carcinomas.

Thus hER expressed in the bacterial system retained the epitopes expressed by wild-type ER and exhibited similar immunorecognition properties. Previously we used this approach to demonstrate that certain hER isoforms from human breast tumors have variable recognition of the D-547 and H-222 monoclonal antibodies (Sato et al, 1986).

Recognition of Hormone-Response Elements

Extracts were prepared in DNA-binding buffer and incubated with 100 nM unlabeled estradiol (final concentration) to stabilize the receptor protein (Wittliff et al, 1990). The estrogen response element (ERE) from *Xenopus* vitellogenin A2 gene was used as a wildtype ERE with the sequence shown:

'5 GTCCAAAGTCAGGTCACAGTGACCTGATCAAAGTT 3'
'3 CAGGTTTCAGTCCAGTGTCACTGGACTAGTTTCAA 5'

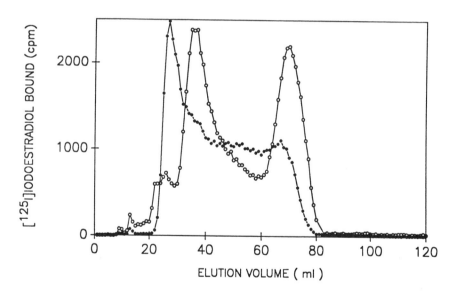

FIGURE 7. *HPHIC of hER Isoforms in Extracts of S. cerevisiae Induced by Copper Sulphate.* Yeast expressed hER was purified by HPHIC using a SynchroPak propyl-300 column (SynChrom, Inc.) where it resolved into a major peak eluting at 27 ml and additional components (●). A profile of hER isoforms from human breast carcinomas which were separated under identical conditions by HPHIC is shown (○).

Synthetically prepared ERE was filled in with the Klenow-fragment in the presence of dGTP, dATP, dTTP and [α-^{32}P]dCTP (3000 Ci/mmol, Amersham) and purified on NICK^{TM}Spin Columns (Pharmacia) as described by Butt et al, (1989). Gel-mobility shift assays containing extract, poly-dIdC and labeled oligonucleotide in buffer were incubated and treated as described earlier (Wittliff et al, 1990). Samples containing hER and labeled-ERE were detected by autoradiography after separation by PAGE. An example of a DNA band-shift assay utilizing hER expressed in bacteria is shown in Figure 8. Note the high specificity for association with the synthetic ERE again confirming its retention of properties similar to naturally-occurring hER.

Western Blot Analyses of hER Expression

The hER gene expression in yeast was initiated by the addition of copper sulphate to growing cultures. As a function of time after induction, samples of yeast cells were harvested and the expression of hER was measured by Western blot analysis (Figure 9). Note that a single species of the 66 kDa isoform was identified in all cultures using the H-222 monoclonal antibody prepared against human estrogen receptor (Greene et al, 1986). These data indicate that the fusion product between the N-terminus of hER and the C-terminus of ubiquitin was cleaved. Eukaryotes including yeast cleave ubiquitin from fused proteins after translation releasing the authentic recombinant protein (Ecker et al, 1989).

Influence of Ligand Association with hER on Binding to ERE

Because of our interest in characterizing various aspects of ligand-receptor and receptor-ERE associations, we conducted experiments to evaluate the influence of ligand binding on these reactions. Figure 10 provides an example of an experiment in which *E. coli*-expressed hER was incubated with various concentrations of either estradiol-17β, tamoxifen or 4OH-tamoxifen and association of these ligand-receptor complexes with ERE was evaluated by DNA bandshift assays. Note that there were no significant differences observed in the mobility of hER-ERE regardless of the type or concentration of estrogenic ligand bound in the steroid-binding domain.

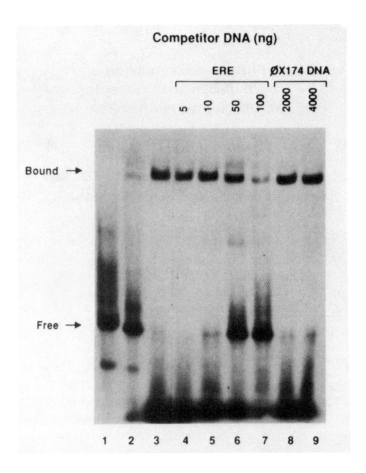

FIGURE 8. *Specific Binding of E. coli-Expressed hER to Estrogen-Response Elements.* *E. coli* cell extracts were prepared as described by Wittliff et al (1990) and incubated with [^{32}P]labeled estrogen-response elements. Unbound labeled ERE and receptor-associated ERE were separated on a 7% gel by PAGE and visualized by autoradiography. Lane 1 contained labeled ERE only; lane 2, ERE incubated with protein extract from uninduced *E. coli*; lane 3, ERE incubated with 5 μg of protein extract from induced cells; lanes 4, 5, 6, and 7 contained labeled ERE incubated with induced-cell extract and increasing amounts (ng) of unlabeled ERE; lanes 8 and 9 contained protein extracts from induced cells incubated with labeled ERE in the presence of 2,000 and 4,000 ng of φX 174, *Hae*III-digested DNA.

488

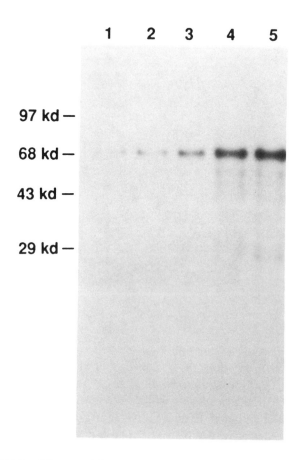

Regulated Expression of
Human Estrogen Receptor
as Ubiquitin-fusion in Yeast

FIGURE 9. *Western Blot Analyses of hER Expression in Yeast.* Yeast containing the YEpE10 expression vector were treated with copper sulphate for various times as described by Lyttle et al (1992). An uninduced culture served as a control. Western blots were performed using H222 monoclonal antibody on samples induced for 10, 30, 60 and 120 min (lanes 2-5).

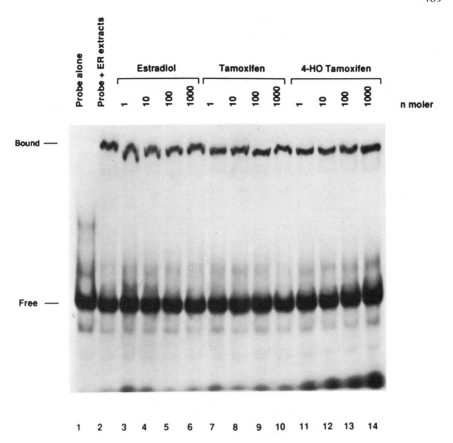

FIGURE 10. *Lack of Influence of Ligand-Binding to hER When Associated With Estrogen-Response Elements.* *E. coli* extracts were prepared from hER expressing cells, 5 μg of extracted protein were incubated with [^{32}P]labeled ERE and the analysis was performed on PAGE using a band-shift assay as described previously (Wittliff et al, 1990). Protein samples were incubated with increasing concentrations of 1 to 1000 nM estradiol-17β (lanes 3-6). Lanes 7-10 contained the samples treated with tamoxifen while lanes 11-14 contained the samples treated with 4HO-tamoxifen. Lanes 1 and 2 illustrate the analysis of probe alone or of probe incubated with hER in the absence of ligand. No significant difference was observed in the mobility of hER-ERE complexes when different concentrations of either estradiol-17β or of the other ligands were present.

Estrogen Regulation of Gene Transcription

The yeast system provides an excellent avenue for rapidly evaluating the biological activity of hormone analogs and chemically unrelated substances on gene transcription. To demonstrate this, yeast cultures

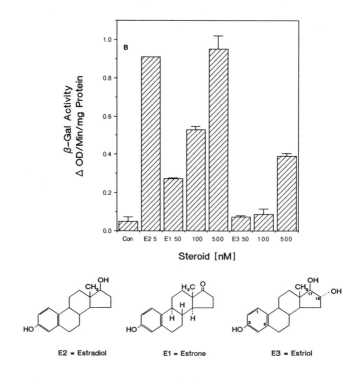

FIGURE 11. *Influence of Ligand Type and Concentration on hER Function in S. cerevisiae is Similar to the Receptor Function in Humans.* Yeast cultures containing the hER expression vector and the reporter gene were incubated with different concentrations of estradiol-17β (E2, 5 nM), estrone (E1 at 50, 100 or 500 nM) and estriol (E3 at 50, 100 or 500 nM). Hormone-dependent increase in β-galactosidase activity was monitored as described previously by Lyttle, et al (1992). The potency and structure/activity relationships of various ligands are duplicated faithfully in the system.

were developed in which the same cells contained both the hER expression vector and a vector containing a reporter vector expressing β-galactosidase (Lyttle et al, 1992). When the yeast, which had been co-transformed with the two vectors, were grown in the presence of various concentrations of naturally-occurring estrogenic ligands (Figure 11), a hormone-dependent increase in β-galactosidase was observed. The response observed in the yeast test system was similar to that observed in mammalian systems when ligand potency and structure activity relationships were considered.

POLYMORPHISM OF STEROID HORMONE RECEPTORS

The origin and physiological significance of the multiple forms of receptors which occur in human breast and endometrial carcinomas have been major foci of our investigations (Wittliff et al, 1981, 1987, 1989; Wittliff, 1984, 1987). Although certain of these components may represent distinct physiologic species, a few may arise due to proteolytic cleavage. Presumably, the latter possibility may occur during homogenization, prolonged incubation, and overnight centrifugation (Wittliff 1985; Wittliff et al, 1988). To our knowledge, no one has provided conclusive evidence of the composition of the "native state" of either ER and PR in breast and endometrial carcinomas.

To circumvent the problem of prolonged manipulation of receptor preparations, we developed the use of HPSEC, HPIEC, HPCF and HPHIC modes for the rapid separation of isoforms. Receptor isoforms are defined as protein components in a target organ which exhibit a high ligand binding affinity and specificity for a single class of steroid hormones (estrogens) and may be identified based upon characteristics of size, shape, and surface ionic properties (Wittliff et al, 1981, 1989).

The sources of wild-type ER and PR were residual samples of human breast tumors and uteri which had been utilized for clinical determinations (Wittliff, 1987), collected and stored at -80°C as described previously (Hyder and Wittliff, 1988). Properties of the isoforms from human tumors were compared with those of rabbit endometrial carcinoma, rat lactating mammary gland and MCF-7 human breast cancer cells grown in tissue culture (Table 2). In general many of the hER isoforms detected by the HPLC modes of separation had properties in common with those receptors isolated from the animal and tissue culture cells.

TABLE 2. Comparison of Properties of Estrogen Receptors Separated by HPLC

Tissue	HPSEC (TSK-3000 SW)	HPIEC (AX-1000)	HPCF (AX-500)
MCF-7 Cells	68A	55 & 180 mM phosphate	pH 6.3 & 4.3
Human Breast Carcinoma	>61 A & 29-32 A	52, 100 & 190 mM phosphate	pH 6.3, 5.3 & 4.8-3.5 shoulder
Human Uterus	>61 A & 29-32 A	115 & 203 mM phosphate	pH 7.1, 6.6 6.15 & 5.3
Rabbit Endometrial Carcinoma	71 A	50 & 175 mM phosphate	pH 6.8, 6.4 & 5.6
Rat Lactating Mammary Gland	>61 A	90 & 205 mM phosphate	pH 6.8 & 6.5

Adapted from Wittliff et al, 1989 with permission.

TABLE 3. Summary of Properties of Progestin Receptors in Human Uterus

Ligand	HPSEC (TSK3000 SW)	HPIEC (AX-1000)	HPCF (AX500)
R5020	>80 A & 35 A	100 mM phosphate	pH 5.7
ORG-2058	70 A & 20 A	100 mM phosphate	pH 5.7

Adapted from Wittliff et al, 1989 with permission.

Table 3 provides a summary of our results evaluating the chromatographic behavior of PR from human uterus using various modes of HPLC. Similar results were obtained when either [³H]R5020 or [³H]ORG-2058, two progestomimetic steroids, were used as ligands.

CLINICAL SIGNIFICANCE OF RECEPTOR POLYMORPHISM

We have proposed that some of the isoforms in human breast carcinoma represent receptor variants that compromise the endocrine responsiveness of patients treated by endocrine therapy (Wittliff, 1984; Wittliff et al, 1981, 1988). Sex-hormone receptors are required for breast cancer patients to respond to anti-hormone therapies, such as treatment with tamoxifen (Fisher et al, 1983). However, 30-40% of breast cancer patients with tumor biopsies containing estrogen and progestin receptor do not respond to endocrine therapy. With the GR mouse mammary carcinoma as a model, we evaluated the influence of ER isoform variants present in tumor cytosols on the reliability of the receptor assay as a prognostic index (Sluyser and Wittliff, 1992). Using high performance ion-exchange and size-exclusion chromatography, ER isoforms were determined in hormone-responsive and non-responsive mammary tumors of GR mice. Isoform distribution and immunorecognition by monoclonal antibodies indicated that there was a difference in receptor expression in the hormone-responsive tumors compared to that of the non-responsive mammary cancers (Sluyser and Wittliff, 1992).

Complementary to our studies of microbial expression systems, we have developed a rapid means of simultaneously identifying isoforms of hER and hPR using a double-isotope assay (Folk et al, 1992). This double-isotope assay utilizes various ligand combinations of [¹²⁵I]iodoestradiol-17β, [³H]estradiol-17β, [³H]R5020 and [¹²⁵I]iodovinyl-nortestosterone and separation of the receptor isoforms by HPLC. In this way one may either examine the hER or hPR isoforms in a single sample. As shown in Figure 12, using two labeled ligands of the same hormonal activity allows analysis of a single receptor type under two conditions, e.g. with and without sodium molybdate. We are using this procedure to study polymorphic variations of ligand-binding isoforms of hER and hPR in human breast cancer (Table 4). Note that human breast cancers exhibit different expression profiles of these isoforms.

TABLE 4. Polymorphic Variations of ER and PR Isoform Expression Identified by HPHIC Using 50 Specimens of Human Breast Cancer

Receptor (n)	Parameter	Range Observed	75th percentile	Mean
ER (n = 50)	Conc. (fmol/mg):	65-986	98-326	196
	K_d Values (M):	1.1×10^{-11}-6.4×10^{-10}	2.8×10^{-11}-9.1×10^{-11}	4.5×10^{-11}
	Distribution as (% of total):			
	MI	0-35	6-19	15
	I	21-62	42-58	47
	II	20-76	29-47	37
PR (n = 19)	Conc. (fmol/mg):	21-1456	72-412	245
	K_d Values (M):	6.1×10^{-11}-9.3×10^{-10}	8.6×10^{-11}-6.7×10^{-10}	2.1×10^{-10}
	Distribution as (% of total):			
	MI	20-44	22-36	31
	I	56-80	63-78	69

Results were taken from Folk et al (1992) and the specimens were treated and chromatography was performed as described. ER and PR concentrations and the respective K_d values were estimated by a ligand-titration assay. The relative distribution of ER and PR into their respective isoforms is given. The 75th percentile interval contains 75% of the data.

SUMMARY

Sex-hormones promote their characteristic response in normal target organs such as the breast and uterus by first binding with high affinity to intracellular receptor proteins which then associate with hormone-response elements near endocrine-induced genes in chromatin. No biologic response is possible unless both hormone and receptor protein are present

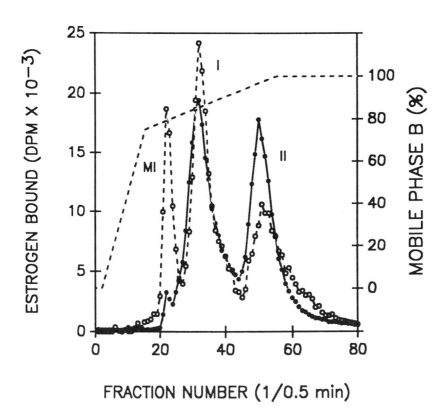

FIGURE 12. *Polymorphic Variation in ER Isoforms in Specimens of Human Breast Cancer.* Cytosols were prepared from each of two specimens and labeled separately with either [^{125}I]iodoestradiol-17β (●) or with [^{3}H]estradiol-17β (○) as described by Folk et al (1992). After removal of the unbound ligand, the cytosols were mixed, HPHIC was performed and the protein-bound radioactivity in the fractions eluted was measured using a double-isotope program.

in required quantities. It has been known for decades that many breast cancers require female sex hormones for their continued growth, i.e. they are hormonally dependent cancers. Some breast cancers, while not dependent upon estrogen for growth, may be inhibited by giving the patient high doses of estrogen-like compounds such as Tamoxifen. Therefore, ER and PR may be considered cellular markers of a breast cancer which is potentially endocrine-responsive. Breast cancer patients most likely to respond to additive (administration of antihormones) or ablative endocrine manipulation (surgical removal of hormone-producing organs such as the ovary, adrenal and pituitary gland) may be identified from analyses of sex-hormone receptors.

As many as 30-40% of human breast cancers are resistant to endocrine therapy in spite of the fact that they contain ER. Earlier we demonstrated that both ER and PR exhibit polymorphism (presence of isoforms) and suggested that their distribution in breast carcinomas may be related to endocrine responsiveness (Wittliff 1974; Wittliff and Savlov, 1975). Recent studies from other laboratories have now confirmed our earlier proposal (Wittliff et al, 1981) that there are variant estrogen receptor molecules in certain human breast cancers.

To better understand the structure and biological role of ER, we expressed the human ER gene in *E. coli* and now have expressed it in yeast as a ubiquitin-fusion construct under the control of a metallothionein promoter. Evaluation of ligand and antibody-binding activities and specificity revealed that the expressed hER retained properties of the naturally occurring receptor. Assessment of molecular properties by various modes of HPLC revealed the presence of isoforms with characteristics similar to those of estrogen receptor species expressed by human breast cancer. Additional confirmation of retention of native properties was shown by specific association of expressed hER with estrogen-response DNA elements. This approach allows the generation of large quantities of recombinant hER for crystallographic studies of both native and mutated receptor isoforms in the presence and absence of a variety of ligands.

REFERENCES

Beato M (1989): Gene regulation by steroid hormones. *Cell* 56: 335-344.

Berger FG and Watson G (1989): Androgen-regulated gene expression. *Annu Rev Physiol* 51: 51-65.

Butt TR, Jonnalagadda S, Monia BP, Sternberg EJ, Marsh J, Stadel J, Ecker DJ and Crooke ST (1989): Ubiquitin fusion augments the yield of cloned gene products in *E. coli*. *Proc Natl Acad Sci USA* 86: 2540-2544.

Butt TR, Khan MI, Marsh J, Ecker DJ and Crooke ST (1988): Ubiquitin-metallothionein fusion protein expression in yeast. *J Biol Chem* 263: 16, 364-16, 371.

Conneely OM, Sullivan WP, Toft DO, Birnbaumer M, Cook RG, Maxwell BL, Zarucki-Schulz T, Greene GL, Schrader WT and O'Malley BW (1988): Molecular cloning of the chicken progesterone receptor. *Science* 233: 767-770.

Dobson ADW, Conneely DM, Beattie W, Maxwell BL, Mak P, Tsai M-J, Schrader WT and O'Malley BW (1989): Mutation analysis of the chicken progesterone receptor. *J Biol Chem* 264: 4207-4211.

Ecker DJ, Stadel JM, Butt TR, Marsh JA, Monia BP, Powers DA, Gorman JA, Clark PE, Shatzman A and Crooke ST (1989): Increasing gene expression in yeast by fusion to ubiquitin. *J Biol Chem* 264: 7715-7719.

Eul J, Meyer ME, Tora L, Bocquel MT, Quirin-Striker C, Chambon P and Gronemeyer H (1989): Expression of active hormone and DNA-binding domains of the chicken progesterone receptor in *E. coli*. *EMBO J* 8: 83-90.

Evans RM (1988): The steroid and thyroid hormone receptor superfamily. *Science* 240: 889-895.

Farrell SE, Lees JA, White R and Parker MG (1990): Characterization and colocalization of steroid binding and dimerization activities in the mouse estrogen receptor. *Cell* 60: 953-962.

Finley D, Bartel B and Varshavsky A (1989): The tails of ubiquitin precursors are ribosomal proteins whose fusion to ubiquitin facilitates ribosomal biogenesis. *Nature* 338: 394-401.

Fisher B, Redmond C, Brown A, Wickerham DL, Wolmark N, Allegra JC, Escher G, Lippman M, Savlov E, Wittliff JL and Fisher ER et al (1983): Influence of tumor estrogen and progesterone receptor levels on the response to tamoxifen and chemotherapy in primary breast cancer. *J Clin Oncol* 1: 227-241.

Folk P, Dong J and Wittliff JL (1992): Simultaneous identification of estrogen progesterone receptors by HPLC using a double isotope assay. *J Steroid Biochem Mol Biol* 42: 141-150.

Fuller PJ (1991): The steroid receptor superfamily: mechanism of diversity. *FASEB J* 5: 3092-3099.

Green S and Chambon P (1988): Nuclear receptors enhance our understanding of transcription regulation. Trends Genet 4: 309-314.

Green S, Walter P, Kumar V, Krust A, Bornert JM, Argos P and Chambon P (1986): Human oestrogen receptor DNA sequence expression and homology to v-erb-A. *Nature* 320: 134-139.

Greene GL, Gilna P, Waterfield M, Baker V, Hort V and Shine J (1986): Sequence and expression of human estrogen receptor complementary DNA. *Science* 231: 1150-1154.

Guichon-Mantel A, Loosfelt H, Lescap P, Sar S, Atger M, Perrot-Applanat M and Milgrom E (1989): Mechanisms of nuclear localization of the progesterone receptor: evidence for interaction of monomers. *Cell* 57: 1147-1154.

Hershko A and Ciechanover A (1986): The ubiquitin pathway for the degradation of intracellular proteins. *Prog Nucleic Acid Res Mol Biol* 33: 19-56.

Howard KJ, Holley SJ, Yamamoto KR and Distelhorst CW (1990): Mapping of the HSP90 binding region of the glucocorticoid receptor. *J Biol Chem* 265: 11928-11935.

Hyder SM and Wittliff JL (1987): High-performance hydrophobic interaction chromatography of a labile regulatory protein: The estrogen receptor. *Biochromatogr*, Vol 2, No 3: 121-130.

Hyder SM and Wittliff JL (1988): High performance hydrophobic interaction chromatography as a means of identifying estrogen receptors expressing different binding domains. *J Chromatogr* 444: 225-237.

Klein-Hitpass L, Ryffel GU, Heitlinger E, Cato AC (1988): A 13 bp palindrome is a functional estrogen responsive element and interacts specifically with estrogen receptors. *Nucleic Acids Res* 16: 647-663.

Lyttle CR, Damian-Matsumura P, Huul H and Butt TR (1992): Human estrogen receptor regulation in a yeast model system and studies on receptor agonists and antagonists. *J Steroid Biochem Mol Biol* 42: 677-685.

Mak P, McDonnell DP, Weigel NL, Schrader WT and O'Malley BW (1989): Expression of functional chicken oviduct progesterone receptors in yeast (*Saccharomyces cerevisiae*). *J Biol Chem* 264: 21613-21618.

McDonnell DP, Pike JW and O'Malley BWJ (1988): The vitamin D receptor: A primitive steroid receptor related to thyroid hormone receptor. *J Steroid Biochem* 29: 459-464.

McDonnell DP, Pike JW, Drutz DD, Butt TR and O'Malley BW (1989): Reconstitution of the vitamin D-responsive osteocalcin transcription unit in *Saccharomyces cerevisiae. Mol Cell Biol* 9: 3517-3523.

McDonnell DP, Nawaz Z, Densmore C, Weigel NL, Pham TA, Clark J H and O'Malley BW (1991): High level expression of biologically active estrogen receptor in *Saccharomyces cerevisiae. J Steroid Biochem Molec Biol* 39: 291-297.

Metzger D, White JH and Chambon P (1988): The human oestrogen receptor functions in yeast. *Nature* 334: 31-36.

Miesfeld RL (1989): The structure and function of steroid receptor proteins. *Critical Rev Biochem Mol Biol* 24: 101-117.

E, Grant R, Ho Y and Platt T (1985): Maximizing gene expression from plasmid vectors containing the λ P_L promoter: Strategies for overproducing transcription termination factor. *Proc Natl Acad Sci USA* 82: 88-92.

O'Malley B (1990): The steroid receptor superfamily: More excitement predicted for the future. *Mol Endocrinol* 4: 363-369.

Power RF, Conneely OM, McDonnell DP, Clark JH, Butt TR, Schrader WT and O'Malley BW (1990): High level expression of a truncated chicken progesterone receptor in *E. coli. J Biol Chem* 265: 1419-1424.

Rechsteiner M (1987): Ubiquitin mediated pathway for intracellular proteolysis. *Annu Rev Cell Biol* 3: 1-30.

Rusconi S and Yamomoto KR (1987): Functional dissection of the hormone and DNA binding activities of the glucocorticoid receptor. *EMBO J* 6: 1309-1315.

Sato N, Hyder SM, Chang L, Thais A and Wittliff JL (1986): Interaction of estrogen receptor isoforms with immobilized monoclonal antibodies. *J Chromatogr* 359: 475-487.

Schaupp C, Folk P, Butt T and Wittliff JL (1992): The Endocrine Society, 74th Annual Meeting, San Antonio, Texas, June 24-27, 1992.

Schena M and Yamamoto KR (1988): Mammalian glucocorticoid receptor derivatives enhance transcription in yeast. *Science* 241: 965-967.

Shatzman A and Rosenberg M (1986): Efficient expression of heterologous genes in *E. coli*: The PAS vector system and its applications. *Ann N Y Acad Sci* 478: 233-248.

Sluyser M and Wittliff JL (1992): Influence of estrogen receptor variants in mammary carcinomas on the prognostic reliability of the receptor assay. *Mol Cell Endocrinol* 85: 83-88.

Tora L, Mullick A, Metzger D, Ponglikitmongkol M, Park I and Chambon P (1989b): The cloned human estrogen receptor contains a mutation which alters its hormone binding properties. *EMBO J* 8: 1981-1986.

Tora L, White J, Brou C, Tasset D, Webster N, Scheer E and Chambon P (1989a): The human estrogen receptor has two independent nonacidic transcriptional activation functions. *Cell* 59: 477-487.

Tsai SY, Tsai M-J and O'Malley BW (1988): Molecular interactions of steroid hormone receptor with its enhancer element: evidence for receptor dimer formation. *Cell* 55: 361-369.

Vijay-Kumar S, Bugg CE, Wilkinson KD and Cooke WJ (1987): Structure of ubiquitin refined at 1.8Å resolution. *J Mol Biol* 194: 531-544.

Wahli W and Mertinez E (1991): Superfamily of steroid nuclear receptors: positive and negative regulators of gene expression. *FASEB J* 5: 2243-2249.

Weber PL, Brown SC and Mueller L (1987): ^{1}H-NMR resonance assignment and secondary structure identification of human ubiquitin. *Biochemistry* 26: 7287-7290.

Wiehle RD, Hofmann GE, Fuchs A and Wittliff JL (1984): High-performance size exclusion chromatography as a rapid method for the separation of steroid hormone receptors. *J Chromatogr* 307: 39-51.

Wittliff JL (1974): Specific receptors of the steroid hormones in breast cancer. *Seminars in Oncol* 1: 109-118.

Wittliff JL (1984): Steroid hormone receptors in breast cancer. *Cancer* 53: 630-643.

Wittliff JL (1985): Separation and characterization of isoforms of steroid hormone receptors using high performance liquid chromatography. In: *Molecular Mechanisms of Steroid Hormone Action*, Moudgil V K, ed. pp. 791-813, Berlin, Germany: Walter de Gruyter and Co.

Wittliff JL (1987): Steroid hormone receptors. In: *Methods in Clinical Chemistry*, Pesce AJ and Kaplan LA, eds. Chapter 99, pp 767-795, St. Louis, Missouri: The CV Mosby Co.

Wittliff JL and Savlov ED (1975): Estrogen-binding capacity of cytoplasmic forms of the estrogen receptors in human breast cancer. In: *Estrogen receptors in human breast cancer*, McGuire WL, Carbone PP and Vollmer EP, eds. pp 73-91, New York: Raven Press.

Wittliff JL, Allegra JC, Day TG Jr and Hyder SM (1988): Structural features and clinical significance of estrogen receptors. In: *Steroid Receptors in Health and Disease*, Moudgil V K, ed. pp 287-312, New York: Plenum Publishing Corp.

Wittliff JL, Feldhoff PM, Fuchs A and Wiehle RD (1981): Polymorphism of estrogen receptors in human breast cancer. In: *Physiopathology of Endocrine Diseases and Mechanisms of Hormone Action*, Soto R, DeNicola AF and Blaquier JA, eds. pp 375-396, New York: Alan R Liss, Inc.

Wittliff JL, Wiehle RD and Hyder SM (1989): HPLC as a means of characterizing the polymorphism of steroid hormone receptors. In: *The Use of HPLC in Receptor Biochemistry*, Kerlavage AR, ed. pp 155-199, New York: Alan R. Liss, Inc.

Wittliff JL, Pasic R and Bland KI (1990): Steroid and peptide hormone receptors identified in breast tissue. In: *The Breast: Comprehensive Management of Benign and Malignant Diseases*, Bland K I and Copeland E M, III, eds. Chapter 43, pp 900-936, Philadelphia, PA: WB Saunders Co.

Wittliff JL, Wenz LL, Dong J, Nawaz Z and Butt TR (1990): Expression and characterization of an active human estrogen receptor as a ubiquitin fusion protein from *Escherichia coli*. *J Biol Chem* 265: 22016-22022.

THE AH LOCUS: A REVIEW

Alan Poland
McArdle Laboratory for Cancer Research
University of Wisconsin-Madison

Chris Bradfield
Department of Pharmacology
Northwestern University Medical School

AH LOCUS: ORIGINAL DESCRIPTION

The Ah locus was first described as a difference among inbred strains of mice when challenged with polycyclic aromatic hydrocarbons. Schmid et al (1) noted that when 7,12-dimethylbenzanthracene was applied to the skin of C57BL/6J they developed skin ulceration, while AKR/J did not. Nebert et al (2) noted that intraperitoneal administration of 3-methylcholanthrene (3-MC) induced hepatic cytochrome P-450IA1 and the associated monooxygenase activity—aryl hydrocarbon hydroxylase (AHH) in C57BL/6 mice but not in DBA/2 mice. In crosses and backcrosses between these strains, the trait of aromatic hydrocarbon responsiveness (hence the Ah locus) is inherited in a simple autosomal dominant mode.

STEROID HORMONE RECEPTORS
V. K. Moudgil, Editor
© 1993 Birkhäuser Boston

2,3,7,8-Tetrachlorodibenzo-p-dioxin

2,3,7,8-Tetrachlorodibenzo-p-dioxin (TCDD) arises as a trace contaminant in the synthesis of 2,4,5-trichlorophenol. TCDD was found to produce a similar pattern of induction as 3-MC and other polycyclic aromatic hydrocarbons; including cytochrome P-450IA1, τ-aldehyde dehydrogenase, glucuronosyl transferase, glutathione S-transferase and DT-diaphorase.

A comparison of TCDD and 3-MC on the induction of hepatic AHH activity in the rat, revealed the compounds elicit parallel dose response curves with the same maximum response, administration of maximally effective doses of both compounds together produced no greater response, and that TCDD was ≈30,000 more potent than 3-MC (3). For congeners of TCDD, there was a well-defined structure-activity relationship. Chloro- and bromo-substituted dibenzo-p-dioxins which induced AHH activity (in chick embryo liver) had halogen substitution in at least 3 of the lateral 4 ring positions (positions 2,3,7, or 8) (4).

The potency of TCDD and the well-defined structure-activity relationship suggested that it bound to some specific cellular recognition site, a receptor, to initiate the induction of AHH activity. The potency of TCDD relative to that of 3MC, prompted us to examine the effects of TCDD in mice unresponsive to 3-MC. TCDD induced hepatic AHH activity in all strains of mice tested, those responsive and unresponsive to 3-MC (5). The half-maximal inducing of dose of TCDD (ED_{50}) is approximately 10-fold greater in the latter strains; i.e., the nonresponsive mice show an absolute insensitivity to a weak ligand (3-MC) and a relative insensitivity to a more potent agonist (TCDD).

Using high specific activity $[^3H]$-TCDD, it was possible to identify a high affinity binding species in mouse liver cytosol that had the *in vitro* properties expected for the postulated receptor: (a) for TCDD, K_D ≈0.27 nM ≈ED_{50}, (b) structure-activity relationship for induction of AHH activity for various dibenzo-p-dioxin, dibenzofuran, and polycyclic aromatic hydrocarbon congeners matched the structure-activity for their receptor binding, and (c) the specific binding of $[^3H]$-TCDD was much greater in hepatic cytosol from C57BL/6J mice (responsive to 3-MC) compared to DBA/2 mice (3-MC nonresponsive) (6).

The Ah receptor and the pleiotropic response

TCDD and related congeners produce a broad range of biochemicals, morphologic and toxic responses (some of which are specific for one or a few species of animals). These include (a) induction of specific enzymes such as cytochromes P450IA1 and IA2, τ-aldehyde dehydrogenase, glucuronosyl transferase, (b) proliferation and/or altered differentiation in various epithelial tissues (e.g., skin in human, monkey, rabbit ear, hairless (hr/hr) mouse, gastrointestinal tract and bladder in guinea pig and monkey), (c) a wasting syndrome, (d) immune suppression, and (e) tumor promotion. Two independent approaches were used to implicate the Ah receptor in mediating the induction of AHH activity: (a) Among a group of congeners their rank-ordered binding to the Ah receptor corresponded to their rank-ordered potency to induce AHH activity and, (b) exploitation of the genetic polymorphism among mice, i.e., in genetic crosses segregation of 3-MC induction of enzyme activity with mice carrying the Ah[b] allele. The same two approaches, have been used to indicate all, or nearly all of the responses evoked by TCDD and congeners are mediated through the Ah receptor (7).

This does not mean that all responses result from direct transcriptional activation as in the case of cytochrome P450IA1 (see below), but that binding to the Ah receptor is the initial event, even when the details of the signal transduction pathway are unknown, as it is in most cases. A third independent approach has been used more recently. Several compounds, e.g., 6-methyl-1,3,8-trichlorodibenzofuran (8), 1-amino, 3,7,8-trichlorodibenzo-p-dioxin (9) and α-naphthoflavone have been shown to be partial agonists for the receptor, i.e., they bind to the receptor and produce a response that is less than that of a full agonist (e.g., TCDD) and when administered with a full agonist diminish its response. Thus, all three approaches (a) structure-activity correlations of biological potency and receptor binding, (b) genetic segregation, and (c) antagonism by partial agonists can be used to determine if new compounds act through the receptor and that newly discovered responses are mediated through the receptor.

Biochemistry of the Ah receptor

The Ah receptor bears many similarities with the steroid hormone receptors. The unliganded Ah receptor is found in the 100,000 xg fraction of disrupted cells as a 280 kD complex which includes the 90 kD heat shock protein (10). The Ah receptor is stabilized, (i.e., retains greater ligand binding) if it is prepared in the presence of sodium molybdate. After treatment with [^{3}H]-TCDD or another agonist, the liganded receptor is tightly associated with the nucleus, and extractable with a high salt concentration (≈0.4 M NaCl).

When cytochrome P-450IA1 was cloned, it became possible to examine the 5' regulatory sequences. The upstream region was cloned into a reporter plasmid containing the chloramphenicol acetyltransferase gene and transfected into cells that were then treated for TCDD. The expression assay identified dioxin response elements which behaved like classical enhancer elements. Footprinting and gel shift studies identified 4 enhancer elements with the minimal core consensus sequence 5'-T(A/T)GCGTG-3' (11,12).

A model system that has been extensively studied is the well-differentiated hepatoma cell line, Hepalc7c7, which has a high concentration of the Ah receptor. The wild-type cells when challenged with 3-MC, display appreciable induction of cytochrome P-450IA1 and metabolize the 3-MC to electrophilic metabolites that are toxic. This provided a selection system for noninducible or poorly inducible cells which fell into 3 complementation groups (and a dominant noninducible phenotype): (A) mutation in cytochrome P450IA1 gene, (B) low level of Ah receptor binding activity with normal nuclear "translocation", (C) normal Ah receptor-ligand binding, but no "translocation" of the ligand-receptor complex to the nucleus. This suggested genes other than the Ah receptor (presumably the B class mutants) are involved in P450IA1 induction (i.e., dominant negative and C class mutants) (13). Elferink et al (14) provided evidence in gel-shift experiments — TCDD incubated rat liver cytosol, and a double-stranded oligonucleotide which had bromodeoxyuridine in the consensus sequence — that two proteins were covalently linked by UV light to the responsive element: (a) a 100 kD protein which bound the [^{125}I]-dioxin ligand, and (b) a 110 kD nonligand-bound protein. Hoffman et al (15) transfected class C Hepa 1 mutant cells with human genomic DNA and a selectable marker, and corrected the noninducible phenotype. They cloned this cDNA which encoded a protein, Arnt (Ah receptor-nuclear translocator) which from the deduced amino acid sequence had a molecular weight of 87.5 kD with a basic-helix-loop-helix motif.

Using a [^{125}I]-labeled photoaffinity ligand ([^{125}I]-2-azido-3-iodo-7,8-dibromodibenzo-p-dioxin) that covalently labels the Ah receptor (thus permitting isolation and identification under denaturing conditions), the Ah receptor from liver cytosol from C57BL/6J mice was purified to homogeneity ($\approx$150,000 fold), and N-terminal amino acid analysis obtained (16).

A degenerate oligonucleotide based on this sequence was used as a probe and the murine Ah receptor cDNA was cloned and sequenced. The deduced amino acid sequence of the encoded protein was 805 amino acids with a spliced leader sequence of 9 amino acids, giving a mature protein of 796 amino acids, with a molecular weight of 89.4 kD (17).

From the deduced sequences, the Ah receptor and Arnt protein share three motifs: (a) At the amino-terminal end of both is a basic helix-loop-helix (BR/HLH) region. (b) Immediately adjacent to the BR/HLH region is an imperfect 51 amino acid repeat spaced by about 100 amino acids. From a data base search this region is found in only two other proteins, the Drosophila "single-minded" (Sim) protein, and circadian rhythm protein (Per). The Ah receptor, Arnt and Sim, (but not Per) also have a similar BR/HLH region. (c) At the carboxyl terminus, Ah receptor, Arnt, and Sim share a glutamine rich sequence (15,17).

The liganded Ah receptor binds to the Arnt protein to form a heterodimer which binds to the enhancer sequence. Like other BR/HLH proteins, e.g., Myo-D and E1A, Myc and Max it is presumed that the proteins bind to each other through their helices and to DNA through their basic amino acid regions. From photoaffinity-labeling of the Ah receptor, the ligand-binding site is presumed to involve the second (carboxyl terminal) 51 amino acid repeat, and possibly sequences amino-, and carboxyl-terminal to it.

Summary and Perspective

Cloning of the Ah receptor and ARNT proteins defines a novel motif for a transcriptional activator: two BR/HLH proteins, one of which, the Ah receptor binds ligand to initiate the formation of the active heterodimeric complex.

This is in contrast to other ligand-binding transcription activators which have been members of the erb-A superfamily (or steroid hormone receptor superfamily) and which recognize specific DNA enhancer sites by 2-zinc fingers. It remains to be determined if other xenobiotics which produce distinct pleiotropic responses in the liver (e.g., phenobarbital, pregnenolone-16α-carbonitrile) act on putative receptors that are (a) of the BR/HLH motif, like TCDD or (b) the erbA family, like the peroxisome proliferators.

With the advent of cloned cDNA for the Ah receptor and Arnt proteins, it should be possible to answer a whole series of questions: for instance (a) the phylogenetic and developmental appearance of these proteins (b) the localization of these proteins in various cell types, (c) the critical structural domains requisite for Ah receptor binding to a ligand and to heat shock protein 90 kD, and (d) the molecular details of activation of this transcription factor. We further need to know more about the specific genes whose expression is activated (or turnoff) and the cascade of pathways and signals. Perhaps, the most obvious and glaring unknown is what this system exists for. Does the Ah receptor and the pleiotropic response it controls have a physiologic purpose? Is there a physiologic or endogenous ligand that activates this system? The alternative view of this system, which is conserved in vertebrate evolution, is that it responds only to foreign planar aromatic ligands to hasten their elimination from the organism. These questions seem more elusive, but with powerful new methodology, hopefully they will be addressed in the near future.

ACKNOWLEDGMENTS

Supported in part by grants from the National Institute of Environmental Health Sciences ES-01884 and ES-05703 and the National Cancer Institute Core Grant 07175.

REFERENCES

Astroff B, Zacharewski T, Safe S, Arlotto MP, Parkinson A, Thomas P and Levin W (1988): 6-Methyl-1,3,8-trichlorodibenzofuran as a 2,3,7,8-tetrachlorodibenzo-p-dioxin antagonist: inhibition of the induction of rat cytochrome P-450 isozymes and related monooxygenase activities. *Mol Pharmacol* 33: 231-236.

Bradfield CA, Glover E and Poland A (1991): Purification and N-terminal amino acid sequence of the Ah receptor from C57BL/6J mouse. *Mol Pharmacol* 39: 13-19.

Burbach KM, Poland A and Bradfield CA (1992): Cloning of the Ah-receptor cDNA reveals a distinctive ligand-activated transcription factor. *Proc Natl Acad Sci USA* 89: 8185-8189.

Denison MS, Fisher JM and Whitlock JP Jr (1988): The DNA-recognition site for dioxin-Ah receptor complex. *J Biol Chem* 263: 17221-17224.

Denison MS, Fisher JM and Whitlock JP Jr (1988): Inducible, receptor-dependent protein-DNA interactions at a dioxin-responsive transcriptional enhancer. *Proc Natl Acad Sci USA* 85: 2528-2532.

Elferink CJ, Gasiewicz TA and Whitlock JP Jr (1990): Protein-DNA interactions at a dioxin-responsive enhancer. *J Biol Chem* 265: 20708-20712.

Hoffman EC, Reyes H, Chu F-F, Sanders F, Conley LH, Brooks BA and Hankinson O (1991): Cloning of a factor required for activity of the Ah (dioxin) receptor. *Science* 252: 954-958.

Luster MI, Hong LH, Osbourne R, Blank JA, Clark G, Silver M T, Boorman GA and Greenlee WF (1986): 1-Amino-3,7,8-trichlorodibenzo-p-dioxin: a specific antagonist for TCDD-induced toxicity. *Biochem Biophys Res Commun* 139: 747-756.

Nebert DW, Goujon FM and Gielen JE (1972): Aryl hydrocarbon hydroxylase. Induction by polycyclic hydrocarbons: simple autosomal dominant trait in the mouse. *Nature New Biol* 235: 107-110.

Perdew GH (1988): Association of the Ah receptor with 90 kDa heat shock protein. *J Biol Chem* 263: 13802-13805.

Poland A and Glover E (1974): Comparison of 2,3,7,8-tetrachlorodibenzo-p-dioxin, a potent inhibitor of aryl hydrocarbon hydroxylase, with 3-methylcholanthrene. *Mol Pharmacol* 10: 349-359.

Poland A and Glover E (1973): Chlorinated dibenzo-p-dioxins: potent inducers of δ-aminolevulinic acid synthetase and aryl hydrocarbon hydroxylase. *Mol Pharmacol* 9: 736-747.

Poland A and Glover E (1975): Genetic expression of aryl hydrocarbon hydroxylase by 2,3,7,8-tetrachlorodibenzo-p-dioxin: evidence for a receptor mutation in genetically non-responsive mice. *Mol Pharmacol* 11: 389-398.

Poland A, Glover E and Kende AS (1976): Stereospecificity, high affinity-binding of 2,3,7,8-tetrachlorodibenzo-p-dioxin by hepatic cytosol. *J Biol Chem* 251: 4936-4946.

Poland A and Knutson JC (1982): 2,3,7,8-Tetrachlorodibenzo-p-dioxin and related halogenated aromatic hydrocarbons: examination of the mechanism of toxicity. *Annu Rev Pharmacol Toxicol* 22: 517-554.

Schmid A, Elmer I and Tarnowski GS (1969): Genetic determination of differential inflammatory reactivity and subcutaneous tumor susceptibility of AKR/J and C57BL/6J Mice to 7,12-dimethylbenz(a)anthracene. *Cancer Res* 29: 1585-1589.

Van Gurp JR and Hankinson O (1984): Isolation and characterization of revertants from four different classes of aryl hydrocarbon hydroxylase-deficient Hepa-1 mutants. *Mol Cell Biol* 4: 1597-1604.

AUTHOR INDEX

SUBJECT INDEX